Metallurgy and Technology of Steel Castings

Authored by

Jan Głownia
AGH – University of Science and Technology, Cracow, Poland

Metallurgy and Technology of Steel Castings

Author: Jan Głownia

ISBN (Online): 978-1-68108-570-8

ISBN (Print): 978-1-68108-571-5

First published in 2017.

Bentham Science Publishers Ltd.
Executive Suite Y - 2
PO Box 7917, Saif Zone
Sharjah, U.A.E.
Email: subscriptions@benthamscience.org

BENTHAM
SCIENCE

CONTENTS

PREFACE i

ACKNOWLEDGEMENTS ii

CHAPTER 1 INTRODUCTION 1
- **1. SHORT HISTORY OF STEEL AND STEEL CASTING** 1
- **2. DEVELOPMENT OF STEEL TECHNOLOGIES** 6
- **CONFLICT OF INTEREST** 9
- **REFERENCES** 9

CHAPTER 2 FUNDAMENTALS OF MELTING PROCESSES 11
- **1. INTRODUCTION** 11
- **2. MELTING OF STEEL IN AN ELECTRIC ARC FURNACE** 12
  - 2.1. Repair of the Furnace Lining 12
  - 2.2. Loading the Charge 13
  - 2.3. Smelting the Charge 15
  - 2.4. Oxidation Period 16
    - *Oxidation of Silicon* 18
    - *Oxidation of Manganese* 19
    - *Oxidation of Phosphorus - Dephosphorization* 20
    - *Reversion of Phosphorus into the Molten Steel* 24
    - *Oxidation of Carbon* 25
    - *Oxidation of Chromium* 28
  - 2.5. Refining Period 31
    - *Desulphurisation* 31
- **CONFLICT OF INTEREST** 36
- **REFERENCES** 36

CHAPTER 3 SLAGS 38
- **1. INTRODUCTION** 38
  - Basicity of Slags B 41
- **2. STRUCTURE OF SLAGS** 42
- **3. SLAG PHASE SYSTEMS** 45
- **4. CHEMICAL COMPOSITIONS OF STEEL SLAGS** 46
  - Slags in Basic Electric Arc Furnace 47
  - Slags for the Ladle Treatment (Synthetic Slags) 50
- **CONFLICT OF INTEREST** 51
- **REFERENCES** 51

CHAPTER 4 GASES AND NON-METALLIC INCLUSIONS IN STEEL CASTINGS 53
- **1. INTRODUCTION** 53
- **2. THE SOLUBILITY OF GASES IN MOLTEN STEEL** 55
- **3. THE EFFECT OF TEMPERATURE ON THE GAS CONTENT IN THE SOLID STATE** 60
- **4. SOLUBILITY OF HYDROGEN AND NITROGEN IN IRON AND STEEL** 61
  - Carbon oxide {CO} 68
- **5. INCLUSIONS IN STEEL CASTINGS** 69
  - 5.1. Conditions for the Formation of Sulphides 74
  - 5.2. Modification of Sulphides 78
    - *Sulphides of Ti and Zr* 81
    - *Influence of Calcium* 83

The Effect of Rare Earth (REs) Treatment 85
5.3. The Influence of Non-Metallic Inclusions on the Properties of Cast Steels 85
CONFLICT OF INTEREST 89
REFERENCES 89

CHAPTER 5 DEOXIDATION 92
1. INTRODUCTION 92
2. TYPES OF DEOXIDATION 93
Diffusion Deoxidation 93
Precipitation Deoxidation of Steel 94
3. DEOXIDATION WITH ALUMINIUM AND OTHER ELEMENTS 95
Deoxidation with Silicon 96
Deoxidation with Manganese 97
Deoxidation by Titanium 98
Deoxidation by Rare Earths REs 98
Deoxidation by Complex Deoxidisers 99
Deoxidation by Ca-Additions 100
4. REMOVAL OF THE NON-METALLIC INCLUSIONS FROM THE MOLTEN STEEL 101
5. DEOXIDATION IN VACUUM CONDITIONS 106
6. EFFECT OF DEOXIDISERS ON THE MECHANICAL PROPERTIES OF CAST STEELS 108
CONFLICT OF INTEREST 110
REFERENCES 110

CHAPTER 6 SECONDARY METALLURGY FOR SMALL LADLES 112
1. INTRODUCTION 112
2. CRITERIA FOR THE SELECTION OF SECONDARY METALLURGY TECHNOLOGIES 114
I. Metallurgical 114
II. Technological 114
3. THE CHOICE OF SECONDARY METALLURGY PROCESS 115
4. PRINCIPLES OF THE PROCESSES AND LAYOUT 117
4.1. AP process 117
4.2. IP – Injection Processes 120
Wire Applications 121
5. SECONDARY METALLURGICAL UNITS WITHOUT VACUUM TREATMENT 122
6. VACUUM PROCESSES 127
Evaporation of impurities 128
VD Process 128
Technology of the VD Process 129
The VOD Process 130
Results Obtained 133
7. VACUUM INDUCTION FURNACES 134
Melting and Casting in Vacuum 136
CONFLICT OF INTEREST 137
REFERENCES 137

CHAPTER 7 CAST STEELS FOR SERVICE CONDITIONS 139
1. INTRODUCTION 139
2. CASTINGS OF WALL THICKNESS UP TO 100 MM 147
Low-Alloy Manganese Cast Steels 149
Manganese-Nickel Cast Steel 151

Manganese-Molybdenum Cast Steels 154
Chromium and Chromium-Molybdenum Cast Steels 155
3. CAST STEEL FOR THICK-WALLED CASTINGS 157
4. HIGH STRENGTH CAST STEEL 160
4.1. High Strength Low-Alloy Cast Steels 163
CONFLICT OF INTEREST 163
REFERENCES 163

CHAPTER 8 STEEL CASTINGS FOR LOW TEMPERATURE APPLICATION 165
1. INTRODUCTION 165
2. THE EFFECT OF THE CHEMICAL COMPOSITION ON THE IMPACT TOUGHNESS 172
3. THE EFFECT OF DEOXIDATION OF STEEL 180
4. CHEMICAL COMPOSITION OF CAST STEELS FOR OPERATION AT LOW TEMPERATURES 182
5. THE INFLUENCE OF STRUCTURE AND HEAT TREATMENT 185
6. THE INFLUENCE OF WALL THICKNESS OF THE CASTING 187
CONFLICT OF INTEREST 190
REFERENCES 190

CHAPTER 9 CORROSION – RESISTANT CAST STEELS 192
1. THERMODYNAMICS OF AQUEOUS CORROSION 192
2. TYPES OF CORROSION 201
3. THE EFFECT OF ALLOYING ELEMENTS 210
4. MICROSTRUCTURE OF CAST STEEL 212
5. FERRITIC – AUSTENITIC, F-A (DUPLEX) CAST STEELS 215
6. HEAT TREATMENT 216
CONFLICT OF INTEREST 216
REFERENCES 216

CHAPTER 10 HEAT – RESISTANT CAST STEELS 219
1. INTRODUCTION 219
2. OXIDATION - SELECTION OF ALLOYS AND MICROSTRUCTURE 220
2.1. Kinetics of Oxidation 221
2.2. The Mechanism of Scale Formation 225
2.3. The Rate of Oxidation 228
Linear (Fig. 4) 229
Parabolic 230
3. HEAT-RESISTANT CAST STEELS 232
4. MICROSTRUCTURE OF HEAT-RESISTANT CAST STEELS 236
5. THE EFFECT OF TECHNOLOGICAL FACTORS ON THE IMPROVEMENT OF HEAT RESISTANCE 238
6. SULPHUR CORROSION 240
7. CASTINGS WORKING UNDER CARBURISING CONDITIONS 243
CONFLICT OF INTEREST 245
REFERENCES 245

CHAPTER 11 WEAR – RESISTANT CAST STEELS 247
1. INTRODUCTION 247
2. DEFINITION AND GENERAL CHARACTERISTICS OF WEAR 247
2.1. Structural Factors 249
2.2. Operating Conditions 252
2.3. Factors Characterising the Mechanism of Interaction 252

2.4. Types of Tribology Wear ... 252
3. WEAR-RESISTANT CAST STEELS ... 258
Carbon Cast Steel - Group 1 ... 258
Low-Alloy Cast Steel - Group 2 ... 258
Cast Steel Containing 12% Chromium ... 269
4. HIGH MANGANESE CAST STEELS ... 271
NOTES ... 273
CONFLICT OF INTEREST ... 274
REFERENCES ... 274

CHAPTER 12 POURING SYSTEMS AND RISING ... 276
1. INTRODUCTION ... 276
Construction of Gating Systems ... 279
2. CALCULATION OF GATING SYSTEM ... 280
Calculation of the Diameter of the Tap Hole in the Ladle ... 283
Calculation of Area of Ingates Fingate ... 284
3. TYPES OF RISERS AND THEIR CALCULATION ... 285
3.1. Feeding of Hot Spots ... 286
4. RISER CALCULATIONS ... 287
The Modulus Method ... 290
5. COMPUTER METHODS ... 291
CONFLICT OF INTEREST ... 292
REFERENCES ... 292

CHAPTER 13 HEAT TREATMENT OF STEEL CASTINGS ... 294
1. INTRODUCTION ... 294
2. MICROSTRUCTURES OF CARBON CAST STEELS IN AS-CAST CONDITIONS ... 295
Normalisation ... 297
Annealing ... 298
Stress Relief Annealing ... 299
Spheroidisation Annealing ... 299
Solution Annealing (Treatment) ... 299
3. QUENCHING OF CAST STEELS ... 301
Stress-Relief Annealing ... 308
4. HARDENING PROCESSES ... 308
CONFLICT OF INTEREST ... 310
REFERENCES ... 310

SUBJECT INDEX ... 312

PREFACE

This work is intended for engineers operating in industries, which apply steel castings in their products. This does not only concern pure castings, but also: machine building (transportation, earthmoving equipment, mining, military machinery), power industry (gas and oil exploration, power generation), chemical industry (refineries, chemical plants), food and medical processing. Foundries produce part of the machines intended for these applications considering their working conditions: temperature (from minus 120 °C to plus 700 – 800 °C), stresses, strength or yield strength and resistance to the environment (corrosion, abrasion). To meet the requirements of casting users, their production process should ensure the appropriate quality of liquid metal, correctness of the technological process (among others: smelting, pouring, solidification and heat treatment) and technical control of each sub-process.

Engineers who produce steel castings for the mentioned working conditions receive the basis in smelting of carbon and alloy steel, significance of the deoxidation process and modification of steel for castings (compared to wrought steel) and the properties of various types of cast steel operating *e.g.* in areas of sub-zero temperatures or corrosion. This knowledge is also useful for engineers in other specialties and working in the field of material engineering, metallurgy and manufacturing processes.

This presentation is based on publications, own co-operations with foundries and institutes, and industrial practices provided in the last decades. The intention of this book is to draw the reader's attention to the possible difficulties in obtaining an appropriate quality of castings as well as to indicate sources where more information on each subject of foundry engineering can be found.

This book is also addressed to students who wish to obtain an overall knowledge of the steel castings technology.

Jan Głownia
AGH – University of Science and Technology,
Cracow, Poland

ACKNOWLEDGEMENTS

It is a great pleasure for me and I am particularly grateful to my wife Jane for her patience, care and indulgence during these long months of writing and staying outside our home. I am very grateful to my son Maciej, daughter Joanna (who spent a lot hours with the pictures) and grandchildren: Justine and Jędrzej.

Many thanks to my colleagues at my Alma Mater – AGH-University of Science and Technology in Cracow:

- Dr Barbara Kalandyk, Dr Renata Zapała, Dr Sebastian Sobula, Dr Grzegorz Tęcza, Antoni Dzieja for many years of co-operation, for their help and valuable suggestions;
- many friends, co-workers from Foundry Engineering Faculty and other Faculties of my University and to the following foundries and factories for the numerous, collective works:
- Metalodlew S.A. Cracow,
- Huta Małapanew in Ozimek,
- Pioma S.A. in Piotrków Trybunalski,
- Zabrzańska Fabryka Maszyn in Zabrze, AS Ferro-Term in Łódź,
- National Centre for Research and Development in Poland,
- Foundry Research Institute in Cracow,

I would like to specially thanks to the Translation Teams: JUNIQUE Agnieszka and Justin Nnorom and Dogadamycie for the translation of the entire text. I feel grateful MSc Grażyna Konieczko and Marcin Wasiak for preparation all figures to their proper conditions before printing.

CHAPTER 1

Introduction

Abstract: The history of steel castings is associated with previously developed technologies for obtaining castings of bronze, copper and cast iron. These formed the basis for the improvement and development of casting technology of metal with a higher melting temperature and higher mechanical properties, and then - better utility properties (corrosion resistance, impact toughness at low temperatures and finally - smelting in a vacuum). The development of steel smelting technology and methods of obtaining steel castings have been presented against the background of extremely extensive literature on bronze casting in ancient times.

Keywords: Ancient castings, Converters, Electric arc furnace, History of steel castings, Lost wax technique, Vacuum technologies.

1. SHORT HISTORY OF STEEL AND STEEL CASTING

Production of metal castings has played an important role in developing societies until these days. Cast elements comprise a significant part of tools and appliances, and currently of: cars, planes or machines (*e.g.* wagons and locomotives, ships, drilling platforms, pumps, excavators *etc.*). Founders have always been a group of innovators who are respected in any society.

Metal casting is a very old technology. The first metal products were probably made of lead [1]. Metal parts made of copper by forging (more than 9000 years BC) are better known.

Only about 5 000 years BC they were cast from copper in the Near East [2, 3], or 3000 BC according to [4, 5]. The following centuries cover the introduction of copper alloys (Cu-As and tin bronzes) during the Bronze Age (3000 to 1500 BC), in Central Europe (2200 to 940 BC [6]). During that period, the technology of lost wax casting was used in the Near East and statues were cast from copper (Pharoah Pepi I in Egypt) [2]. First copper alloy castings did not appear until around 1500 BC. These were mainly Cu-As (4-12%) and Cu-Sn (5-10%) [2] alloys cast into stone moulds, mainly as a weapon (swords, axes), (Fig. **1**). Flat stone forms were carved so sophistically that after casting objects in them, and then bending them, they created artistic objects (these were most often bracelets

Jan Głownia

and other jewellery articles) [1]. They are visible in a lot of museums of the Near and Far East as well as Europe.

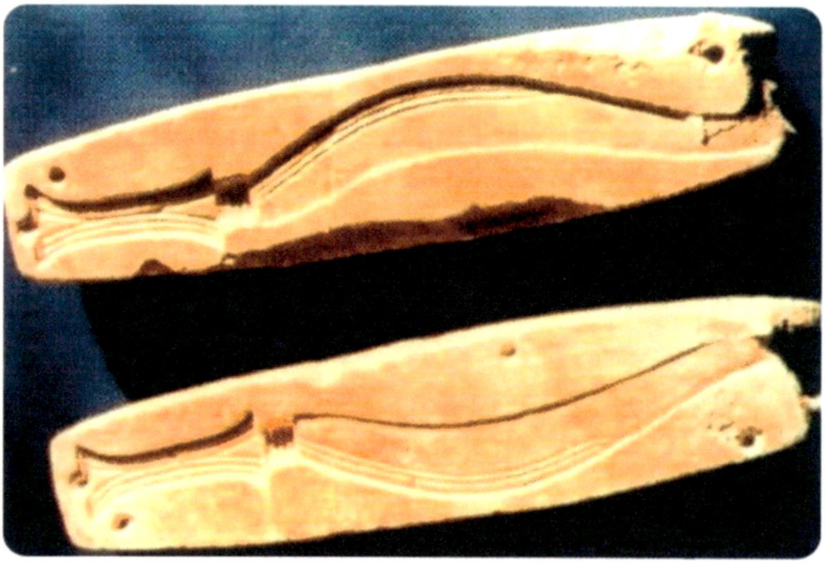

Fig. (1). Sand stone mould from 4500 years BC [11].

The introduction of tin into copper allowed to lower the melting temperature of the alloy, bath deoxidation and to increase the alloy's strength. Lowering the melting temperatures also allowed to make more complex casting shapes.

The development direction of foundry followed from the Near to the Far East (a difference of about a thousand years) [2, 7]. Traces of bronze castings were discovered from 4000 - 3000 BC (*e.g.* "frog" from Nahal Mismar in Mesopotamia 3200 BC [8]) and in Far East, a thousand years later [2]. There is no unambiguous confirmation of the primary source of their performance [3] and the technology of casting moulds (if the castings were made by the lost wax method or in ceramic moulds). The first moulds consisted of clay bound sands, then of sand stone (Fig. **1**) and finally – bronze moulds. There is very detailed literature relating to the first castings made of non-ferrous metals [2 - 10]. However, due to the subject discussed here, it is more reasonable to document the development of iron alloy castings.

Iron alloy castings appeared much later, around 600 BC in China [2, 8], (cast iron boiler from 512 BC in China) [5, 12, 13]. Due to the high content of phosphorus and sulphur, the melting temperature of these alloys was low, close to the melting temperature of bronze.

According to J. Needham [14], the first cast iron castings in China were agricultural tools (axes, pickaxes, wedges), vases and pitchers. Due to the high carbon content (white cast iron with a low content of silicon, phosphorus and sulphur), these tools were brittle; therefore, the raw castings were subjected to graphitization annealing at the temperature of 900 - 1000^{0}C [12, 15], and then forged. There are a number of examples of discovered cast iron castings, which were made in China in the years 500 to 200 BC. These are vases and moulds [14], tombstones, and in later years - monuments of dragons and lions, weighing even up to 48 tons [3, 14]. Moulds for these castings were made of clay (and also baked clay).

The first iron alloy castings appeared in Europe much later than in China: only around AD 1400, there were references on iron melting [16]. There is a consensus for the formation of the first cast iron castings in Europe: M. Goodway [2] indicates the fifteenth century, and G. Engels and H. Wübbenhorst [3] about AD 1400. Although it is unknown which country this technology comes from, but the direction "from the East" is undeniable. The first castings were bullets and water pipes [3, 9], which replaced bronze and stone pipes. In Germany, cast water pipes were already made in the fifteenth century (around 1450). They were 1 meter long and 30 mm in diameter and were joined with muffle connections.

In the second half of the fifteenth century, gratings, plates and openwork elements for fireplaces [17,11 - Germany], bells [3 -Switzerland AD 1435], and cooking pots were also cast from cast iron [18]. In the United States, the first castings were made in 1642 [19]. In plants Saugus Iron Works, a cooking pot with a weight of 1.4 kg was cast from ground.

In Coalbrookdale (England), A. Derby introduced smelting of cast iron from ores with the use of carbon [2]. This method enabled the production of cast iron on an industrial scale. The introduction of blowing air in furnaces for a reduction of iron ore allowed to increase combustion temperatures and thereby to obtain a temperature of cast iron of about 1400^{0}C [2, 3].

The particular development of foundry occurred in the eighteenth century as a result of a lot of technological developments introduced and high demand on cast iron castings. Replacement of moulding in clay for the use of moulds made in sand was such an improvement [18]. This way, moulding with a template was converted into flask moulding [A. Darby's patent from 1707 - 20].

Further development of casting of iron alloys occurred in the second half of the eighteenth century as a result of the industrialisation entering a lot of countries, and the associated need for cast parts to newly constructed machines (cylinders of steam engines [Scotland 1766 -21], cylinders and shafts for pumps [5, 21, 22] or bridge elements [England 1777-1779 2,5,23]; Germany 1796 Lower Silesia [5] and Upper Silesia 1794, (Fig. **2**).

Fig. (2). The bridge, cast from cast iron elements in Ozimek, Poland,1827 year; Courtesy Marcin Widera, (Local Administration of a Commune Council in Ozimek).

The use of elements made of cast iron instead of wood was preferred for safety reasons (fires) and compressive strength (for stone and wood). Hence, a lot of elements of industrial halls are cast iron castings (load-bearing columns, girders, roof supports and in mining - shaft linings). In many cases, these castings have ornaments and also serve as decorations (of windows, balconies, especially as decorative elements in architecture). Artistic castings are of particular importance in the development of cast iron castings. The end of the eighteenth century brought important examples of the art of casting:

- tombstones [3, 5], Germany 1800,
- vases, bowls and tableware,
- all kinds of jewellery, often with sophisticated designs.

Such a rapid development of foundry was possible only in combination with new technological solutions, introduced directly into industrial applications. These are primarily:

- replacement of clay moulding with sand mass moulding [18, 20]. In case of casting pots, the technological process was thereby improved by replacing moulding with a template with flask moulding with sand mass. The use of the model reduced labour input.

- introduction of blast into shaft furnaces, Fig. (**3**) in Western Europe in the second half of the fifteenth century [5, 24].

Initially, "iron beads" (iron lumps) obtained from Rennfeuer furnaces (through the reduction of iron ore with carbon) were smelted for further forging. Later, these iron lumps, with the addition of charcoal, were smelted in a crucible or shaft furnaces [4, 12].

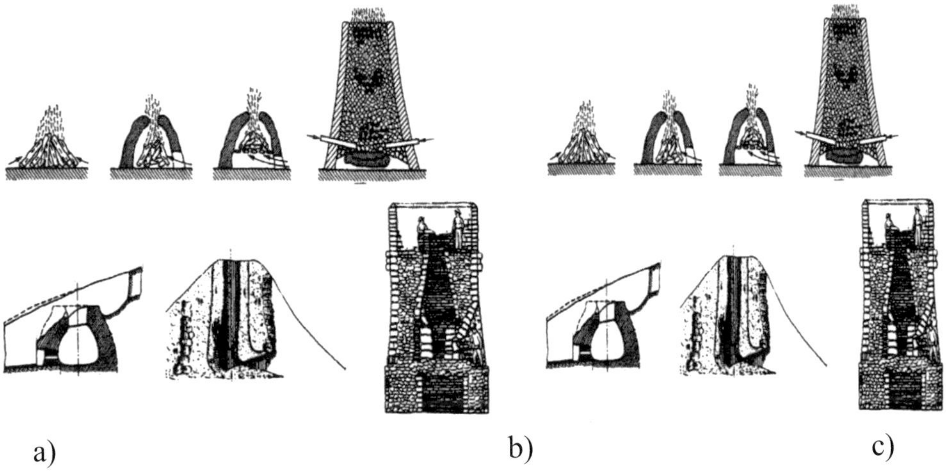

Fig. (3). Original furnace (a) and shaft furnaces (**b,c**) [11].

In the seventeenth century, iron smelting in reverberatory furnaces with natural draught was started.

- Introduction of coke instead of charcoal for melting iron in "blast furnaces". A. Darby started this process in England in 1709 [3, 20], and it was further used in Upper Silesia in 1796 and in the Ruhr Region in 1845 [3].
- The invention of the steam engine by J. Watt in 1765 [5] made it possible to implement water drive to introduce air blast.
- An important moment of progress of foundry is also the development of new alloys: malleable cast iron, spheroidal cast iron, and currently a lot of types of ADI cast iron.

The driving force for these changes is seeking to achieve better mechanical properties and especially now - utility properties: fatigue, wear, corrosion (cast steels with less than 0.03%C), crack resistance or favourable strength values in combination with plastic values – HSLA steels). More plastic materials from brittle cast iron were already known in China [25, 26].

2. DEVELOPMENT OF STEEL TECHNOLOGIES

Primary tools from iron alloys are forged “lumps” smelted in Rennfeuer furnaces [5]. As mentioned earlier, these elements were also annealed in China in order to increase their plasticity and resistance to cracking.

Before the first castings were made of steel, liquid steel was smelted in crucibles: in India about AD 500 [1], (according [25] in southern India around 300 BC), in the United States in 1818 (in Valley Forge Foundry [2]). In 1740 [2, 3] or 1750 [8, 21], the Englishman B. Huntsman smelted steel in a crucible made of clay, and later of graphite. Unfortunately, it was cast in small ingots (blocks) that were later forged. Performing shapes, cast from liquid steel to ceramic moulds, occurred much later, no earlier than in the 19th century in Germany. J. Meyer [3, 25, 26], (1841 bell castings in Bochum) and C. Fischer (1845 castings of small gears in Schaffhausen) started this process. In 1847, the first steel cannons were cast in Krupp Works [1, 27]. In the first two cases, steel was smelted in a shaft or reverberatory furnaces, in which graphite and chamotte crucibles were used. Introduction of coke instead of charcoal allowed to increase the combustion temperature and to reduce the costs of smelting [5]. However, smelting steel in crucibles was characterised by some drawbacks: manually pouring the steel into moulds limited the capacity of the crucible, and the crucible acidic material did not allow to obtain low contents of sulphur and phosphorus, resulting in low plasticity.

Subsequent improvements and inventions in the melting technology have contributed to the rapid spread of steel castings. The most important events were: launching the first Bessemer converter with acid lining (1856 [25]; 1851 [2]), Thomas basic converter (1879), (1878 – [28]) and open-hearth furnace (E. and P. Martin 1863 [25]). The second half of the nineteenth century was very rich in inventions that were significant for metallurgy:

- in 1878 [2, 25], W. Siemens constructed an electric arc furnace (installed in the USA in 1906 at Holcomb Steel Co. Syracuse);
- in 1913, the first induction furnace of low frequency was constructed and in 1916 E. Northrup invented a coreless induction furnace [2];
- a coreless induction furnace of high frequency was installed for the first time in 1930 in Lebanon Steel Foundry [2].

Firstly, steel smelted in crucible furnaces was expensive, it made it possible to obtain castings of light weight, and its chemical composition contained too much sulphur, phosphorus and carbon (as high temperature of molten steel with low carbon content couldn't be obtained). These furnaces are fired with coke (as well as gas or oil to raise the temperature) and blown (heated) with compressed air. Since the capacity of crucibles is low, and the heat loss is large, there are constructions of crucible furnaces in which steel is melted simultaneously in a lot of crucibles. This solution, however, is rarely used due to possibilities of obtaining liquid steel in other (induction, arc) furnaces, which have higher capacities and lead to a lower cost of steel production.

With the development of a new method of steel smelting by H. Bessemer (1856), (1851- [8]), liquid steel for larger castings could be obtained. However, there were limitations due to the fact that this steel was not deoxidised and products made of it were not very plastic (high contents of sulphur and phosphorus). Improvements in the process of steel smelting in the converter consisted in:

- introduction of manganese oxide and carbon by R. Musket (1801- [23]) and thereby in reduction of carbon from a cast iron melt. With variable carbon content, this allowed to deoxidise the metal bath;
- conversion of air blowing from the bottom of the converter into the side air supply (S.J.Tropenas converter). By air supply at 5-120mm under the metal surface, the rate of oxidation of carbon and silicon was increased, and therefore liquid steel of higher temperature (up to about 1700^0C) was obtained. This also allowed to increase the amount of smelted steel in the converter.

Steel smelting in Tropenas converter (1891) consists in [23, 29]:

- filling the converter with liquid cast iron from the cupola of the chemical composition (in wt %): C = 2.7 – 3.5; Si = 1.3 – 2.0; Mn = 0.5 – 0.6; S and P each max. 0.04,
- immediately after filling the converter with liquid steel, an air is blown. Then, oxidation of C (up to 0.05%), Si (up to 0.02-0.05%, and Mn (up to 0.03%) occurs, after deoxidising the bath, the chemical composition is completed to the required content and tapping of steel into a ladle is started,
- increasing the amount of air blast (G. Goeranson),
- conversion of the acid into basic lining.

Propaedeutics of improvements to the Bessemer process by G. Göransson (increasing the amount of blast) and R. Musket, who introduced deoxidation of steel with pig-iron, eliminated brittle fracture and increased the plasticity of steel, which in turn allowed for plastic deformation [3].

In 1899, P. Héroult received a patent for heating metal with an electric arc. His joint works with R. Eichhoff led to the first tool steel smelting in a 500 kg arc furnace (1906). Simultaneously introduced solutions of induction melting of metals by F. Kjellin (1891,1905 [4],) allowed for the application of network induction furnaces and furnaces of medium frequency for melting steel since 1930 in Sweden [4].

All these technical innovations concerned the production of castings mainly for the railway sector, and over the years, for the following industries: machine construction, automotive, aviation and astronautics.

The modern industry uses a small percentage of steel for castings; this constitutes from 2 to 10% of the total amount of steel produced. However, owing to the development of better technology of smelting, including oxygen blowing, calcium injection, introduction of secondary metallurgy, processes AOD, VOD, VAD, OBM, Vacuum Degassing, LF (ladle furnace) or DETEM, as well as forming - for example, Shaw processes, Replicast and evaporated pattern - castings with a wall thickness of 2.5-3 mm with even better quality, *i.e.* with lower content of gases (H;N;O) and non-metallic inclusions, as well as lower content of unwanted elements (not only phosphorus and sulphur but also such as Pb, As, Bi, Sn, Sb) and of more stable chemical composition (carbon, silicon, manganese and aluminium) can be obtained from low-alloy steel and ductile cast iron.

Introduction of CAD/CAM methods, allowing for guiding the process of solidification in hot spots and an elimination of microshrinkage, improved the quality of steel castings. On the other hand, modern non-destructive tests, *e.g.* with the use of ultrasound, form a safe level of production by disclosure of defects and enforcing high-quality products. The possibility of welding steel castings also allows for:

- combining castings of simple shapes into a complex final product, which reduces expenditures when making a mould, cores and instrumentation,
- making cast - forged parts,
- combining castings of various requirements in their individual parts, *e.g.* good plasticity with the appropriate wear resistance in certain areas;

All these possibilities make up a set of features of steel castings, used in various industries. Then, the machine building industry requires these cast steel grades, which have good strength, plasticity, fatigue and corrosion resistance; petrochemical industry requires corrosion and creep resistance; turbine building industry requires creep and cavitation resistance; cosmonautics requires strength and plasticity at low temperatures.

Meeting the above mentioned requirements results in the steel castings having an established position, being important in the steel industry. Examples of the use of castings working in difficult, often extreme conditions of machines and equipment, are described in each of the following chapters.

CONFLICT OF INTEREST

The author (editor) declares no conflict of interest, financial or otherwise.

REFERENCES

[1] H. Schmidt, and H. Dickmann, *Bronze-und Eisenguss.*Düsseldorf, 1958.

[2] W.W. Krysko, *Lead in History and Art,* Stuttgart, 1980.

[3] *History of Casting.* vol. 15. ASM Intern, 1988.M. Goodway, ASM Handbook

[4] G. Engels, and H. Wübbenhorst, *5000 Jahre Giessen von Metallen.* Giesserei-Verlag: Düsseldorf, 1994.

[5] P.F. Wiser, *Steel Casting Handbook* Steel Founder`s Soc.of America: Ohio, 1980.

[6] P.R. Sahm, *Casting Bronze, Studies in Simulation of Ancient Foundry Processes* Foundry Institute, Aachen University.

[7] M.C. Flemings, *Invention and the Foundry Industry* WFO Technical Forum: Istanbul, Turkey, 2004.

[8] M.J. Lessiter, and E.L. Kotzin, *Timeline of Casting Technology.* Modern Casting`s Guide to CASTEXPO, 1996.

[9] A. Wittmoser, "Einige untersuchungen an historischen gusseisernen Rohren", *Giesserei,* vol. 44, no. 20, pp. 557-563, 1957.

[10] J. Sobczak, *Odlewnictwo w Rozwoju Cywilizacji.* Foundry Research Institute: Cracow, Poland, 2011. (in Polish)

[11] C. Böhne, *Rev.d'Historie de la Siderurgie*, 1967, no. 8, p. 237.

[12] R.F. Tylecote, *A History of Metallurgy.* The Metals Society: London, 1976.

[13] N. Bernard, and S. Tamotsu, *Metallurgical Remains of Ancient China.*Tokyo, 1975.

[14] J. Needham, *The Development of Iron and Steel Technology in China.*London, 1958.

[15] W.G. Henger, "The metallography and chemical analysis of iron-base samples dating from antiquity to modern times", *Bulletin of the Historical Metallurgical Group,* vol. 4, no. 2, pp. 45-72, 1970.

[16] G. Agricola, *Zwölf Bücher vom Berg-und Hüttenwesen, 5.* Auflage: Düsseldorf, 1978.

[17] A. Kippenberger, Die Kunst der Ofenplatten, Dargestellt an der Sammlung des Vereins Deutscher Giessereifachleute.

[18] O. Johannsen, Geschichte des Eisens, Hrsg.Verein Deutscher Eisenhüttenleute, 3.Auflage, Düsseldorf, 1953.

[19] E. Kotzin, "Development of Foundry Technology in the United States", *Metals Handbook,* vol. 15, pp. 24-37, 1988.

[20] J.J. Robinson, "Introducing the Nominees for the Greatest Materials Moments in History", *JOM,* no. Nov, pp. 51-55, 2006. [http://dx.doi.org/10.1007/s11837-006-0227-1]

[21] E. Werner, Die Eisernen Brücken. Einige aspekte ihrer Entwicklung.*Eisen Architektur (1).* Icomos-

Colloquium: Hannover, 1958.

[22] K.C. Barraclough, Steelmaking Before Bessemer.*Crucible Steel: The Growth of a Technology* The Metals Society: London, 1984.

[23] N. Vukovich, "A Short History of the Steel Foundry", *AFS Trans.,* pp. 643-649, .

[24] R.F. Tylecote, "The early history of metallurgy in Europe", In: *Bulletin HMG,* Longman, Ed., Royal Numismatic Soc.: London, 1972, no. 10, pp. 261-278.,

[25] J. Mayer, *Schriftenreihe der Arbeitsgemeinschaft für Technikgeschichte des VDI.* vol. 11. Berlin, 1938.

[26] R. Roesch, and K. Zimmermann, *Stahlguss.* Verlag Stahleisen GmbH Düsseldorf, 1982.

[27] D. Diderot, and J. d'Alembert, *Encyclopedie Methodique*, 1976.

[28] A. Schulenburg, *Giesserei Lexikon.* Fachverlag Schiele und Schön GmbH: Berlin, 1958.

[29] R.W. Heine, C.R. Loper, and Ph.C. Rosenthal, *Principles of Metal Casting, Tata McGrow-Hill Publ.* Comp., New Dehli, 1983.

CHAPTER 2

Fundamentals of Melting Processes

Abstract: The primary technology of steel making in the foundry is a basic step to obtain a good quality of molten steel: with proper deoxidation, with low gases quantity and the control of non-metallic inclusions. These processes are discussed in detail through oxidation of elements and then during deoxidation in reduction period and in the ladle.

Keywords: Basic arc furnace, Ellingham-Richardson diagram, Furnaces in steel foundry, Melting processes, Reactions in oxidation and refining periods.

1. INTRODUCTION

Smelting steel for castings is based on small furnace units, compared to strictly metallurgical devices used in steel works. The most commonly used furnaces for melting steel in foundries are:

- electric arc furnaces EAF (basic and acid),
- induction furnaces (acid),
- modern units:
  - AOD (Argon Oxygen Decarburisation);
  - VOD (Vacuum Oxygen Decarburisation);
  - VAD (Vacuum Arc Degassing);
  - VD (Vacuum Degassing);
  - LF (Ladle Furnace),

and a number of modifications of these processes, combining their possibilities.

In practice, arc and induction furnaces are applied. Their advantage in the foundry is their smaller capacity (35 - 50 tons) than in the case of typical steel furnaces (150 - 350 tons), and thereby - better adaption of castings to the production cycle (it is difficult to make so many moulds to pour 150 tons of liquid steel every hour, *e.g.* from the VOD converter). The possibility of smelting steel of various grades (carbon, low-alloy Cr-Ni-Mo) during one - two shifts, in the same furnace, is also not without significance. Electric arc furnaces with basic lining also provide good

Jan Głownia

desulphurisation, dephosphorization and deoxidation of steel, necessary to obtain high-quality castings.

The choice of smelting technology and the furnace for smelting steel for castings depends on:

- foundry productivity (the possibility of performing the proper number of moulds, the possibility of cleaning the produced castings, their heat treatment and quality control),
- weight, size and complexity of produced castings, cost of production resulting from the possessed energy sources and available raw materials,
- types of cast steel grades produced (low carbon for work at minus 40°C, high carbon for wear conditions),
- wear resistant or high-alloy, *e.g.* Hadfield (12-14% Mn), high-alloy 26% Cr, corrosion-resistant 18% Cr - 11% Ni with carbon content below 0.03%).

The main unit for melting steel in the foundry is an electric arc furnace with basic lining. Therefore, smelting technology is based on the application of this type of furnace.

2. MELTING OF STEEL IN AN ELECTRIC ARC FURNACE

The whole process of steel smelting is divided into the following periods:

1. post-tapping repair,
2. loading the charge,
3. melting,
4. oxidation,
5. slagging-off,
6. refining,
7. tapping the steel into the ladle.

2.1. Repair of the Furnace Lining

After each time that the metal is tapped into the ladle, it is necessary to review the furnace devices (including the tilt mechanism, drive of electrodes, their cooling, power supply) and the lining. Current repair most often concerns ceramics (hearth, furnace side walls, roof and tapping spout). The main causes of the wear of furnace elements are:

- radiation of the electric arc (arc temperature of 4000 - 8000°C [1]),
- large temperature changes in liquid steel - lining boundary,
- reactions of liquid steel - lining and liquid steel - slag,
- (ZnO) oxides generated during evaporation as a result of oxidation of pure

metals (Zn, Bi, As, Sb, Pb, Sn),
- bath movement during the whole smelting process.

Lining losses are most visible on the side walls, in place of metal - slag impact. Existing rapid temperature changes (on the walls and the roof) lead to a reduction in the durability of ceramics. Oxides included in the slag composition (*e.g.* SiO_2, MnO, Cr_2O_3, Al_2O_3) flow into the connections of bricks form chemical compounds, and they lower the melting temperature of ceramics. While volatile oxides and elements attack the furnace roof the most, they also create low-melting chemicals.

Staff (master with smelters) consists of controlling the consumption of ceramics after each smelting process. Repair is especially needed for:

- losses in liquid steel - slag boundary,
- drain opening and tapping spout,
- removing metal and slag residues from the hearth surface.

In the case of arc furnaces with basic lining, magnesia and dolomite bricks are materials used to repair the furnace hearth and walls. These works are performed manually in furnaces with a small capacity (7 - 30 tons) or with special devices, allowing to throw or spray powdered mass (spray guns).

Post-tapping repair usually takes 10 - 15 minutes for furnaces with a capacity of up to 30 tons.

2.2. Loading the Charge

This process is performed mechanically. After raising and half-opening the roof, a basket suspended on a crane moves above the furnace hearth, lowers its position and from the low height - pours out the entire portion of the charge onto the furnace hearth. A single portion of charge is weighed and prepared earlier in the charging shop, in which sorted grades of the foundry scrap (risers, gate systems), pig iron, ferroalloys and commercial scrap (purchased from the outside) are stored. The charge preparation process is an important technological element. In large, it is responsible for subsequent consumption of electricity during the melting and smelting time.

The preparation and the selection of the charge are based on the following principles:

- each heat is another steel grade; so for castings of carbon cast steel, the charge cannot include alloyed scrap (Cr, Ni, Mo, *etc.*);
- the mass of the charge corresponds to furnace capacity and the predicted amount

of prepared moulds (additional charges take place only when the used scrap is fine and comprises a large volume of the furnace);

- before loading the furnace, limestone is introduced on the hearth as beginning of slag (in the amount of 1-2% of the charge weight);
- after melting, charge selection should ensure a carbon content higher by 0.2 - 0.4% than the upper carbon content required in the smelted steel grade;
- the manner of placing the charge is important for the length of both the melting period and the whole smelting process. Therefore, fine scrap is placed on the bottom of the hearth (of the charging basket), while large and alloyed scrap is placed in the middle of the furnace (under the electrodes). In the case of smelting alloyed steel, it is necessary to:
  - avoid charge containing elements with high affinity for oxygen (chromium, cerium, titanium - silicon and manganese are present in each scrap),
  - use these charge elements that do not oxidise during melting and oxidation (nickel, cobalt, copper, molybdenum - see Ellingham-Richardson diagram),
- modern methods of charge preparation are based on the shredding and packaging of the charge. This method increases charge density (0.8 - 1.1 t/m^3 [1]), improves the operation of the electric arc, and thus increases the rate of melting and reduces the total time of smelting. Production technologies of this type of charge are used for shredding old cars. Therefore, a certain amount of copper and tin can be found within it. They are detrimental to steel castings, as they reduce plasticity parameters at low temperatures (reduction of area, impact toughness) and provide to hot tears in castings (Cu),
- in melting shop, where steel is smelted in arc furnaces with a large capacity (150 - 350 tons), the charge is heated in the air counter-current from the furnace chamber. This technology reduces the time of melting and lowers the costs of smelting. This process has limited possibilities in foundries,
- the use of foundry scrap in foundries very often leads to an increase in the content of nitrogen and hydrogen in the castings. In the end, the formation of gas porosity, most often in hot spots, is possible, These defects are caused by the presence of the burnt moulding sand on elements of the foundry scrap, which is the source of gases from resins used as clay binder in the moulding sand. Similar effects are caused by scrap contaminated with oils, enamels and lacquers,
- the introduction of damp materials (iron ore, limestone, oxidized scrap) along with scrap also introduces hydrogen into the liquid steel (an addition of 500 kg of iron ore with moisture of 2-4% introduces 10-20 dcm^3 of water [2]).

While the charge loading process lasts for minutes, its proper preparation requires special attention.

2.3. Smelting the Charge

This smelting period has the highest consumption of electricity, (Fig. **1**). For furnaces with a capacity of 8 - 15 tons, the total consumption of energy in melting period is over 50% of total energy and depends mainly on the power of used transformers.

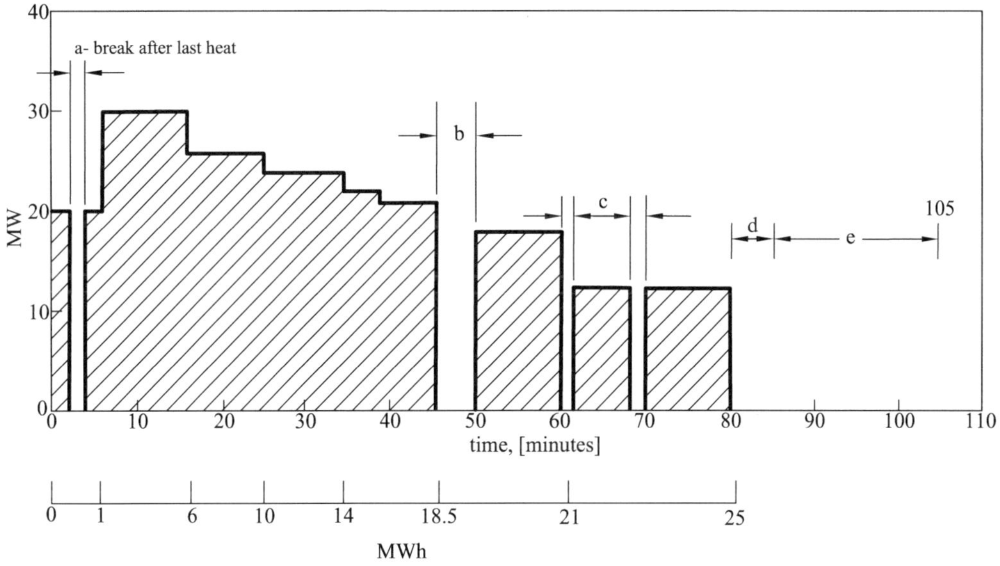

Fig. (1). Consumption of electric energy for melting steel in EAF [1, 2, 13] a-b melting and oxidation periods; b-slag-off; b-d refining period; d-tapping; e-pouring.

In the initial melting period, the electric arc is unstable. It is characterised by changes of strong current and secondary voltage. During this time, the furnace should work at a lower voltage, and after the work is stabilised, it should be switched to a higher voltage. Short electric arc leads to local melting of the charge (under the electrodes) and to lowering of the electrodes in relation to the surrounding scrap pieces. Then, it may lead to scrap slide and electrode fracture. Solid scrap pieces on the furnace walls should be collected into the bath when the operation of the furnace is switched off. The end of melting is checked by stirring the bath and by the presence of solid scrap pieces on the furnace hearth. When the charge is completely melted, the first sample for steel chemical composition is taken.

In modern steel plants, at the end of melting, the method of foaming slag is applied [1, 3]. Blowing of powdered carbon (with lance and compressed air) beneath the slag (at the depth of 5-10 cm to the bath) leads to the increase in its volume by forming bubbles of {CO}. Simultaneous blowing of oxygen facilitates the creation of foamed slag [1]. The beneficial effects of this method include:

- reduction of heat transfer from the bath,
- decarburisation of the bath,
- protection of the furnace lining,
- softening the operation of the furnace.

The process of melting steel for castings can be performed in accordance with the following two methods:

- with one slag (without the slagging-off after oxidation period),
- using two slags during smelting.

The first method allows for:

- a reduction in temperature fluctuations, which protects the walls and roof of the furnace against wear,
- oxidation of the bath with slag (lower heat loss),
- temperature measurements and sampling for the chemical composition is performed with slag (also lower heat loss),
- using slag with high phosphorus content for other technological purposes,
- more regular and stable operation of the electric arc.

In contrast, smelting with two slags is beneficial due to:

direct (preliminary) deoxidation of the bath,
possibility of introducing reduction slag, which allows for diffusion deoxidation and desulphurisation of the bath,
oxidation with gas oxygen.

2.4. Oxidation Period

After obtaining the 1st sample for the chemical composition, iron ore or pure oxygen (with the lance) is introduced into the bath. The main purpose of oxidation is the reduction of phosphorus to the lowest value (possibly below 0.015 - 0.020%). Intensive oxidation also brings the removal of hydrogen and nitrogen from the metal bath.

For dephosphorization of the bath, the following conditions are required:

- the presence of oxidising slag,
- high basicity of slag,

- a small content of (P_2O_5) in slag,
- a lower temperature of the bath.

However, under these conditions, other elements beyond phosphorus also oxidise. First of all, those whose affinity to oxygen is higher than for iron (red line in Fig. **2**). Ellingham - Richardson diagram indicates that they are in descending order: Ca, Mg, Zr, Ce, Ti, Si, B, V, Mn, Cr, P, Fe. From these data, it appears that silicon and manganese oxidise prior to phosphorus.

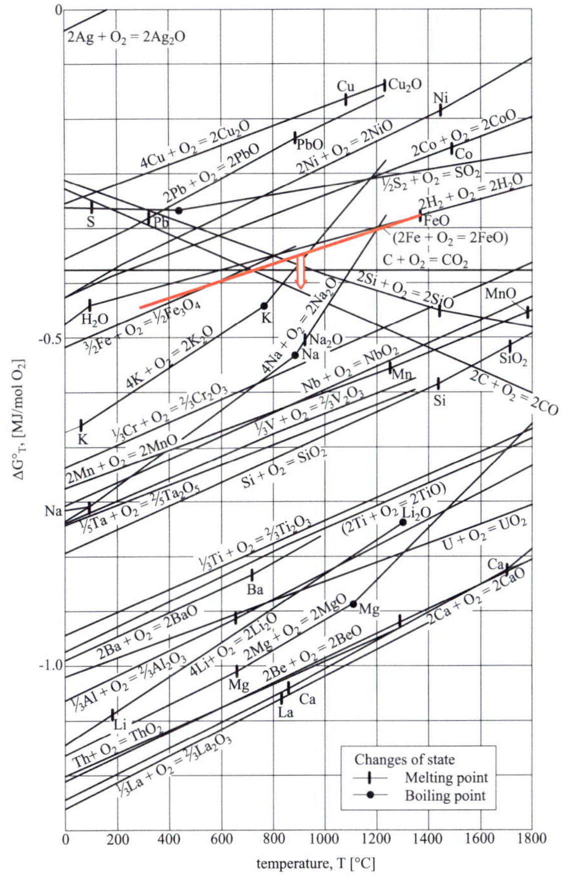

Fig. (2). Ellingham- Richardson diagram [3 - 5].

These elements are present in each steel (scrap in the furnace), others are in steel in such low concentrations and have such a high affinity for oxygen that they have already passed into the slag during melting.

Oxidation of Silicon

Silicon is completely soluble in liquid iron, according to the reaction:

$$Si_{liq} = [Si]\ \Delta G^{o} = (5.54\ T{-}28\ 000)\cdot 4.184\ [J/mol]\ [3, 7] \quad \textbf{(1)}$$

which generates significant amount of heat. In the alloys of Fe-Si, silicon activity has a negative deviation from Raoult`s law, which indicates a tendency to form compounds. Silicon dissolved in liquid steel reacts both with components of slag and oxygen, according to the reaction:

$$[Si] + 2(FeO) = (SiO_2)+2[Fe] \quad \textbf{(2)}$$

$\Delta G^{o} = 98{,}13\cdot T{-}299\ 960$ [J/mol]

$[Si] + 2[O] = SiO_2\ \Delta G^{o} = 218.2\cdot T{-}576\ 440$ [J/mol] [7],

$lgK_{Si} = lg\ a_{SiO2} / a_{[Si]} \cdot a_{[O]}^{2} = 30\ 110/T{-}11.40$

$lgK_{Si} = 30\ 410/T{-}11.59$ [12a]

[Si] means that silicon is in the melt

(SiO_2) silicon oxide is in the slag

{CO} carbon oxide is in the gas phase

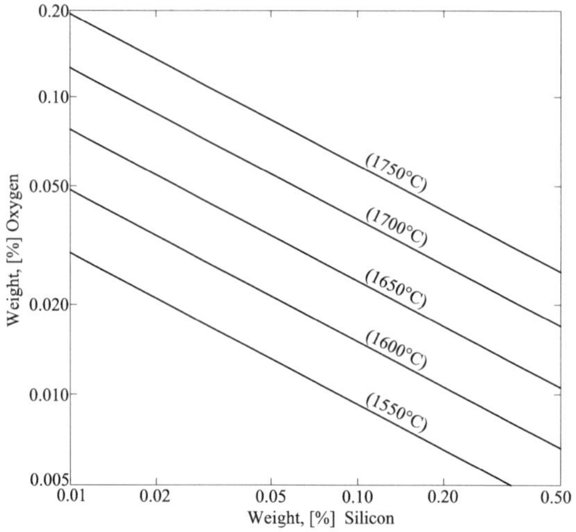

Fig. (3). Effect of temperature on the dependence of silicon content from oxygen content in liquid iron at 1600°C [6].

Knowing the activity coefficients of oxygen and silicon, the equilibrium concentration of oxygen with silicon for the elected temperatures in the Si-O system is shown on Fig. (**3**). According to [7], at 1600°C and at a concentration of 0.020% oxygen, the silicon concentration equals about 0.12%. In fact, at the end of oxidation, there is a trace content of silicon (less than 0.05%) in the metal bath. In the bath, there is so much oxygen that during the following smelting period, its reduction (deoxidation) must be performed, since in steel casting, the total oxygen content (dissolved in metal and in the form of non-metallic inclusions) should not be more than 40 -50 ppm (0.005%).

Oxidation of Manganese

The Fe - Mn system is an ideal solution and in the whole concentration range, the coefficient of manganese activity γ_{Mn} equals 1.

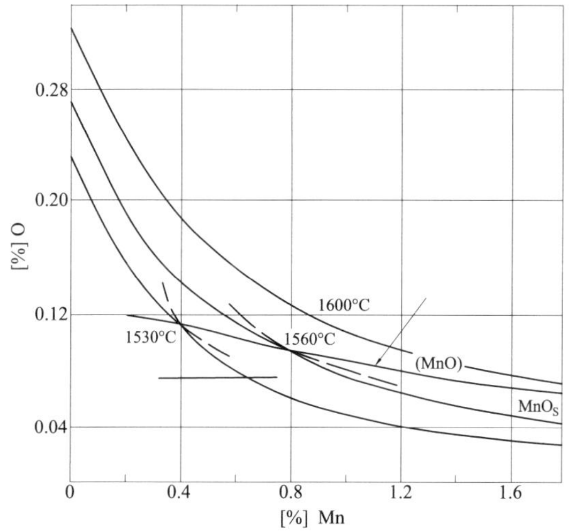

Fig. (4). Effect of temperature on the oxidation of manganese.

For the solution of Fe -1%Mn:

[Mn] = [%Mn] the value $\Delta G^o = (-9.12 \cdot T) \cdot 4.184$ [J/mol] [3, 7]

and for the reaction at the metal - slag boundary

$$[Mn] + (FeO) = (MnO) + [Fe] \qquad \textbf{(3)}$$

$\Delta G^o = 56.1 \cdot T - 123\ 300$ [J/mol] [7a]

while inside the bath (*e.g.* during precipitation deoxidation):

$$[Mn] + [O] = MnO_s \quad \textbf{(4)}$$

$\Delta G^o = 124.0 \cdot T - 282\ 000$ [J/mol] [7a]

Manganese oxide exists in liquid state (MnO) or in stable form MnO_s at lower temperature (Fig. **4**).

The course of these reactions means that they produce less heat than the oxidation reaction of silicon. Both, however, lead to an increase in bath temperature and silicon - to a reduction in slag basicity (B). Therefore, during the oxidation period, new portions of lime should be thrown into the slag.

The course of manganese oxidation can be estimated from the ratio of (MnO)/(FeO) in slag and the temperature value. From the dependence given by [3].

$[\%Mn] = 1/K_{Mn} \cdot(\%MnO)/(\%FeO)$ (4`)

It appears that the higher the content of (FeO) in the slag, the lower the content of [Mn] in the bath. This results in the transition of manganese into the slag.

Technical slags do not correspond to the ideal FeO-MnO system. Deviations are characterised by the activity coefficients γ_{MnO} and γ_{FeO} presented in the equation: $K`_{Mn} = K_{Mn}\, \gamma_{FeO} / \gamma_{MnO}$. Changes in apparent equilibrium constant for basic and acid slags indicate that with basic slags, the equilibrium manganese content is higher than with acid slags.

Although pure Fe-Mn system are the ideal solutions, activity of manganese is subject to significant changes in the presence of carbon. According to H.Schenck and F.Neumann [8] with increasing carbon concentration, there is a negative deviation from the ideal solutions.

Oxidation of Phosphorus - Dephosphorization

The content of phosphorus in each cast steel grade should be as small as possible, although most frequently, up to 0,025 - 0.030% is allowed. In castings of the highest quality, the phosphorus content is typically below 0.015 - 0.020%. In the charge for arc furnaces, the initial phosphorus content typically ranges between 0.04 and 0.08%. Phosphorus is responsible for a reduction in the plasticity of cast steel, especially at minus temperatures (Fig. **5**).

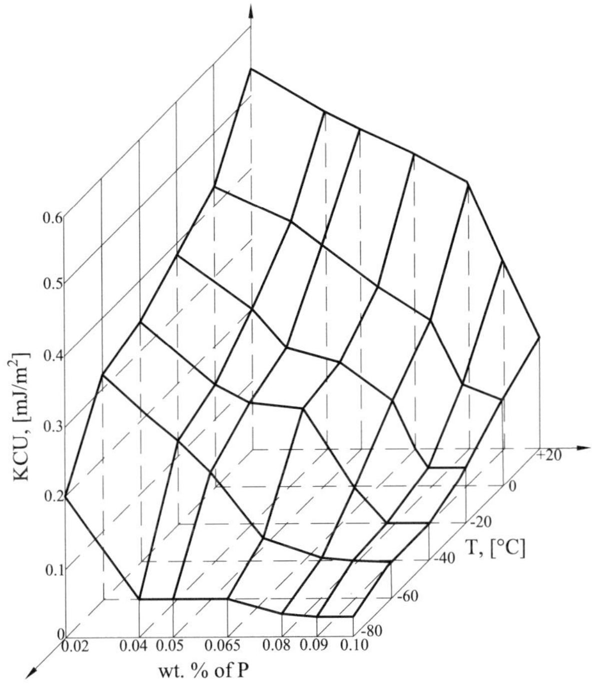

Fig. (5). Effect of P – contents on the impact toughness of carbon cast steel (0.35% C) after normalisation at 900°C and tempering at 600°C [9].

In the solid state, the following phosphides exist in steel: Fe_3P, Fe_2P, FeP; while Fe_2P is the most stable phosphide at elevated temperatures. In liquid iron, phosphorous occurs in the form of Fe_2P or in a dissolved form [P] (Fig. **6**) [9]. The maximum solubility of phosphorus in γFe equals ≈ 0.8% at 1150°C. The maximum solubility of phosphorus at ambient temperature equals to 0.015%.

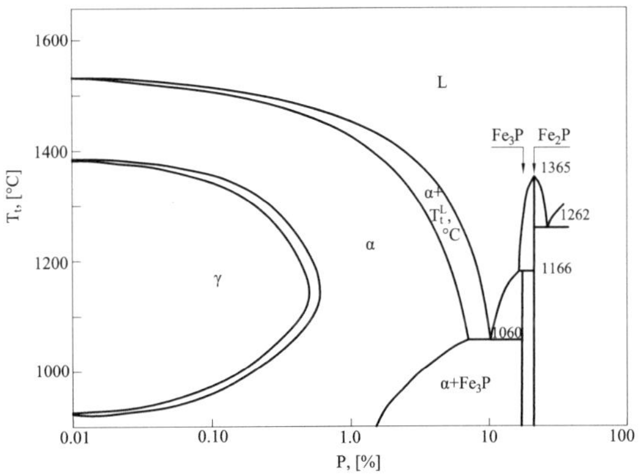

Fig. (6). The solubility of phosphorus in iron [9]

The highest tendency to form phosphides in iron is presented in:

Mg (ΔH^{o}_{298K} = 535.91 [kJ/mol]), Ca (502.42), Ti (265.44), Fe (121.41), Si (69.08), [kJ/mol] [9].

The solubility of phosphorus in liquid iron only depends on the partial pressure. Phosphorus has a low evaporation temperature ($T_{vol.}$ = 280°C [10]), which leads to the low concentration of phosphorous in liquid iron: under pure evaporation conditions - at least 0.030%, under conditions of evaporation and crucible reaction - 0.025% [11].

The removal of phosphorus from liquid steel is only possible in furnaces with basic lining.

According to the theory of molecular structure of slags, prior reactions of phosphorus oxidation in molten steel are described with the following reactions:

$$2[P] + 5[O] = (P_2O_5) \quad \textbf{(5)}$$

$K_P = a_{(P2O5)} / a^2_{[P]}\, a^5_{[O]}$

$\lg K_P = \lg(a_{P2O5)})/[a_p]^2 \cdot [a_O]^5 = 35\,880/T - 30.31$

$\Delta G^o = -164\,175 + 138.63 \cdot T$ [J/mol] [5]

$\Delta G^o = (-168\,600 + 133 \cdot T) \cdot 4.184$ [J/mol] [22]

$$2[P] + 5(FeO) = (P_2O_5) + 5[Fe] \quad \textbf{(6)}$$

$K_P = a_{(P2)} / a^2_{[P]}\, a^5_{(FeO)}$

Oxide (P_2O_5) is not stable at high temperatures and it interacts easily with the oxide (FeO), and then with the oxide (CaO). Therefore, the process of the dephosphorization reaction constitutes the following sequence:

$3(FeO) + (P_2O_5) = (3FeO \cdot P_2O_5)$

$3(CaO) + (P_2O_5) = (3CaO \cdot P_2O_5)$ $\Delta G^o = (18.16 \cdot T - 172\,800) \cdot 4.184$ [J/mol] [22]

$$4(CaO) + (P_2O_5) = (4CaO \cdot P_2O_5) \quad \textbf{(7)}$$

4(CaO P_2O_5) is the most stable compound. In order to perform an optimum dephosphorization, a proper amount of free (CaO) in the slag is required. The point here is not only about the share of (CaO) in binding the oxide (P_2O_5), but also reactions with (SiO_2) and (MnO), which are always present in the slag.

The final dephosphorization reaction is typically presented as:

$$2[P] + 5(FeO) + 4(CaO) = (4CaO \cdot P_2O_5) + 5[Fe] \quad \textbf{(8)}$$

and the reaction constant as:

$K_P = a_{(4CaO \cdot P2O5)} / a^2_{[P]} \, a^5_{(FeO)} \, a^4_{(CaO)}$

lg K = 40 067/T−15.06 [21, 23]

This equation shows that the reaction of phosphorus oxidation occurs when the following conditions are fulfilled:

1. a low temperature (see Ellingham - Richardson diagram),
2. the slag is oxidised (contains a lot of (FeO)),
3. the slag is strongly basic (B > 2.5-3.0),
4. the concentration of the reaction product is low (small activity (4CaO P_2O_5) in slag),
5. a large bath-slag interface.

These conditions are well-illustrated in Fig. (**7**). The highest degree of dephosphorization $\eta_P = (P_2O_5)/[P]$ is obtained by the basicity B > 3.5-4.0, and by the content of (FeO) in the range of 14-16% [12, 13]. These conditions are obtained during oxidation by continuous putting of limestone and iron ore.

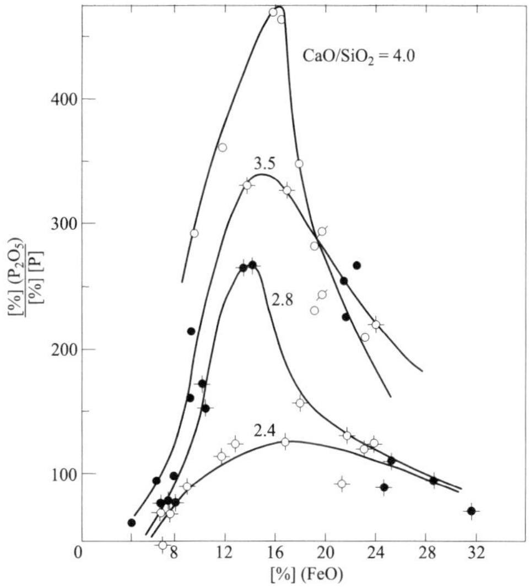

Fig. (7). Effect of basicity on the degree of dephosphorization [5, 8, 13].

The reduction of the ($4CaO\ P_2O_5$) content in slag takes place automatically when the slag flows out through the furnace's threshold during intensive oxidation. An introduction of new portions of limestone is then necessary and obvious.

According to recent studies [12, 13], the course of dephosphorization is described on the basis of the ionic theory of slag construction, as follows:

$$2[P] + 5[O] + 3(O^{2-}) = 2(PO_4^{3-}) \qquad \textbf{(9)}$$

$K_p' = (PO_4^{3-})^2 / [\%P]^2 \cdot [\%O]^5 \cdot (O^{2-})^3$

According to H. Flood and K. Grjotheim [11], the constant

$K_p' = \Sigma N_M^{2+} \cdot lgK_M^{2+}$

where: N_M^{2+} - ionic fraction of a basic oxide cations,

K_M^{2+} - equilibrium constant for reaction phosphorus – cations

and is dependent on the share of cations in basic slag:

$$\lg K_p' = 21N_{Ca}^{2+} + 18N_{Mg}^{2+} + 13N_{Mn}^{2+} + 12N_{Fe}^{2+} \qquad \textbf{(10)}$$

This equation indicates the correct order of the activity of basic cations during dephosphorization of molten iron. E.T. Turkdogan and J.Pearson [14] introduced a similar equation, taking into account the effect of silicate anions:

$$\lg K_p'' = 22N_{CaO} + 15N_{MgO} + 13N_{MnO} + 12N_{FeO} - 2N_{SiO2} \qquad \textbf{(11)}$$

This equation is closer to the experimental studies [7], and by taking silicon anions into account, it indicates the share of free basic ions in the course of the phosphorus oxidation reaction.

Reversion of Phosphorus into the Molten Steel

In practice, there are several cases of an increase in the phosphorus content in steel after tapping into the ladle. This most often concerns the following cases:

• the slag was not completely slag-off after the oxidation period, leaving some amounts of $4(CaO\ P_2O_5)$ from the oxidising slag in the slag of the refining period.

In the case of an incomplete slag-off, the conditions of phosphorus oxidation change according to the Ellingham - Richardson diagram.

In this regard, after introduction of the mixture of CaO + C+ Si (FeSi) + CaF_2 (in the ratio 4:1:1:1) into the new slag, the following reactions may occur:

$$2(P_2O_5) + 5Si = 5(SiO_2) + 4[P] \quad \textbf{(12)}$$

$$(P_2O_5) + 5C = 5\{CO\} + 2[P] \quad \textbf{(13)}$$

$$3(P_2O_5) + 10[Cr] = 5(Cr_2O_3) + 6[P] \quad \textbf{(14)}$$

The latter reaction takes place during smelting of corrosion-resistant steel *e.g.* Cr18-Ni9; Ni25-Cr20-Mo or after introducing scrap of these cast steel grades into the charge.

- there was an increase in the temperature of liquid steel in the reducing period,
- when tapping steel from the furnace into the ladle with large amounts of slags, another increase in the concentration of phosphorus is also possible,
- precise tapping of steel and the use of synthetic slags into the ladle reduce the risk of phosphorus increase.

Oxidation of Carbon

This is one of the most important reactions during the oxidation period, although it takes place after oxidation of silicon and manganese. The Ellingham - Richardson diagram shows that it occurs at higher temperatures than the oxidation of Si and Mn. Hence, there are conditions in which no oxidation of carbon occurs after throwing ore into the bath. If there is scrap with very high silicon content in the charge, the silicon oxidises according to the reaction (2) until it reaches the amount of about 0.05% Si. Only then does the boiling of the bath occur, *i.e.* reaction:

$$[C] + [O] = \{CO\} \quad \textbf{(15)}$$

$$K_{C\text{-}O} = p_{\{CO\}} / [a_{[C]}]\,[a_{O]}] = p_{\{CO\}} / [\%C]\cdot f_C\cdot[\%O]\cdot f_O$$

The free energy change of this reaction has the number of values, *e.g*:

$\Delta G^o = (27.400 - 27.07\cdot T)\cdot 4.184$ [J/mol] $\lg K_{C\text{-}O} = -\,5.986/T + 5.917$ [14]

$\Delta G^o = (33.300 - 30.40\cdot T)\cdot 4.184$ [J/mol] $\lg K_{C\text{-}O} = -\,7.280/T + 5.917$ [15]

The differences between these and others values [7] are small near the melting temperature of iron, and relatively higher at temperatures of 1650-1700°C.

The carbon activity of liquid iron is proportional to its mole fractions, so at $p_{\{CO\}} = 1$ atm., it can be noted:

$$K' = m = 1/ [\%C]\cdot[\%O] \qquad \textbf{(16)}$$

At 1600°C, value m equals 0.0025 - 0.0020 [16] and is presented like in Fig. (**8**).

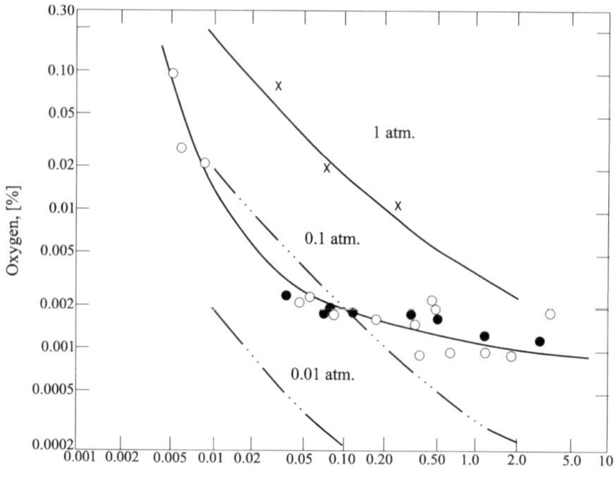

Fig. (8). Oxygen-carbon equilibrium (o-o) and effect of vacuum pressure (atm.) [13]

The solubility of both carbon and oxygen in liquid iron increases with increasing pressure. This effect is used during the process of smelting steel in a vacuum because a removal of the reaction product {CO} leads to a continuous reduction of the oxygen concentration (deoxidation in a vacuum).

A removal of the carbon oxidation reaction product in electric arc furnaces is associated with difficulties in the nucleation of gas bubbles in liquid steel. A lack of interfacial tension and atmospheric pressures of the metal bath forces the nucleation of {CO} in the existing pores of the furnace walls and the hearth lining. An outflow of {CO} bubbles is possible only after exceeding a certain pressure $p_{\{CO\}}$

$$p_{\{CO\}} = p_{atm} + p_{fer} + 9.87 \times 10^{-7} (2\sigma/r) \qquad \textbf{(17)}$$

where: p_{atm} - atmospheric pressure,[Pa]

p_{fer} - ferrostatic pressure above the bubble

$p_{fer} = h_m \cdot \gamma_m + h_s \cdot \gamma_s$

h_m and h_s are the depth of the melt and slag, [m]

γ_m and γ_s – metal and slag densities, respectively, [kg/m^3]

r – radius of the bubble, [m]

σ – surface tension between bubble and the refractory material, [mN/m]

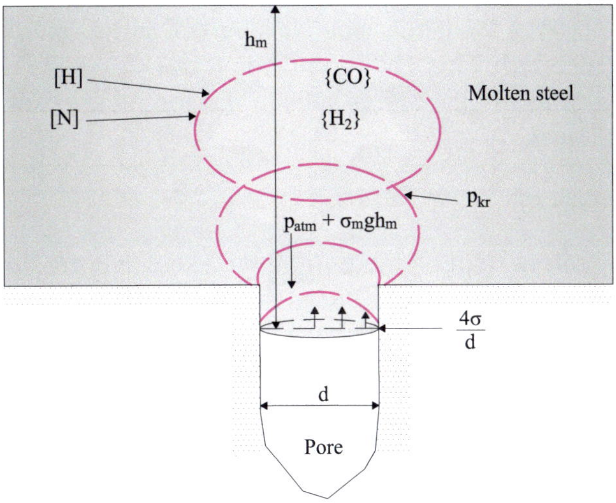

Fig. (9). Increasing of pore diameter at the lining-molten steel interphase.

This means that these bubbles are only capable of leaving the pore and rise up in the melt of which pressure exceeds the critical radius. The bubbles increase their radius by the diffusion of oxygen or carbon into the bubble, or by collecting other bubbles.

A refractory surface favours the nucleation of {CO} bubbles. If the furnace hearth is coated with a layer of viscous slag, the rate of decarburisation is very low.

The oxidation rate of carbon during bath oxidation is described by the following equation:

$$-d[\%C]/dt = D\,(12[\%O_{eqil.}]-[\%O])/16\cdot l\cdot \Delta x) \qquad \textbf{(18)}$$

where: [%O] – actual oxygen content,

[$O_{equil.}$] – oxygen content in the melt at equilibrium with iron oxide in the slag,

D – diffusivity of oxygen in liquid iron,

l - depth of the melt,

Δx – film thickness,

t - time.

The oxidation process ends when in the sample taken for chemical composition the phosphorus content are much lower than the value required by the standard (usually lower by 0.01% P). Then, slag is slag-off and a new one consisting of CaO and CaF_2 is inserted. This change means a transition from the oxidation to the reduction conditions of slag and is substantial for subsequent desulphurization and deoxidation of steel.

Oxidation of Chromium

Chromium is a very important element in alloyed steels: corrosion-resistant, creep resistant and heat resistant steels. Hence, there is a possibility of a presence of certain amounts of chromium in the charge from these cast steel grades. For this reason, chromium is also expected in molten steel. The liquid alloys Fe - Cr create ideal solutions [17] and for 1% of chromium concentration, it can be written:

$$Cr_{liq.} = [\%Cr] \qquad \textbf{(19)}$$

or $Cr_s = [\%Cr]$ $\Delta G^o = 4\ 600 - 11.13 \cdot T$ [J/mol] [3]

During melting and oxidation periods, chromium is oxidised according to the reaction:

$$Fe + 2[Cr] + 4[O] = FeO \cdot Cr_2O_{3(s)} \qquad \textbf{(20)}$$

$\Delta G^o = -\ 102\ 700 + 438.8 \cdot T$ [J/mol] [5]

$lgK = 53\ 420/T - 22.92$

$$2[Cr] + 3[O] = Cr_2O_{3(s)} \qquad \textbf{(21)}$$

$\Delta G^o = -\ 843\ 100 + 371.8 \cdot T$ [[J/mol] [12a]

$lgK = 44\ 040/T - 19.42$

$$3[Cr] + 4[O] = Cr_3O_{4(s)} \qquad \textbf{(22)}$$

$\Delta G^o = -\ 1\ 025\ 712 + 459.22 \cdot T$ [J/mol]

$lgK = 53\ 521/T - 23.96$

The arising chromium oxides exist in the form of suspension of solid oxides due to their limited solubility in the slag. Hence, these slags have a high density, reducing the kinetics of the oxidation reaction.

The scope of occurrence of chromium oxides in slag, depending on chromium, oxygen and the temperature, is presented in Fig. (**10**).

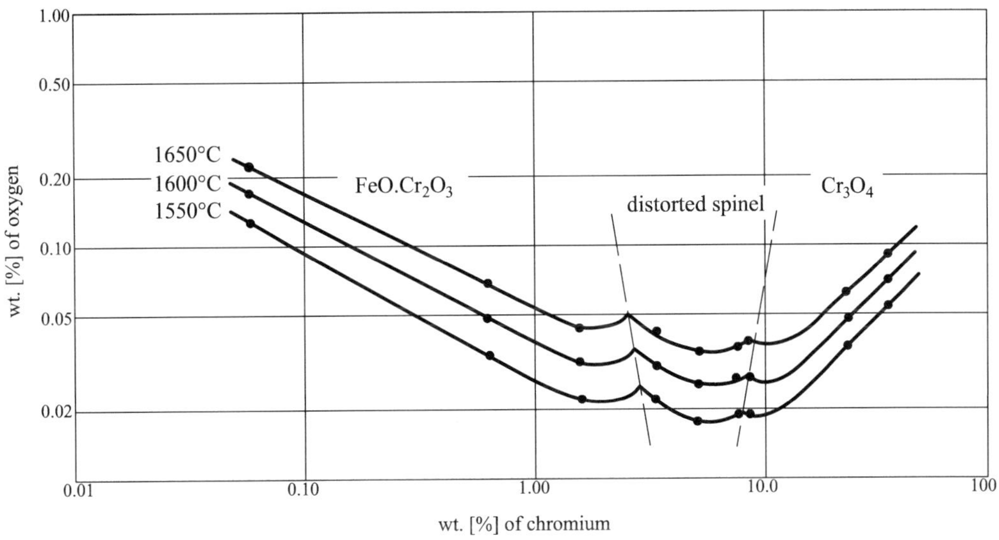

Fig. (10). Equilibrium of chromium and oxygen in the alloy Fe-Cr [18].

Oxidation of chromium with basic slags also takes place according to the equation:

$$2[Cr] + 3(FeO) = (Cr_2 O_3) + 3[Fe] \quad \textbf{(23)}$$

$K_{Cr} = a_{(Cr2O3)} / a^2_{[Cr]} \cdot a^3_{(FeO)}$

The course of this reaction depends on the slag basicity, and temperature according to $lgK`_{Cr} = 14\ 500/T - 3.57$. The higher the temperature, the higher the oxidation of chromium, Fig. (**10**). A low content of (FeO) in slag and high basicity act in the same direction.

The activity of chromium in the liquid alloys Fe - Cr depends on the presence of other elements in them. First of all, an important role is played by carbon, which increases the deviation of chromium from the ideal solution.

The activity of carbon in the presence of chromium is described by the following equation [5]:

$lg\gamma_c^{(Cr)} = -5.5\ lg(1 + 0.8 \cdot N_{Cr} / 1 - N_c)$

which shows that carbon activity decreases with an increasing chromium content, Fig. (**11**).

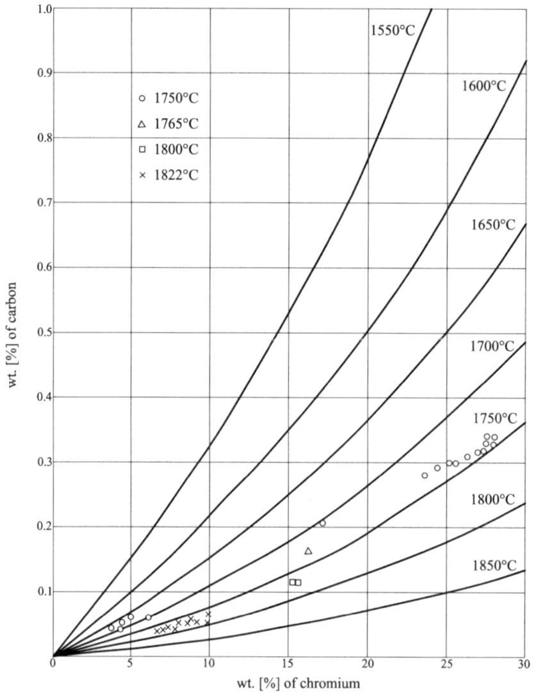

Fig. (11). Activity of chromium in molten Fe-Cr-C alloys at 1540°C [19].

For practical reasons, it is important to keep low chromium contents in alloys containing carbon. In liquid alloys Fe - Cr – C, the oxidation of chromium can be avoided if the bath contains less than 0.5%C (at the temperature of 1600°C) or less than 0.13%C (at the temperature of 1800°C).

The degree of chromium oxidation ($Cr_{start} - Cr_{finish}$), before and after oxidation, depends on the degree of decarburisation of the bath under conditions of oxidation with oxygen in the electric arc furnace. At higher chromium contents before oxidation, the degree of chromium oxidation is lower. This means that during melting corrosion-resistant steels (18-20%Cr + 9 -11% Ni and low carbon content <0.03%), in order to prevent large losses of chromium, the process of carbon oxidation should be performed at high temperature (from 1750 to 1800°C) and with slags of high basicity. Similar conditions are fulfilled when a mixture of oxygen with argon is blowing, or when vacuum is applied.

2.5. Refining Period

The refining period begins with applying new slag ($CaO + CaF_2$). Initially, it is the slag that protects the liquid steel against oxidation and heat loss. During refining, its task is modified after the introduction of the so-called deoxidizing mixture, and it becomes reduction slag, which enables the desulphurization and deoxidation of the bath. The whole refining period includes:

- desulphurisation of steel,
- diffusion and precipitation deoxidation of steel,
- supplementing the chemical composition to the values required by the standard,
- increasing the bath temperature to the value required for tapping steel into the ladle.

Diffusion deoxidation and desulphurization processes take place at the metal - slag boundary, hence the chemical composition of slag is important. Both of these processes depend on the oxygen equilibrium in the bath, with the iron oxide in the slag. Therefore, it is so important to completely slag-off after the oxidising period, in which (FeO) equals 10 - 15%, and at the end of the refining period, it should be below 1%.

Desulphurisation

Sulphur solubility in liquid iron is limited (Fig. **12**), and in the solid state, it is insignificant. In αFe, up to 0.020% S is dissolved (at the temperature of 913°C), and in γFe - up to 0.050% S (at the temperature of 1365°C). Solubility of sulphur in γFe is less than in αFe (at 913°C).

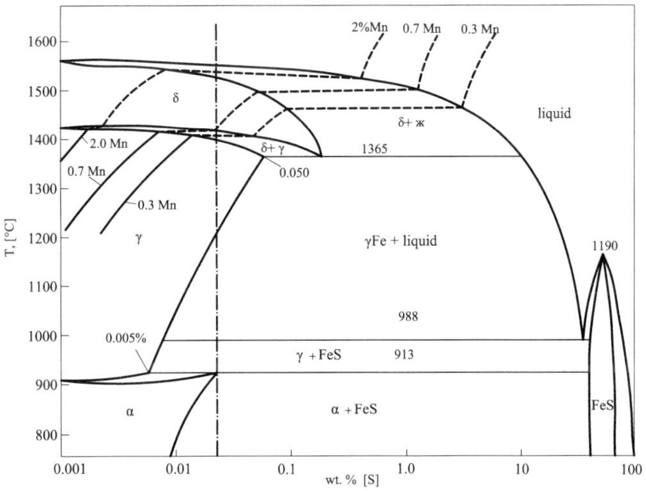

Fig. (12). System Fe – S in half-logarithmic scale; effect of manganese – dash line [9].

FeS sulphide arises at this temperature. The presence of other elements in steel leads to the formation of additional sulphides, which can be divided into three groups, Fig. (**13**):

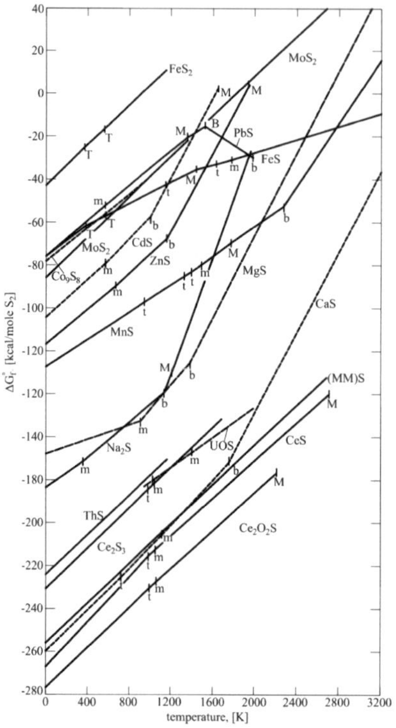

Fig. (13). Free Energy formation of sulphides in temperature [9, 10].

- sulphides less stable than FeS
  ($NiS-T_m = 992°C$; $NiS_2-T_m = 1010°C$; $CoS - T_m = 1180°C$; $Co_9S_8-T_m = 832°C$
  WS_2-T_m above 1800°C; $Cu_2S - T_m = 1129°C$
- elements that form compounds with iron and reduce solubility of FeS in liquid iron (C, Si, P)
- sulphides more stable than FeS
  $CeS-T_m = 2450°C$; $Ce_2S_3-T_m = 1890°C$; $Ce_3S_4 - T_m = 2080°C$;
  $LaS - T_m = 2200°C$; $LaS_2-T_m = 1650°C$; $CaS - T_m = 2525°C$; $MnS-T_m =$ 1610°C

The most stable sulphides form rare earths (mischmetal) elements, especially over 1500°C.

At lower temperatures, the energy of steel desulphurization consists of two steps:

- diffusion of iron sulphide from the liquid steel into the slag

[FeS] → (FeS)

- reaction of iron sulphide in the slag with basic oxides

$$[FeS] + (CaO) = (CaS) + (FeO) \quad \textbf{(24)}$$

or [FeS] + (MnO) = (MnS) + (FeO)

The product of these two reactions (FeO) is an unwanted component of slag because it hinders the desulphurization process. Diffusion deoxidation, proceeding in parallel with desulphurization in the refining process, shifts the desulphurization reaction to the right:

(FeO) + C = {CO}+ [Fe]

This way, some part of iron is recovered and the course of the reaction is visible due to the oxidising and burning bubbles of carbon monoxide:

$\{CO\} + \frac{1}{2} O_2 = \{CO_2\}$

Total desulphurization reactions can be presented as:

$$[FeS] + (CaO) + C = (CaS) + \{CO\} + [Fe] \quad \textbf{(25)}$$

this is the course of the desulphurization reaction after putting the reduction slag CaO + Fe-Si + C

or

$$[FeS] + (CaC_2) + 2(FeO) = (CaS) + 2\{CO\} + 3[Fe] \quad \textbf{(26)}$$

$\Delta G_o = 31\,140 - 34.4 \cdot T$ [J/mol]

The reaction is endothermic and requires a higher bath temperature [13].

The reaction constant $K_{24} = a_{(CaS)} \cdot a_{(FeO)} / a_{[FeS]} \cdot a_{(CaO)}$ indicates that, in order to ensure the course of desulphurization reaction, activities of (CaS) and (FeO) in the slag should be low, while the activity of sulphide [FeS] in the bath and the activity of (CaO) in the slag should be high. These conditions lead to statements that desulphurization of the bath is favoured by:

- high basicity of slag Fig. (**14**), (high activity of oxygen anions in slag),

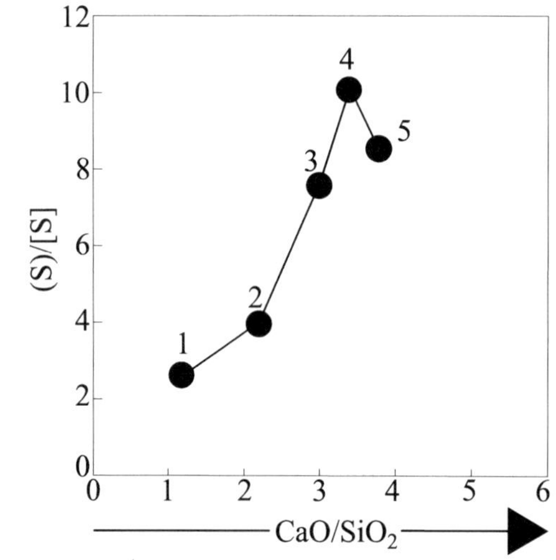

Fig. (14). Effect of basicity B = CaO/SiO_2 on the degree η_s = (S)/[S]

- low content (activity) of iron oxide in the slag,

- high temperature,

- low concentration of the reaction product (CaS) in the slag,

- low viscosity of slag.

High slag basicity (B = 3.5 - 4.0) is obtained by continuous addition of new portions of CaO, along with CaF_2 fluorspar, which reduces the viscosity of slag, and thus, accelerates the diffusion processes of sulphur transfer at the slag-metal interface.

For a proper course of desulphurization, a small activity of (FeO) in the slag should be provided. When introducing new portions of limestone, the activity of CaO increases and the activity of (FeO) decreases. The desulphurization degree, referred to as η_s= (S)/[S], is strictly dependent on the content of (FeO) in the slag. At the content of (FeO) below 1%, conditions for the transfer of sulphur from metal into the slag are more beneficial.

Low activity of (FeO) in the slag corresponds to the low oxygen activity in metal. According to the ionic theory of slag, the desulphurization reaction can be written as follows:

$$[S] + (O^{2-}) = [O] + (S^{2-}) \quad \textbf{(27)}$$

$\Delta G^{o} = 17\ 200 - 9.12 \cdot T$ [J/mol] [21]

$\lg K_s = \lg (a_{s2-}) \cdot [a_o] / (a_{o2-}) \cdot [a_s] = -3\ 750/T + 1.996$

and according to [10] constant $K' = N_{(S2-)} / N_{(O2-)} \cdot [\%O]/[\%S]$

whose value of the sulphur separation constant value η_s as a result of (FeO) content in the slag is presented in Fig. (**15**).

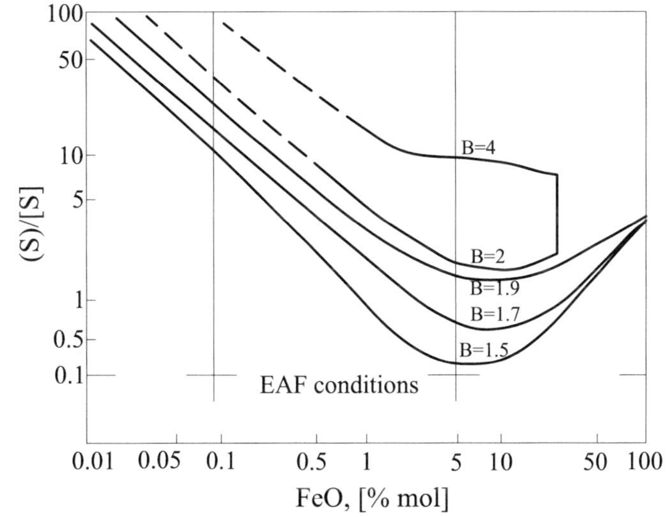

Fig. (15). Relation between (FeO) contents and degree of desulphurization η_s [7, 20]

$\eta_s = (\%S)/[\%S] = 32\ K' \cdot n_{(O2-)} / [\%O]$

In reduction slags of the refining period, low concentrations of (FeO) in the slag are achieved by applying the deoxidising mixture

$CaO + FeSi + C + CaF_2$ in the ratio 4:1:1:1 [3, 13, 15]

or $CaO + CaF_2 + C$ 6:2:1 [3]

or carbide slag

$CaO + CaF_2 + C$ in the ratio 6:4:2 [3]

The significance of the slag deoxidation process in the course of desulphurization is expressed, in practice, with the dependence:

$[\%S] = 4\ [\%O]$ [3, 2, 13]

or $[\%S] = 30 \cdot a_{[O]}$ [21]

The desulphurization reaction is controlled by sulphur diffusion from metal into slag, and oxygen atoms through the metal - slag interphase boundary. Increasing the surface area is beneficial for the course of the desulphurization reaction, and is used in modern steel smelting methods. Because these processes are slow, particularly in the arc furnace, there are methods for their acceleration by slag foaming, installation of an induction coil in the furnace walls to force the movement of liquid steel, or the application of a vacuum and blowing of Fe-Ca-Si powder with argon into the liquid steel.

CONFLICT OF INTEREST

The author (editor) declares no conflict of interest, financial or otherwise.

REFERENCES

[1] M. Jellinghaus, "Stahlerzeugung im Lichtbogcnofcn, 3.Auflagc", *Stahl Eisen,* 1994.

[2] S. Sobula, *Metody wytapiania stali na odlewy, Poradnik Odlewnika* II. 6th. Polish Foundrymen`s Association, 2013.

[3] K. Leitner, and E. Ploeckinger, *Die Edelstahlerzeugung.* Springer- Verlag: Wien, 1965.

[4] F.D. Richardson, and J.H. Jeffes, "The Thermodynamics of Substances of Interest in Iron and Steel Making from 0-2400oC, I.Oxides", *J. Iron Steel Inst.,* vol. 160, pp. 261-270, 1948.

[5] F.D. Richardson, *Physical Chemistry of Melts in Metallurgy.* vol. 1-2. Academic Press INC: London, 1974.

[6] J. Chipman, J.C. Fulton, N.A. Gocken, and G.R. Caskey, "Activity of silicon in liquid Fe-Si and Fe--Si alloys", *Acta Metall.,* vol. 2, p. 439, 1954. [http://dx.doi.org/10.1016/0001-6160(54)90064-4]

[7] C. Bodsworth, *Physical Chemistry of Iron and Steel Manufacture.* Longmans: LondonLondon, 1963, 1973.C. (a). Bodsworth, and H.B. Bell, *Physical Chemistry of Iron and Steel Manufacture.* Longmans: LondonLondon, 1963, 1973.

[8] H. Schenck, and G.H. Gerdom, "Equilibria of Carbon and Oxygen in Liquid Iron and the Effect of Low Silicon Contents Upon the Equilibria", *Arch. Eisenhuettenwesen,* vol. 79, no. 8, pp. 451-460, 1959.

[9] W.W. Lunev, and W.W. Averin, *Siera i Fosfor w Stali, Metalurgija.* Moskwa, 1988. (in Russian)

[10] *ASM Materials-Metalworking.* Technology Series: Metal Park, Ohio, 1975.

[11] W.A. Fischer, and A. Hoffmann, "Einfluss von Kohlenstoff und Sauerstoff auf das Verhalten einiger Begleitelemente in Eisenschmelzen unter Hochvacuum", *Arch.Eis,* vol. 31, pp. 411-417, 1960.

[12] E.T. Turkdogan, "Chromium-Oxygen Equilibrium in Liquid Iron", *J. Iron Steel Inst.,* vol. 178, Gordon and Breach Science Publishers, p. 278, 19541988. *The Japan Society for the Promotion of Science.,* vol. 178, Gordon and Breach Science Publishers, p. 278, 19541988.

[13] F.P. Edneral, *Electrometallurgy of Steel and Ferro-alloys* Mir Publishers: Moscov, 1979.

[14] E.T. Turkgodan, and J. Pearson, "Activities of constituents of iron and steelmaking slags.Part I", *J. Iron Steel Inst.,* vol. 173, pp. 217-223, 1953.

[15] A. Rist, and J. Chipman, *Physical Chemistry of Steelmaking.* JohnWiley: N.Y., 1958.

[16] H.C. Vacker, and E.H. Hamilton, "The carbon-oxygen equilibrium in liquid iron", *Amer. Inst. Min. Met. Eng., Iron and Steel Div.,* vol. 95, pp. 124-140, 1931.

[17] M. Hansen, and K. Anderko, *Constitution of Binary Alloys.* 2nd ed. McGraw Book Comp. Inc.: London, 1958.

[18] D.C. Hilty, W.D. Forgeng, and R.C. Folkman, "Oxygen-Solubility and Oxide Phases in Fe-Cr-O System", *J.Metals Trans,* vol. 7, pp. 253-268, 1955.

[19] M. Ohtani, and N.A. Gokcen, "The equilibrium controlling the decarburization of iron-chromiu--carbon-melts", *J. Iron Steel Inst.,* vol. 175, pp. 264-266, 1953.

[20] N.A. Gokcen, "Equilibria in reaction of hydrogen and carbon monoxide with dissolved oxygen in liquid iron, Trans.Amer.Inst.min.(metal.)", *Engrs,* vol. 206, no. 8, pp. 1558-1567, 1956.

[21] J. Senberger, Z. Buzek, A. Zadera, K. Stransky, and V. Kafka, *Metalurgie Oceli na Odlitky.* VUTIUM: Brno, 2008. (in Czech)

[22] L. Coudurier, D.W. Hopkins, and I. Wilkomirsky, *Fundamentals of Metallurgical Processes.* Pergamon Press: Oxford, 1978.

[23] T. Winkler, and J. Chipman, "An Equilibrium study of the distribution of phosphorus between liquid Iiron and basic slags", In: *Trans. AIME* vol. 167. , 1946, pp. 111-133.

CHAPTER 3

Slags

Abstract: Slags play an important role in melting processes (dephosphorization, desulfurization and deoxidation) and during tapping of steel in the ladle. During melting of steel in basic EAF, the slags change the chemical composition in each melt. These aspects govern the quality of molten steel for castings and in consequence – the properties of each grade of cast steel. The structure of multicomponent systems, the structure of slags according to molecular or ionic theories, composition of slags during oxidation and reduction periods of melting and calculation of their basicity are shown.

Keywords: Basicity of slags, EAF slags, Multicomponent systems, Slag properties, Synthetic slags.

1. INTRODUCTION

Knowledge of the chemical composition, properties and structure of slags are of significant importance in steel smelting processes. In the steel smelting process, chemical composition of the slag is continuously changing due to the oxidation of iron and components of the metal bath, the increase in the bath temperature, dissolution of slag-forming additions and erosion of the refractory lining. Forming slag is dependent on the steelmaking process (a converter, arc or induction furnace) and lasts up to 10 minutes in an oxygen converter process, but from 1.5 to 2 hours in an electric arc furnace. Slags play a special role in dephosphorization, desulfurization and diffusion deoxidation of steel. They also have a significant effect in the process of removing non-metallic inclusions in the reduction process, argon treatment of steel in the ladle or in the secondary oxidation of steel during its pouring and filling the mould cavity. In a metal - mould reaction, slag residues in liquid steel also play an important role. In steels intended for castings, their importance is even more crucial due to the lack of possibility of fragmentation of the inclusions after the solidification of steel, as it is possible during rolling or forging steel. For this reason, the proper selection of slag in the smelting process, and the argon treatment of steel, is of particular importance in obtaining correct properties of cast steel. Steel slags comprise multicomponent systems based on oxides such as: MnO, SiO_2, Al_2O_3, Cr_2O_3, FeO, P_2O_5, CaF_2 and CaO. These are the most common components of triple, quadruple

Jan Głownia

and multi-component systems. In steel practice, slags comprise multi-component systems, and as imperfect solutions, it is difficult to calculate their equilibrium state [1 - 5].

The tasks of slag in the steel smelting process include:

- protection of liquid bath against heat loss (quick reduction in temperature) and protection against direct action of the furnace atmosphere,
- protection against oxidation of bath components (mostly silicon, manganese and the elements of which chemical affinity for oxygen is higher than for iron (see Ellingham - Richardson diagram, chapter 2), but also the removal of harmful elements of sulphur and phosphorus by reactions at the metal – slag boundary,
- enabling reactions of liquid steel - slag during the oxidizing period (decarburization and dephosphorization of the bath) and in the reducing period (desulphurisation and diffusion deoxidation of the bath),
- ensuring binds of non-metallic inclusions flowing out of the bath during precipitation deoxidation, argon stirring of the molten steel or other processes of the secondary metallurgy,
- binding of inclusions flowing out from the bath into slag by coagulation or coalescence.

These tasks can be fulfilled only if the slag satisfies relevant properties. These are: density, viscosity, surface tension, melting temperature, activity of individual components. These properties can be characterised as follows:

The melting temperature T_m [°C]- should always be lower than the temperature of liquid steel. Its value can be determined theoretically from the triple or the quadruple system (CaO-SiO_2-MnO-FeO), but it is always an approximate value due to the deviations of the actual chemical composition of slag from the theoretical balance system (in practice, there are more complex, multi-component systems), the effect of fluxing additions (fluorite, bauxite) or chemical composition of liquid steel (*e.g.* when melting chromium steel, slag contains certain amounts of Cr_2O_3).

Viscosity η [Pa.s] - is the fundamental physical property of molten slags. The activity of slag depends mainly just on the viscosity, which is a function of temperature T and chemical composition of the slag.

According to Arrhenius' formula, along with the increasing temperature, the viscosity decreases Fig. (**1**) as specified in the equation:

$$\eta = A \exp E / RT \qquad \textbf{(1)}$$

where: A – constant,

R - gas constant [$J \cdot K^{-1}\ mol^{-1}$]

T - temperature in Kelvin

E - activation energy of slag viscosity [J / mol]

The higher the viscosity, the slower is the rate of reaction between the bath and the slag. In metallurgical processes, a reduction in the viscosity means an increase in the rate of diffusion of slag components, thereby, an increase in the rate of chemical reactions between the slag and the bath. It also means an increase in slag fluidity.

Basic slags have a lower viscosity at higher temperatures, (Fig. **1**). Hence, in acid induction furnaces, slag located above the induction coil has high viscosity and retards reactions between the bath and the slag.

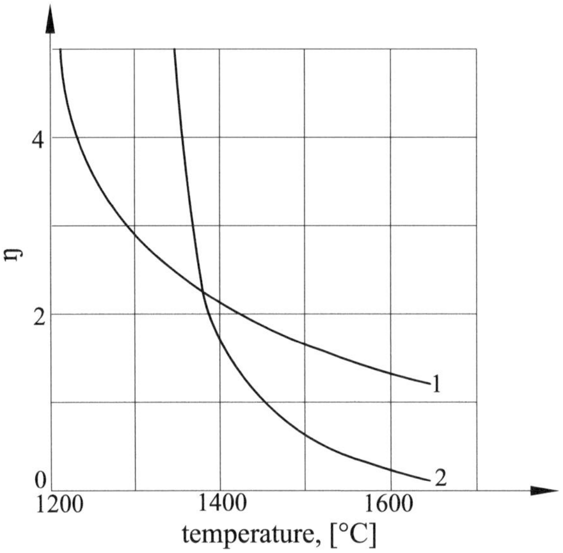

Fig. (1). Effect of temperature on the viscosity of slags [6; 1- for acid slags, 2 - for basic slags].

The viscosity depends on the structural properties of the bath, and especially the structure of liquid silicates [7, 8]. They create bonds of two ions SiO_4^{4+} connected by bridging with oxygen O^0 [8]. The addition of basic oxides (*e.g.* CaO) causes a change of their bonds by oxygen ions, which form free ions O^{2-}, Fig. (**2**). These changes cause a reduction in the viscosity of the slag. Acidic oxides SiO_2 and P_2O_5 contribute to the polymerisation of slags (increase viscosity) and are called network formers; and basic ions (CaO, MgO, Na_2O) - are called network modifiers. Since 2000, a lot of viscosity changes models, associated with modifications in the structure of slag, have been developed [9, 10].

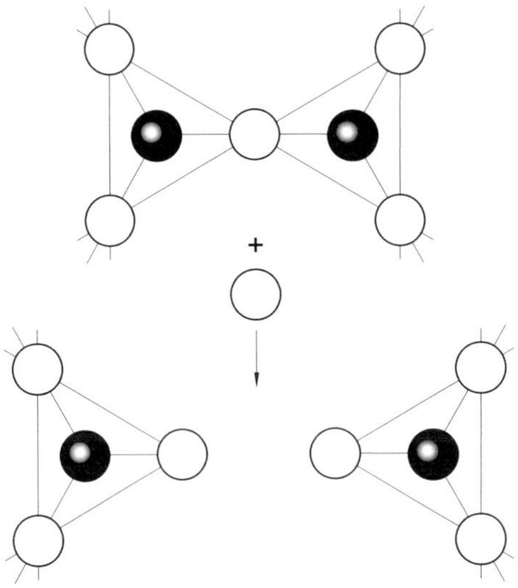

Fig. (2). Effect of CaO additions on the network structure of silicate slags [8].

Thermal conductivity q [W / m·°K] – is defined as

$$q = - k \cdot dT/dx \tag{1a}$$

where: q – heat flux [W/m^2],

k – thermal conductivity [W/m·K]

dT/dx – temperature gradient.

Heat transfer in liquid slag takes place mainly by convection. Metals generally have high thermal conductivity, contrary to ceramics and plastics. Carbon content and chemical composition of cast steel have high influence on the value of thermal conductivity. Temperature has also an important effect on thermal conductivity, spatially in heat and corrosion resistant cast steels [11].

Basicity of Slags B

Metal oxides are the basic ingredients of steel slags. At certain smelting periods, the chemical composition of slags varies substantially, leading to a change in their basicity. Depending on the chemical characteristics, the following oxides are distinguished:

basic: CaO, MgO, MnO, FeO, CrO, Na_2O, K_2O

acid: SiO_2, P_2O_5, TiO_2, MoO_3, V_2O_5, WO_3

amorphous: Al_2O_3, Fe_2O_3, Cr_2O_3

In addition to these simple oxides, more complex compounds can be found in slags:

silicates: $FeO{\cdot}SiO_2$, $MnO{\cdot}SiO_2$, $CaO{\cdot}SiO_2$, $MgO{\cdot}SiO_2$, $Al_2O_3{\cdot}SiO_2$

phosphates: $(FeO)_3{\cdot}P_2O_5$, $(CaO)_3{\cdot}P_2O_5$, $(CaO)_4{\cdot}P_2O_5$

aluminates: $FeO{\cdot}Al_2O_3$, $CaO{\cdot}Al_2O_3$

ferrites: $CaO{\cdot}Fe_2O_3$, $FeO{\cdot}Fe_2O_3$, $(CaO)_3{\cdot}Fe_2O_3$

The shares of each of these compounds in the slag depends on the chemical composition of smelted steel, ferroalloys (FeCr, FeMn, FeMnSi, FeCaSi) introduced, the furnace lining, the kind of ore used for oxidation of the bath and slag-forming materials introduced during smelting. A combination of some of them in the slag causes complex influence on the nature of the slag. With the existing advantage of one of these groups over the other, there are basic and acid slags distinguished. Slag is defined as [12]:

$$\text{basic when: } B => 1.3 \qquad \textbf{(2)}$$

acid when: $B < 1$

Due to the conditions of conducting a lot of steelmaking processes, in practice, the concept of the so-called slag basicity is applied, defined as:

$B = \%CaO/\ \%SiO_2$; $B` = CaO/SiO_2\ P_2O_5$ [13]; $B`` = CaO\ 1.4MgO/\ SiO_2\ 0.84\ P_2O_5$ [14, 22]

This definition assumes that the slag is composed of particles of oxides, and for practical reasons, it is applied by metallurgists, since it reflects metal - slag chemical reactions. Assuming the ionic structure of slags, the definition is replaced by the expression per the number of moles of free oxygen ions in the slag (described in the next sub-chapter 2).

2. STRUCTURE OF SLAGS

In the literature, there are two basic theories of slag structure: molecular and ionic. According to the former, in slags, there are more or less dissociated compounds, which form "free oxides" (CaO, FeO, SiO_2 or P_2O_5) [7]. Only such oxides participate in bath - slag reactions. In more recent expressions of this theory [6, 8], concentrations of "free oxides" are replaced by their activities.

According to the ionic theory, slags comprise a mixture of ions (positive, negative, simple and complex) and reactions between the liquid steel and liquid slag take place in accordance with ion reactions; *e.g*:

$$[Mn] + (Fe^{2+}) = [Fe] + (Mn^{2+}) \quad \textbf{(3)}$$

thus, basic oxides are dissociated into

$$(CaO) \rightarrow (Ca^{2+}) + (O^{2-}) \quad \textbf{(4)}$$

$$(FeO) \rightarrow (Fe^{2+}) + (O^{2-}) \quad \textbf{(5)}$$

and are donors of oxygen ions. In contrast, acidic ions are the result of dissociation of acidic oxides:

$$(SiO_2) \rightarrow (Si^{4+}) + (O^{2-}) \quad \textbf{(6)}$$

$$\Sigma(MO)/(SiO_2) > 2 \quad \textbf{(7)}$$

and neutral oxides:

$$(Al_2O_3) \rightarrow 2(Al^{3+}) + 3(O^{2-}) \quad \textbf{(8)}$$

In liquid slags, these oxides join oxygen anions (are oxygen ion acceptors). They form complex anions, according to the reactions:

$$(SiO_2) + 2(O^{2-}) \rightarrow (SiO_4^{4-}) \quad \textbf{(9)}$$

$$(P_2O_5)\ (O^{2-}) \rightarrow 2(PO_4^{3-}) \quad \textbf{(10)}$$

and

$$(Al_2\ O_3) + 3(O^{3-}) \rightarrow 2(AlO_3^{3-}) \quad \textbf{(11)}$$

In ionic theory, basic slag contains excessive oxygen anions (O^{2-}) and the number of free oxygen anions n_{o2-} is the basis for determining the basicity:

$$no^{2-} = n_{(CaO)} + n_{(MgO)} + n_{(FeO)} + n_{(MnO)} - 3n(Al_2O_3) - 3n(P_2O_5) - 2n(SiO_2) \quad \textbf{(12)}$$

where: n - the number of moles of the oxide.

At $n_{(o^{2-})}$ > the slag is basic; at $n_{(o^{2-})}$ < it is acid, and at $n_{(o^{2-})} = 0$ it is neutral.

The ionic theory of slag structure was developed in 1945 by M.Tiemkin [15] and modified in 1952 by H. Flood and K. Grjotheim [16]. According to M.Tiemkin, slag is an ideal solution that meets the following assumptions:

- the number of anions is equivalent to the number of cations,
- anions are surrounded only by cations, and *vice versa*,
- there is no interaction between unipolar ions,
- unipolar ions behave similarly in reactions with neighbours.

Concentrations (x) of anions A^- and cations B^+ of the compound are defined as:

$$X_{(A^-)} = n_A^- / \Sigma_n^{anions} \text{ and } X_{(B+)} = n_B^+ / \Sigma_n^{cations} \quad \textbf{(13)}$$

M.Tiemkin has also shown that the activity of ions is equal to their concentrations:

$$a_{(A+)} = x_{(A+)} \; a_{(B-)} = x_{(B-)} \quad \textbf{(14)}$$

and hence the **Tiemkin's right:** activity of the AB compound equals:

$$a_{AB} = x_{(A+)} \cdot x_{(B-)}$$

However, it is difficult to implement the Tiemkin's theory because:

- it is difficult to determine the ionic fraction of oxygen,
- as in steel practice, slags are not applied to perfectly ideal ionic solutions.

Modification of M.Tiemkin`s theory by K. Flood and K. Grjotheim [16] consists of the introduction of valence of ions to the formulas for ion fractions. From these assumptions, it can be concluded that, in the slag, there are interactions not only between ions of the opposite sign, but also between ions of the same sign, according to the equations:

ionic fraction of an anion B^-:

$$a_{(B-)} \approx x_{(B-)} = n_{(B^-)} \cdot z_{(B^-)} / \Sigma n_{(i^-)} \cdot z_{(i^-)} \quad \textbf{(15)}$$

ionic fraction of a cation A^+:

$$a_{(A+)} \approx x_{(A+)} = n_{(A^+)} \cdot z_{(A^+)} / \Sigma n_{(i^+)} \cdot z_{(i^+)} \quad \textbf{(16)}$$

3. SLAG PHASE SYSTEMS

The behaviour of multi-component slags encountered in steel-making processes is difficult for practical assessment. Therefore, more simple systems: double, triple or quaternary systems, are applied. They allow to:

- specify the melting temperature in solutions of oxides, sulphides or oxy-sulphides, depending on their chemical composition; for example, in a triple system CaO - FeO - SiO_2, which is the basis of a lot of metallurgical slags (Fig. **3**), it is possible to determine the melting temperature of the triple systems: $CaO{\cdot}FeO{\cdot}SiO_2$ or $CaO{\cdot}2FeO{\cdot}2\ SiO_2$.
- From Fig. (**3**), it can be therefore determined which compound has the lowest melting temperature. This is important for the practice of conducting metallurgical processes in an electric arc furnace, a converter or an induction furnace. Conducting *e.g.* dephosphorization or desulphurisation at more liquid slag accelerates each of these processes, reducing the cost of smelting.
- determine the range of occurrence of liquid phases,
- create certain chemical compounds and their stability in the temperature range (Fig. **3**)
- calculate solubility of components in the slag solution.

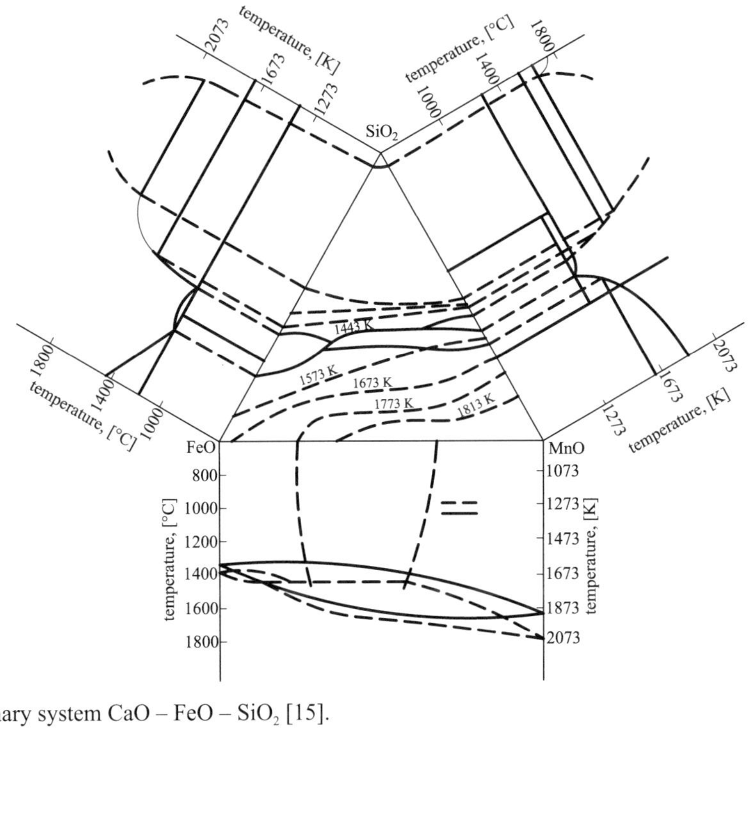

Fig. (3). Ternary system CaO – FeO – SiO_2 [15].

To reduce the temperature of basic slags, the addition of fluorite (CaF_2) was used, but because it decomposes at high temperatures, it is replaced by the addition of bauxite. In the system CaO - SiO_2 -CaF_2, there is eutectic with the melting temperature at 1055^{0}C [12, 14, 17].

It should be emphasized that each of the systems presented refers to the state of equilibrium and is valid only when the slag is in the state of equilibrium.

4. CHEMICAL COMPOSITIONS OF STEEL SLAGS

Starting from the period of melting, through the period of oxidation, refining, the chemical composition of slags is constantly changing in EAF. This composition is deliberately controlled to meet the required conditions in the metal bath and at the bath - liquid slag boundary. The course of the reaction is faster the higher the temperature, the lower the viscosity of the slag, the greater the area between the slag and bath (intensive oxidation), the higher the activity of the specified slag components.

Since there are no equilibrium conditions during intense reactions in technical conditions, the temperature distribution at the phase boundary and activity of the components (concentrations) play the greatest role. Processes to bring reagents to the metal - slag surface, and to transport them from the reaction site, determine diffusion processes. The chemical compositions of slags change in each melt after additions of iron ore or limestone in oxidation period or after addition prepared earlier slag composition. The most often used in practice slags are shown in Table **1**.

Table 1. The chemical compositions of slags used in industry [2, 12, 14, 18].

Basic electric furnace	**CaO**	**SiO_2**	**MnO**	**MgO**	**FeO**	**Fe_2O_3**	**Al_2O_3**	**P_2O_5**	**CaS CaF_2**
	35-50	12-16	3-13		12-25				
	34-44	14-18	6-8		10-14		1-3	2-3	
	40-60	5 – 15	2 - 5		10- 30		-		
At the end of melting period									
Ex1	35-45	18-25	5 - 10	10-14	4-10	1-4	4-7	0.2-0.5	<0.2
Ex2	46.3	10.5	6.1	6.3	16.2	10.3	1.58		-
Ex3	53.38	16.60	0.52	8.48	1.83	3.47	2.96	0.03	-
Ex4	44.2	20.06	8.16	7.28	10.86	3.72	5.14	0.58	<0.3
end of oxidizing									

(Table 1) contd.....

Basic electric furnace	**CaO**	**SiO_2**	**MnO**	**MgO**	**FeO**	**Fe_2O_3**	**Al_2O_3**	**P_2O_5**	**CaS CaF_2**
Ex1	40-50	10-15	6-8	8-10	15-20	2-3	1-2	0.5-1.0	
Ex2	33-44	12-19	6-9	8-12	12-17	4-6	4-6	0.6-1.0	
end of refing									
Ex1	55-65	15-20	<0.4	6-10	<1.0	-	1.5-3	-	<1.5 5-10
Ex2	55 white	22	0.1	10	0.4	-			1.0 10
Acid electric	< 10	57 – 66	15 - 25		12– 20				
		55 – 61	18 - 27		8 - 15		5 – 8		
Basic oxygen LD	37-44	15 – 20	7 - 8		16-24		+		
	45-55	10-20	7-10	2-6	10-25	5-8	1-2 1.5-3		<0.3 <1.5

Slags in Basic Electric Arc Furnace

Throughout the whole smelting process, slags from melting, oxidation and refining periods should be clearly separated. In the first two periods, oxidizing slags are used, and during the refining period - reduction slags are used. Due to high FeO content in oxidizing slags, they are slag-off after oxidation period, so that accurate diffusion deoxidation of the bath can be carried out in the refining process. These slags have completely other compositions (Table **2**), than the slags of reduction period.

Table 2. Chemical composition of oxidising slags after oxidation.

Ref.	**C**	**CaO**	**FeO**	**MnO**	**SiO_2**	**P_2O_5**	**MgO**	**Al_2O_3**	**CaF_2**	**Basicity**
20	0.3	38-53	8-11	5-8	13-16	0.4-0.8	9-15	4-7		2.5-3.5
20	0.1	33-44	12-17	6-9	12-19	0.6-1.0	8-12	4-6		2.1-2.7
21	-	35-45	10-19	4-12	10-18	Max.1	7-13	3-8	Max.0.5	2.5-3.5

Melting of the heat is followed by gradual introduction of ore into the metal bath with additions of limestone and fluxes for carrying out the process of dephosphorization of liquid steel. The course of exothermic reactions of ore into oxidation of carbon and other elements (mainly silicon, manganese and phosphorus) during this period leads to the temperature increase, which creates the possibility of metal losses (loss mainly of iron, manganese and silicon). If in the heat there is chromium (from a certain share of low-alloy steel scrap *e.g.* Cr-Mo or risers and defected castings of *e.g.* corrosion-resistant cast steels), slag contains some amounts of Cr_2O_3, becomes very thick and hinders processes at the bath-slag boundary. The rapid temperature increase, especially local overheating of the bath, is also the cause of evaporation of some elements and dustiness [11, 19]. These phenomena generally occur in the vicinity of the electric arc and concern heavy elements (Zn, Sn, Pb, Bi, As, Sb), (Table **3**).

Table 3. Temperatures of melting and evaporation of heavy elements [2, 12].

Element	Temp. of melting T_m [^oC]	Temp. of evaporation T [^oC]	Density [Mg/m^3]
Zn	420	906	7.1
Sn	232	2750	
Pb	327	1740	11.34
Sb	630.5	1380	6.68
Bi	271.4	634	9.8
As	814.5		
Cd	321	755	8.65
Al	660	2500	2.7
Mg	650	1100	1.74
Fe	1535	3008	7.9
Cr	1875	2773	7.19
Mn	1260	2150	7.2
Ni	1455	3183	8.9
Mo	2610	5823	10.2
W	3410	≈5900	19.3
Ti	1670	>3000 TiC 3152 TiN 2942	4.43
Nb	2468	NbC 3502 NbN 2202	8.57
V	1900	3000, VC 2202	6.1
Zr	1852	ZrC 3527, ZrN 2977	

Losses of metallic elements depend on the manner of smelting (oxidation with ore, gaseous oxygen, melting with one slag, a reversible process of melting, *etc.*). Quality of the applied heat and its positioning in the furnace are very important. Reduction of the final content of harmful elements (P,S,O) can be accomplished by an introduction of different contents of pure carbon cast scrap, lime or low alloyed scrap into the heat. Specific losses of metals refer to smelting process with the use of oxidation with oxygen. Elements with low evaporation temperature oxidize, and their presence can be confirmed by the deposition of white powder on the arc furnace vault (*e.g.* ZnO_2).

Heats of chrome, chrome-nickel and high-manganese steel are characterised by losses of chromium and manganese. In these heats, an important role is played by: the time of oxidation with oxygen, the manner of its introduction, the initial content of the element in the molten charge and the temperature of the bath in the oxidation process. According to E. Pachaly [19] during the melting chrome and chrome-nickel steel in the 7.5-ton arc furnace, the losses of chromium is up to even 25%, (Table **4**).

Table 4. The losses of chromium during melting of Cr and Cr-Ni steels at blowing of oxygen [19]%.

Steel	**% oxidised C**	**% oxidised Cr**	**% Cr losses**	**Bath temp.[^{0}C]**
0.10%C-9%Cr	0.04	22.67	78	-
- „ -	0.07	25.22	83	1890
- „ -	0.07	24.65	88	1905
0.15%C-4%Cr- -21% Ni	0.11	22.69	91	1870
19%Mn-10%Cr-1.5%Ni	-	-	96-97%Cr 63-66% Mn	

In the high-manganese cast steel (pure Hadfield 12÷14%Mn) also large losses of manganese takes place. Especially in heats oxidized by blowing oxygen, manganese losses reach up to 6-8% [17]. Melting losses are mainly due to evaporation after crossing the bath temperature above 1560^{0}C.

The total losses of elements in basic arc furnaces relates to the method of melting. When using two (or more) slags, certain quantities of alloying elements can be recovered by the reduction of oxides. This applies to those elements of which chemical affinity to oxygen is higher than the affinity of iron (Ellingham - Richardson diagram in chapter 2). Such elements as: Ca, Al, Ti, Nb, Mg, Zr, Ce, La oxidize completely in oxidation periods. In contrast, nickel, copper and cobalt pass in 100% to the molten steel in the EAF furnace.

Slags from the refining (reduction) period fulfil two essential tasks: deoxidation and desulphurisation of steel. After slagging-off the first slag (after oxidation period) and pre-deoxidation of the bath by aluminium, the reduction "white" or "carbide" slag is put. In the first case, a mixture of $CaO:CaF_2:C = 6:2:1$ [12, 20], or 2:1:1 [17] is put into the bath. In the latter case, the ratio 3:2:1 is used [12, 21]. The final chemical composition of slags in the refining period (reducing period) is presented in the Table **5**.

Table 5. The compositions of final slags in the reducing period [11, 12, 14, 17].

Slag	CaO	SiO_2	MgO	Al_2O_3	MnO	FeO	CaF_2	CaS	CaC_2	CaO/SiO_2
White	55	22	10	3	0.1	0.4	10	1.0	-	2.75
	≈60	15-20	8-12	2-3	0.3-0.5	<1.0	0.5-1.0	1.0		3-3.5
	50-55	15-18	8-12	0-7	≤2	≤2	6-8	-	≤1	-
Carbide	60	20	9	2	0.1	0.4	6	1.5	2	3.0

Losses of chromium can be decreased in 90% through the reduction of the Cr_2O_3 in the slag with additions of Fe-Si, in the amount of 20kg/t [22].

Slags for the Ladle Treatment (Synthetic Slags)

In steel practice, there is a group of special slags, intended to reduce the final content of oxygen, sulphur and non-metallic inclusions. Processes of accelerated deoxidation and desulphurisation of steel take place during its pouring into the ladle, in which, at the bottom, there is slag of high content of CaO and Al_2O_3 melted in another furnace. A stream of liquid steel falling down from a great height increases the surface of metal - slag contact (by a factor of few hundreds [17]), accelerates reactions of steel deoxidation and desulphurisation. This process was developed and applied to large ladles (with a capacity of about 150 tons), especially for BOF, VOD, VAD technologies and other converter processes.

Table 6. Chemical compositions of typical synthetic slags [in wt-%] [14, 17].

Reference	CaO	Al_2O_3	SiO_2	FeO	MgO	Cr_2O_3	T_m [^{0}C]
[20]	52-55	38-42	2-3	2-3	2-3	2-3	≈ 1400
Eutectic	54.5	45.5					1395

Synthetic slags (Table **6**) are characterised by low melting temperature (≈ 1400^{0}C), low viscosity and high interfacial tension at the metal - slag boundary. These properties of slags promote coagulation of non-metallic inclusions and their number decreases from 0.78 to 0.15% [17]. Very low sulphur content: 0.001 - 0.006% can also be obtained in steels treated with synthetic slags [17].

Increased energy consumption for melting slags in the second furnace is reduced by shortening the reduction period of smelting the steel (even up to 30 minutes). A slightly different technology of desulphurisation with synthetic slags is applied in the foundries: deoxidized steel in EAF is poured into a ladle, onto the bottom of which powdered synthetic slag has been previously introduced. The blowing of argon is started during the taping of the steel to the ladle. After filling the ladle, the slag is already in the liquid state. The blowing of argon increases the interfacial tension of the slag, resulting in a higher degree of desulphurisation. According to [20], the use of these slags in ladles with basic (dolomite) lining allows for obtaining sulphur of even 0.005%, and in acid ladles - 0.0017 - 0.0015%.

CONFLICT OF INTEREST

The author (editor) declares no conflict of interest, financial or otherwise.

REFERENCES

[1] L. Coudurier, D.W. Hopkins, and J. Wilkomirski, *Fundamental of Metallurgical Processes.* Pergamon Press: Oxford, 1978.

[2] J.F. Elliot, M. Gleiser, and V. Ramakrishna, *Thermochemistry for Steelmaking, t.II* Addison-Wesley Publ.Company: Oxford, 1963.

[3] J.J. Moor, *Chemical Metallurgy* 2nd. Butterworth a. Co. Ltd, 1960.

[4] F.D. Richardson, *Physical Chemistry of Melts in Metallurgy.* Pergamon Press: Oxford, 1978.

[5] E.T. Turkdogan, *Fundamentals of Steelmaking.* The Institute of Materials: Cambridge, 1996.

[6] M. Holtzer, *Podstawy metalurgiczne i odlewnicze stopów żelaza. Podstawy fizykochemiczne* Wydawnictwo Naukowe PWN, 1913.

[7] W-Y. Kim, *"Modelling the Viscosity of Silicate Melts Containing Lead Oxide"*, PhD.Thesis, Ecole Politechnique de Montreal, Canada, 2011.

[8] T. Yano, S. Shibata, and T. Maehara, "Structural equilibria in silicate glass melts investigated by Raman spectroscopy", *J. Am. Ceram. Soc.*, vol. 89, pp. 89-95, 2006. [http://dx.doi.org/10.1111/j.1551-2916.2005.00815.x]

[9] K. Leitner, and K. Ploeckinger, *Die Edelstahlerzeugung, Springer Verlag, 2.* Auflage: Wien, N.Y., 1965.

[10] W-Y. Kim, A.D. Pelton, and S.A. Decterov, "Modeling the Viscosity of Silicate Melts Containing Lead Oxide", *Metall. Mater. Trans., B, Process Metall. Mater. Proc. Sci.*, vol. 43, pp. 325-336, 2012. [http://dx.doi.org/10.1007/s11663-011-9610-6]

[11] M. Gajewski, "Structural changes in steel casts applied to power industry and their impact on cracking and corrosion mechanisms", *Monograph of Politechnika Świetokrzyska, Kielce, Poland,* 2003.

[12] T. Matsushita, M. Hayashi, and S. Seetharaman, "Thermochemical and thermophysical property measurements in slag systems", *Int. J. Mater. Prod. Technol.*, vol. 22, p. 351, 2005. [http://dx.doi.org/10.1504/IJMPT.2005.006464]

[13] E. In-Ho Jung, *Jak, J.* vol. Vol. 2. Lehman, Treatise on Process Metallurgy, Elsner Ltd., 2014, pp. 643-674.

[14] A.N. Grundy, H. Lin, I-H. Jung, S.A. Decterov, and A.D. Pelton, "A model to calculate the viscosity of silicate melts", *Int. J. Mater. Res.*, vol. 99, p. 1185, 2008. [http://dx.doi.org/10.3139/146.101752]

[15] H. Schenck, M.G. Frohberg, and W. Rohde, "Viscosität reinen flüssigen Eisenoxyduls im Temperatur zwischen 1380 und 1490oC", *Arch.Eis,* vol. 32, pp. 521-523, 1961.

[16] M. Tiemkin, "Mischungen geloester Salze als Ionenloesungen", *Acta Physicochimica URSS,* vol. 20, pp. 411-420, 1945.

[17] M. Jellinghaus, "Stahlerzeugung im Lichtbogenofen", *Verlag Stahleisen mbH., Düsseldorf, 3.Auflage,* 1994.

[18] H. Flood, and K. Grjotheim, "Thermodynamic Calculation of Slag Equilibria", *JISI,* vol. 171, pp. 64-70, 1952.

[19] W.E. Dennis, and F.D. Richardson, "The Equilibrium Controlling the Decarburization of Iron-Chromium-Carbon-Melts", *J.Iron Steel Inst.,* vol. 175, pp. 264-266, 1953.

[20] F.P. Edneral, *Electrometallurgy of Steel and Ferro-alloys, Izd.Mir.* vol. 1. Moskwa, 1979.

[21] E. Pakulla, and K. Rudnik, "Abbrandverhältnisse in basischen Lichtbogen-Elektrostahl-Öfen", *Stahl Eisen,* vol. 54, p. 621, 1934.

[22] J. Senberger, Z. Buzek, A. Zadera, K. Stransky, and V. Kafka, *Metalurgie Oceli na Odlitky.* Vutium: Brno, 2008. (in Czech)

CHAPTER 4

Gases and Non-Metallic Inclusions in Steel Castings

Abstract: Gases constitute a number of threats to the good quality of steel castings. Therefore, changes in the solubility of oxygen, nitrogen and hydrogen during solidification and defects, caused by rapid reduction of solubility in the liquidus temperature: microshrinkage, microporosity and non-metallic inclusions, have been presented. The types of inclusions and their dependences on the chemical composition of cast steel have been defined. The examples of adverse effect of inclusions on mechanical properties of cast steel have been given. To reduce their impact on the quality of castings, examples of modifications of inclusions and elements leading to changes in the morphology of inclusions, have been identified.

Keywords: Hydrogen, Inclusions morphology, Mechanical properties, Nitrogen, Solubility.

1. INTRODUCTION

The content of oxygen, nitrogen and hydrogen in steel castings is important for their behaviour under operating conditions. Since, very often, when checking castings, there are cracks revealed, caused by too high concentrations of these elements. When testing the fractures, there is porosity or microshrinkage revealed (Fig. **1a**), or both these defects occur with visible non-metallic inclusions on the fracture surface (Fig. **1b**).

The sources of gas origin in castings are: metal charge used for heats (rust or oil on the scrap surface, burnt moulding sand on the surface of circulating scrap in the charge, oxidation of iron during melting metal charge, and also oxidation of Cr, Mn, Si and other elements). The increase in gas content also results from the contact of liquid steel stream with the atmosphere (when pouring steel into a ladle or the mould), between the liquid steel and the lining of the furnace and the ladle or the mould material.

Jan Głownia

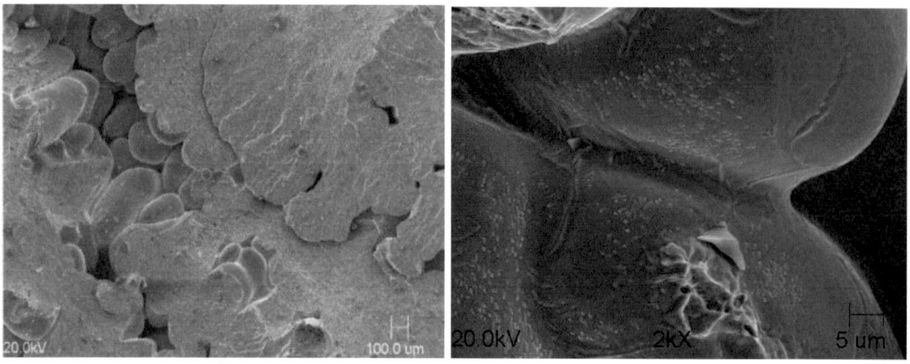

a) b)

Fig. (1). The microshrinkage (**a**) in carbon steel casting (impact toughness test at 20°C); Magn.500X and non-metallic inclusions (**b**) in cast steel (0.2%C-1.4%Mn) after impact test at "-40°C"; Magn 3500X

Yet another source of gases in molten steel, and later in steel casting, is moisture on the input materials (ore, limestone, scrap) and ferro-alloys, introduced as alloying additions during smelting. For this reason, there is an increase in the hydrogen content as a result of dissociation:

$$\{H_2O\} = \{H_2\} + ½\,\{O_2\}\ \Delta G^o = -\,246\,000 + 54.8\cdot T\ [J]\ [1, 2] \qquad \textbf{(1)}$$

$½\{H_2\} = [H]\ lgK_H^{liq} = +2.423 - 1900/T$ [3]

$lgK_H^{liq} = -\,1.591 - 1905/T$ [8b] $\Delta G^o = 36\,460 + 30.46\cdot T\ [J]$ [4]

Therfore, as a result of this reaction in molten steel, oxygen and hydrogen in the atomic form are dissolved.

Hydrogen is a harmful element in steel due to the formation of defects, such as: gas bubbles, microporosity, "fish eye" (Fig. **2**) and decrease in plastic properties of cast steel, especially impact toughness.

Final concentrations of hydrogen in molten steel in basic arc furnaces are equal to 4-6 ppm [5, 6], and in acid induction furnaces: 3.5-5 ppm [5]. The elimination of hydrogen in steel castings is a very difficult, therefore the process of smelting and pouring steel into moulds should be particularly carefully conducted.

Oxygen is also harmful in steel castings. Since steel intended for castings must be well deoxidized, it is recommended that the final concentration after pouring does not exceed 50 - 60 ppm. Unlike steel, which may be produced in the semi-killed form (oxygen content 100 - 120 ppm [7, 8]), or even in the rimmed form (such cases are not permitted in castings). Please note that during the pouring of steel into moulds, there is secondary oxidation of metal and an additional increase in the oxygen concentration.

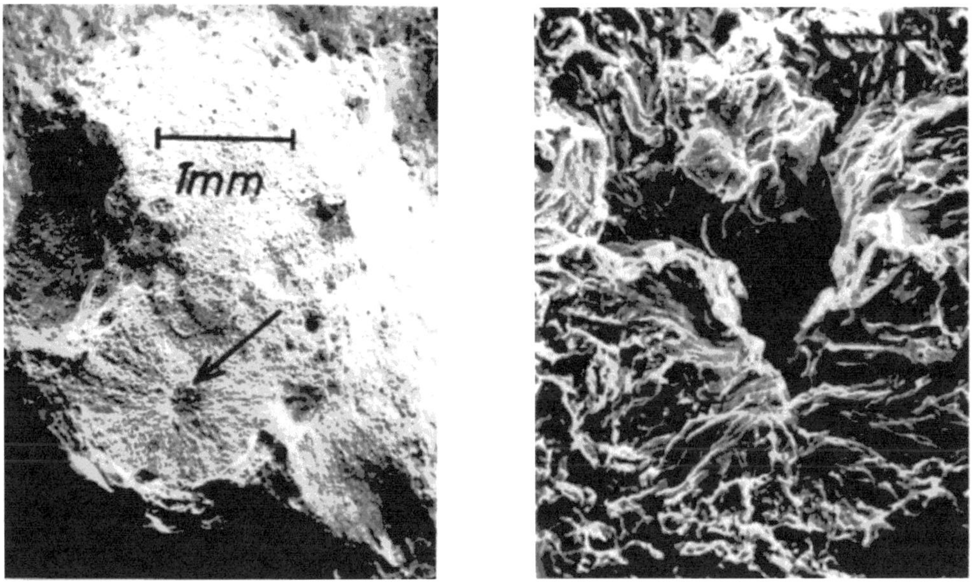

a) b)

Fig. (2). "Fish eye" in massive casting (**a**) and porosity in the centre of this defect (**b**).

2. THE SOLUBILITY OF GASES IN MOLTEN STEEL

Oxygen - it must be stressed that, in steel smelting, in the periods of melting and oxidation, oxygen is the preferred additive and is used to remove excessive quantities of carbon (and hydrogen). In the final stage of smelting - refining - there is a rapid reduction of oxygen through the diffusion and precipitation deoxidation.

There are two major methods of the introduction of oxygen into molten steel:

* by a direct method - as pure gas (over 98.5%), *e.g.* in LD, VOD processes or in an electric arc furnace during steel smelting with recovery method), or as air (formerly Tropenas process). This method of melting results in a temperature increase in molten steel, since the reaction:

$$[Fe] + \frac{1}{2}\{O_2\} = (FeO) \qquad \textbf{(2)}$$

$\Delta G^{o} = 225\,500 - 41.3 \cdot T$ [J] [4]

is an exothermic reaction, and at 1600^{0}C - 58 kcal of heat is generated [8]. This oxide reacts with other oxides of slag components (MnO, SiO_2, CaO), forming complex phase systems.

* by an indirect method through adding iron ore into the bath. In the case of the introduction of hematite ore, the following reaction occurs:

$$[Fe] + (Fe_2O_3) = 3(FeO), \qquad \textbf{(3)}$$

which requires 7.9 kcal of heat at 1600^{0}C. In this case, FeO oxide is the oxygen source.

According to studies of L.S. Darken and R.W.Gurry [9], oxygen does not occur in molten iron in the atomic form, but it does occur in the molecular form $Fe_{1-x}O$ (wüstite) [9]. As a proof, it has been quoted that already at oxygen partial pressure of 10^{-8} at. and temperature at 1600^{0}C, pure iron is in equilibrium with the liquid $Fe_{1-x}O$. Wüstite is stable in the temperature range 560 - 1370^{0}C (Fig. **3**) and passes into the mixture αFe + (Fe_3O_4) below 560^{0}C. Higher oxides Fe_3O_4 (Magnetite) and Fe_2O_3 (Haematite) are formed at higher oxygen concentrations and are stable at lower temperatures. Magnetite (T_m = 1597^{0}C; ρ = 5.1 g/cm^3 [8]) passes into metastable γ-Fe_2O_3 at the temperature below 350^{0}C [4]. Haematite (T_m = 1730^{0}C; ρ = 5.24 g/cm^3 [8] does not exist above 1457^{0}C in air atmosphere.

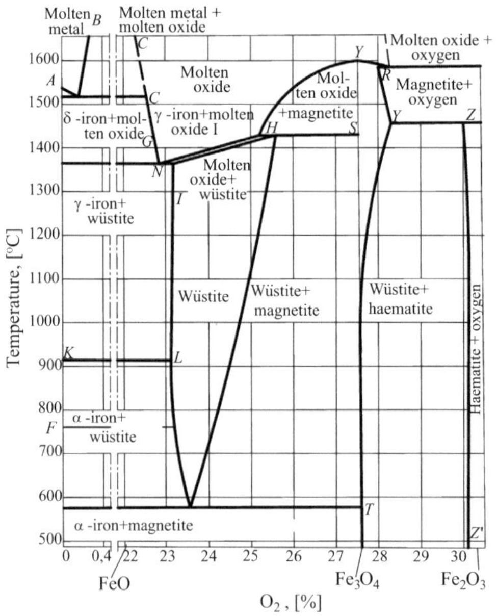

Fig. (3). Iron – oxygen system [13, 14].

According to P. J. Spencer and O. Kubaschewski [10], the free energy equations of forming iron oxides have the following dependence:

$$\text{wüstite: } 0.953[Fe] + ½\{O_2\} = [Fe_{0.953}O];\ \Delta G^0 = -261\ 182 + 62.93\ T\ [J/mol] \quad \textbf{(4)}$$

in the range 777°C – 1277°C

in the range: 527°C – 1327°C;

haematite:

$$2[Fe] + 3/2\ \{O_2\} = [Fe_2O_3];\ \Delta G^0 = -790\ 660 + 231.15\ T\ [J/mol] \quad \textbf{(5)}$$

in the range: 527°C – 1457°C

In the presence of CaO, MnO or SiO_2, iron oxides are subjected to decomposition, depending on the atmospheric pressure, mixture of CO, CO_2 and the temperature. These data include works [15, 16] for additions of CaO and MnO, and works [12] for redox reactions in equilibrium with CaO - FeO - Fe_2O_3.

The issue of solubility of oxygen, in molten iron and steel in particular, is crucial to the smelting process, since such elements like Si, Mn, Cr, C loss by oxidation in next periods of melting, subsequent process of deoxidation of steel is also dependent on the oxygen content in the bath.

Solubility of oxygen in pure liquid iron takes place according to the following reaction:

$$½\{O_2\} = [O]\ \Delta G^0 = -\ 115\ 750 - 4.63 \cdot T\ [J]\ [7] \quad \textbf{(6)}$$

Dependence of oxygen solubility on the temperature is described by the following equation [17, 18]:

$$lg[\%O] = -\ 6\ 320/\ T + 2.734 \quad \textbf{(7)}$$

$lg[\%O] = -6380/T + 2.765$ [7]

The equation concerns oxygen solubility in pure iron, remaining in equilibrium with pure, liquid iron oxide ($a_{(FeO)} = 1$), and indicates that it depends only on the temperature, (Fig. **4**).

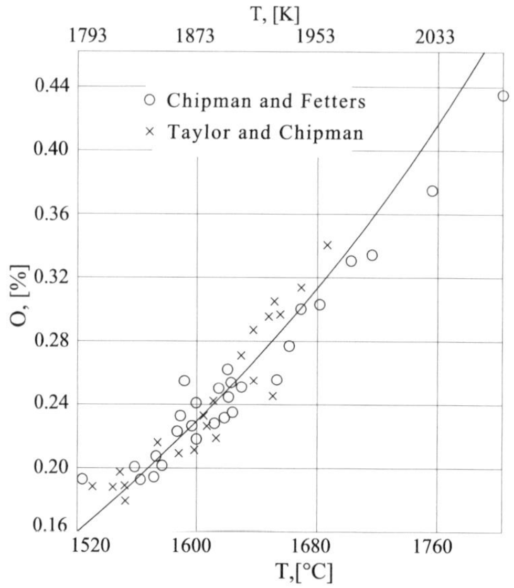

Fig. (4). The solubility of oxygen in molten iron at 1600°C [12].

Temperature [°C]	[%O]
1550	0.185
1600	0.229
1650	0.281
1700	0.340
1750	0.409

During the lowering of the temperature (cooling and solidification), there is a reduction of the solubility of oxygen in iron, and excessive oxygen leads to the formation of a new phase - iron oxide (Fig. **3**).

Line B - B determines the oxygen solubility of oxygen in pure iron (along the line, there is a boundary of saturation of iron with oxide $Fe_{1-x}O$). Line C - C indicates that the chemical composition of the oxide $Fe_{1-x}O$ changes along with the temperature, so that only at an oxygen content of 23.2% and temperature of 1700°C, it corresponds to the stoichiometric compounds.

The solubility of oxygen in molten iron alloys depends on the presence of elements with a high affinity to oxygen (iron has lower affinity to oxygen than silicon, manganese or aluminium). In the presence of these elements in the molten iron, the activity of oxygen is subject to change, Fig. (**5**). The higher the concentration of elements with a high affinity to oxygen, the lower the coefficient of activity.

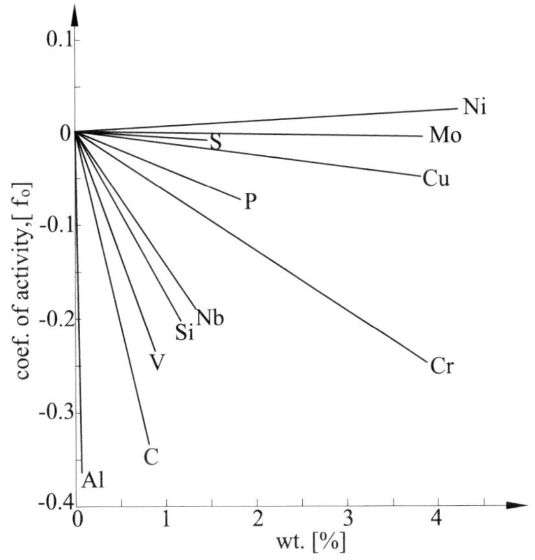

Fig. (5). The influence of alloying elements on the oxygen activity coefficient in iron [11].

As in the case of pure iron, also in the molten steel, a reduction of the concentration of oxygen in the presence of an element with a higher affinity to oxygen than iron causes the formation of insoluble non-metallic inclusions in the steel. Both the quantity and type of inclusions formed depend on the added element. In case of large amounts of oxygen (in steel for castings 100 - 60 ppm), during solidification carbon causes gas precipitations {CO}, silicon, manganese or titanium lead to formation of liquid inclusions, and aluminium or zirconium - also solid inclusions. Both in carbon and low-alloy steels these are usually complex inclusions (*e.g.* $SiO_2 \cdot FeO$, $MnS \cdot FeO$).

The oxygen content in molten steel being in contact with complex slag requires knowledge of the activity (FeO). For basic slags, the content of oxygen is presented by the following equation [15]:

$$\log[\%O] = -5\,200/T + 1.742 \tag{8}$$

and for acid slags by the equation [17]:

$$\log[\%O] = -6\,120/T + 2.212 \tag{9}$$

In the first case, the activity of (FeO) decreases with increasing temperature, and in the other - remains at a constant value.

The presence of the third elements in molten iron affects the activity of oxygen. Most often, this dependence is represented by a change in the value of the oxygen activity coefficient f_o, as a result of concentration of these elements in the molten steel:

$$a_{[\%O]} = f_o^{\Sigma} \cdot [\%O] \quad \lg f_o^{\Sigma} = \lg f_o + \Sigma_{i=3} \lg f_o^{i} \qquad \textbf{(10)}$$

f_o^{i} - coefficient of oxygen activity at the presence of i elements.

Elements that have a higher affinity to oxygen than iron (primarily Al, C, Si, V, Nb, Ti) cause a reduction in the activity coefficient.

Changing the oxygen activity in iron with alloying elements refers both to the process of oxidation of steel and its subsequent deoxidation. This issue is widely characterised in metallurgical processes as a result of the atmosphere effect over the bath, furnace lining materials and temperature. More accurate data on this subject is included in the works [5, 9].

3. THE EFFECT OF TEMPERATURE ON THE GAS CONTENT IN THE SOLID STATE

Knowledge about the solubility of gases in steel, especially in the final products, which are steel castings, is essential for subsequent properties of machine components. This is mainly due to the rapid reduction in gas concentration at the melting temperature of steel. Hence, in the solid state, a lot of defects of a gas origin are observed after the solidification process:

1. gas porosity, Fig. (**1**);
2. microshrinkage, (often with porosity), Fig. (**1**);
3. microcracks induced by hydrogen precipitation, Fig. (**2**).
4. oxides and nitrides segregated along grain boundaries, Figs. (**1b**, **16** and **17**).

The oxygen content in molten steel is removed during its diffusion deoxidation in the furnace before pouring the steel, and in the ladle during precipitation deoxidation, while pouring the steel. A characteristic feature when lowering the temperature is a rapid reduction in the concentration of oxygen (as well as hydrogen and nitrogen) at the temperature of steel solidification. In so far as the content of oxygen at 1600°C equals 0.23%, in αFe it equals only a max. of 0.03%. These contents in steels are even lower: 0.003 - 0.004%. In killed steels, smelted in basic furnaces, they are dependent on the chemical composition of steel, particularly on the carbon and aluminium contents. Oxygen in steel is predominantly in the form of non-metallic inclusions (total oxygen) and in the dissolved form in the metallic matrix.

A maximum of up to 0.03% oxygen is dissolved in αFe. Similarly to the melting temperature of iron (and steel), a rapid change in the oxygen solubility also occurs at the transition temperature γFe →αFe, Fig. (**6**).

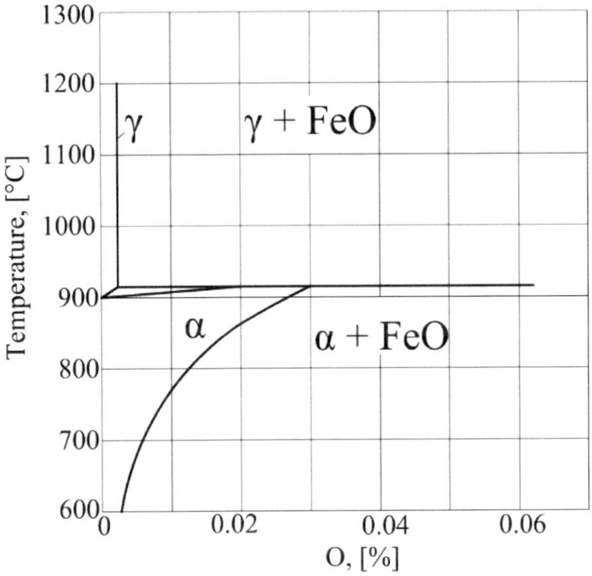

Fig. (6). The solubility of oxygen in solid state of iron [14].

In practical terms, in steels containing 0.5 - 0.6%C, and smelted in basic furnaces, the final concentrations of oxygen amount to 0.003 -0.004%. In cast steels, the concentrations are slightly higher: 0.005 - 0.006%.

4. SOLUBILITY OF HYDROGEN AND NITROGEN IN IRON AND STEEL

Both hydrogen and nitrogen dissolve in molten iron in an atomic form, according to the reaction:

$$\tfrac{1}{2}\{H_2\} = [H] \quad \textbf{(11)}$$

$$\tfrac{1}{2}\{N\} = [N] \quad \textbf{(12)}$$

$K_H = a_{[H]}/p_{H2}\}$ $K_N = a_N/\{p_{N2}\}$

ΔG^o=36 460 +30.46·T [J] [3] ΔG^o=9916 +20.17·T [J] [4]

where: a_H and a_N are the activities of hydrogen and nitrogen in molten iron.

Since, at normal pressure, the gas content in molten iron is less than 1% wt, and Henry`s activity can be assumed at ($a_{[H]} = K_H \cdot \gamma_{[H]} \cdot [\% H]$).

Then:

$$[\%H] = K_H \, (p_{H2})^{1/2} / \gamma_{[H]} = k_s \, p_{H2} \quad \textbf{(13)}$$

where: k_s - Sieverts` constant,

[%H] - hydrogen content in the molten metal,

$\gamma_{[H]}$ - Henry's activity coefficient for hydrogen.

(for pure iron is equal to 1)

Dependence of the constant K_H from the temperature is described by the following equations:

$\log K_{H(l)} = -\ 1\,905/T - 1.591 \ \Delta G_{(l)} = 36\,460 + 30.46\ T$ [J] [4]

$\log K_{H(\gamma)} = -\ 1\,182/T - 2.369 \ \Delta G_{(\gamma)} = 226\,300 + 45.35\ T$ [J]

$\log K_{H(\alpha\delta)} = -\ 1418/T - 2.369 \ \Delta G_{(\alpha\delta)} = 28650 + 45.35\ T$ [J]

$K_{H\,(l)}$ $K_{H\,(\gamma)}$ and $K_{H\,(\alpha\delta)}$ mean respectively the equilibrium constant of hydrogen in the liquid state, for the range of the phase γ and α or δ phases.

H. Schenck and H. Wünsch [19] give the similar equation for the temperature dependence of the equilibrium constant of hydrogen (equation 1).

At a temperature of 1600°C and at a pressure of 10^5 Pa, the hydrogen solubility in molten iron equals 27 ppm, and at a pressure of 10^2 Pa, it only equals 1ppm (under vacuum). According to [9, 20, 21], the solubility of hydrogen at the melting temperature of iron is 0.0025%, and at the temperature of 1700°C - 0.0030%.

Processes of dissociation of H_2 and N_2 are endothermic reactions, which means that, along with the increase in temperature, there is an increase in the solubility, Fig. (7). For this reason, in steel practice, this means that the higher the temperature of the molten steel in the furnace, the higher the content of hydrogen should be expected in steel castings.

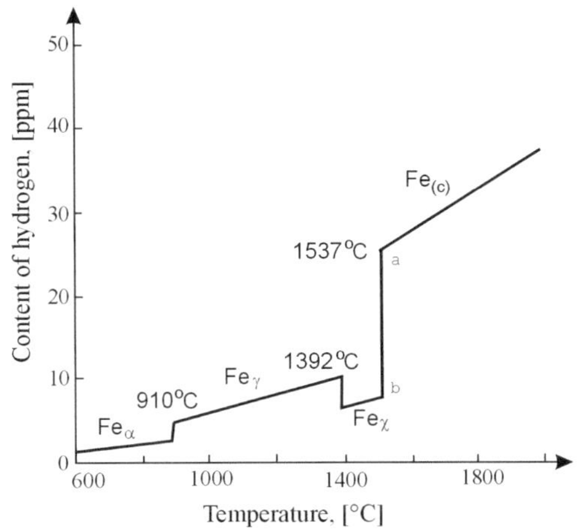

Fig. (7). The solubility of hydrogen in molten iron [15].

Due to the quality of steel castings, the rapid change in the solubility of hydrogen and nitrogen at the melting temperature (point "a" to "b") is important. After exceeding the solubility limit of hydrogen, large quantities of hydrogen are released, causing the occurrence of porosity (Fig. **1**), and in massive castings (more than 5 - 10 tons), the so-called fish eye (Fig. **2**), which initiate cracks and brittle fracture already during the heat treatment, and most often during quenching in oil or in water. Presence of above 4 - 5 ppm of hydrogen in cast steel is the cause of the occurrence of these defects and the so-called hydrogen flakes (thin cracks). In carbon and low-alloyed steel smelted in basic arc furnaces, the hydrogen content ranges from 4-5 ppm [17], 4 ppm [6], or even 40 ppm [18]. Higher contents of hydrogen lead to the occurrence of the mentioned defects.

Changes in the solubility of hydrogen in the solid state are related to the type of crystal lattice of the iron matrix, Fig. (**8**). Due to the size of the hydrogen atom and the radius of the interstitial site "r_i" in relation to that of the host lattice atom "r_h", the relations are as follows:

for the fcc structure - 0.414 (for octahedral symmetry), 0.225 (for tetrahedral symmetry) and for the bcc structure - 0.155 and 0.291.

Hence, for the fcc network, O-sites are predominately filled, while bcc structures for the T-sites are often preferred. However, this dependence explains only the tendencies in the solubility of hydrogen because in this case, processes of entropy and enthalpy must be considered.

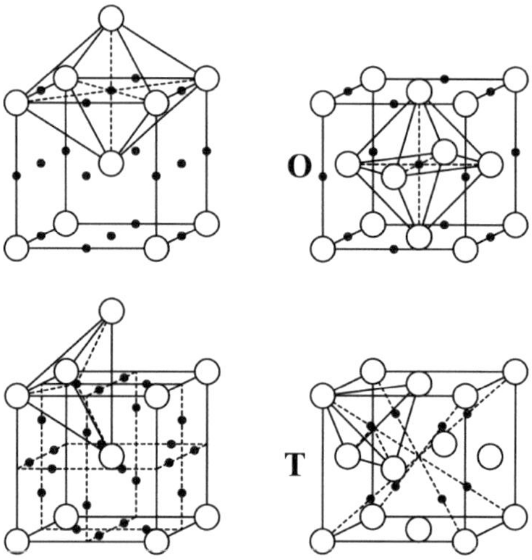

Fig. (8). Interstitial sites (octahedral "O" and tetrahedral "T") in fcc and bcc lattices [20]

The presence of alloying elements in molten iron changes the solubility of hydrogen. Nickel, manganese, titanium and niobium increase, while carbon, aluminium, oxygen and boron reduce the solubility of hydrogen, (Fig. **9**). Regarding the effect of chromium, there are divergent data [5], but chromium steels contain higher concentrations of hydrogen than carbon steels.

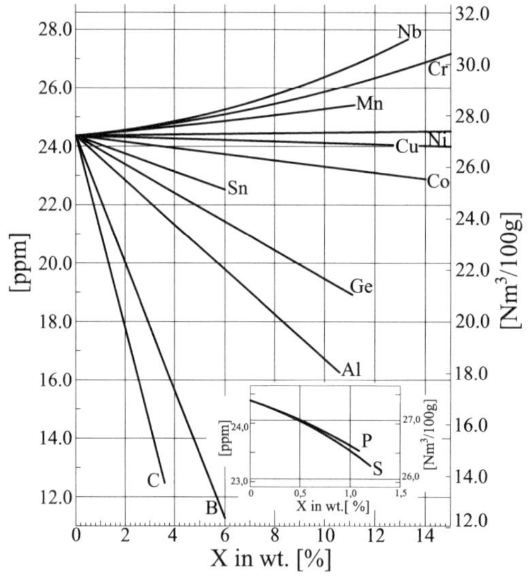

Fig. (9). Effect of alloying additions X on the solubility of hydrogen in the double alloys at T = 1592°C and p = 1 atm [21].

The reasons for the high hydrogen content in cast steel are:

- application of wet charge, rust on the charge parts, damp slag-forming materials (limestone, dolomite) and also damp exothermic insulations. It should be emphasised here that limestone contains chemically bound water that can be eliminated only by its annealing. Large quantities of hydrogen may be from damp ladles, tapping spouts or ceramic fittings, used as gating system.
- reaction of molten steel - material of the mould. The way of application of the coating onto the mould cavity should be mentioned here: using brushes causes tearing off single quartz grains and mixing them with the coating material (quartz as the acid material), combined with an inert material of basic character (coatings containing MgO, MnO), give low-melting compounds (a mixture of 28% SiO_2 +72% FeO has T_m = 1220^{0}C). At 600^{0}C chromite mass (FeO · Cr_2O_3) is disintegrated and increased in volume by 10%. This process is accompanied by cracking and entering of molten steel into the moulding sand. Then, the reaction of molten steel and clay binder of the moulding sand and liberation of large amounts of hydrogen and nitrogen occur.
- method of conducting the smelting: without oxidation (recovery method of melting), without intensive oxidation, long breaks in putting the refining slag, no heating of ferroalloys and chemical composition of the slag lead to increase in hydrogen concentration in the molten steel. Basic slags transfer larger amounts of moisture from the atmosphere than the acid slags,(solubility of hydrogen and diffusion rate are lower) [22].
- in high-alloyed cast steel, hydrogen is introduced into additions of electrolytic nickel. In cast steels 18%Cr - 11%Ni, 25%Cr – 20%Ni and their similar grades, additions of nickel can be an important source of hydrogen. In order to eliminate it (or reduce its final content in the cast steel), it should be introduced together with the charge because it is not oxidised during smelting, and intensive oxidation of the bath is the main parameter of its removal from steel. Granules of nickel are also the cause of an increased hydrogen content because at the time of its production the stream of molten nickel is introduced directly into water.
- type of slag; acid slags are less permeable to transfer moisture from the furnace atmosphere (including 30 – 100 Nm^3/100g of hydrogen) than basic slags (40 – 205 Nm^3/100g) [13].

Nitrogen is dissolved in molten iron similarly to hydrogen; in atomic form and in accordance with the Sieverts` law. The temperature dependence of the equilibrium constant for dissolution of nitrogen in iron is described by the equation:

$$\lg K_{N(l)} = -518/T - 1.063 \quad \Delta G_{(l)} = 9\,916 + 22.17\,T\ [J]\ [4] \qquad \textbf{(14)}$$

$$\lg K_{N(\gamma)} = 450/T - 1.955 \quad \Delta G_{(\gamma)} = -8\,613 + 37.42\,T\ [J]$$

$lgK_{N(\alpha,\delta)} = -1\,520/T - 1.04\ \Delta G_{(\alpha,\delta)} = 29\,090 + 19.91\ T\ [J]$

and the diagram (Fig. **10**). As in the case of hydrogen, at the solidification temperature of iron there is a sudden reduction in the solubility of nitrogen, resulting in the possibility of formation of gas bubbles (particularly in steel castings). To avoid formation of bubbles in carbon cast steel, the nitrogen content should be below 120 - 100 ppm.

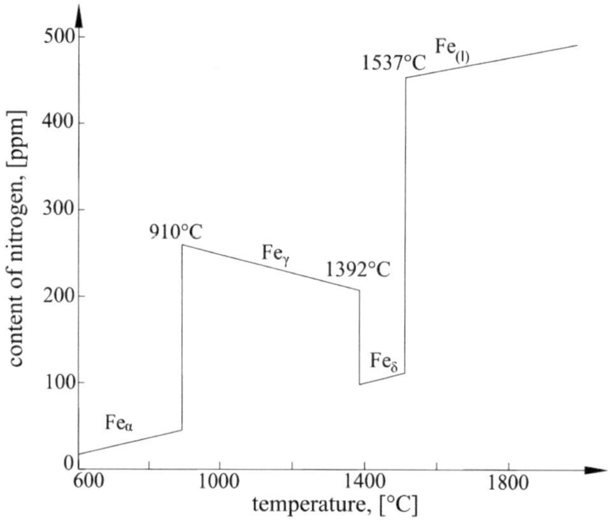

Fig. (10). The solubility of nitrogen in liquid iron at $p_{N2} = 10^5$Pa [15, 23].

In the solid state, structures of the fcc network dissolve more nitrogen than structures of the bcc network. At the melting temperature of iron, nitrogen solubility in iron equals 450 ppm, and in the phase δFe, it only equals 150 ppm. In the phase γFe, nitrogen solubility ranges from 210 to 280 ppm. Alloying elements have a significant effect on the nitrogen solubility in steel, are shown on (Fig. **11**). Titanium, zirconium, vanadium, niobium and chromium (as elements with the highest affinity for nitrogen) have the largest effect, while carbon and silicon have the lowest effect. An addition of 4% of vanadium to iron causes more than double the increase in nitrogen from 450 ppm to about 1000 ppm. Assuming chromium has a solubility coefficient of 1.0, other alloying elements are characterised by the coefficients shown in Fig. (**11**). Alloying elements play an important role in creating properties of steel and corrosion-resistant cast steel. For example, in heat-resistant cast steel, the growth of chromium content to 18% increases nitrogen content even up to 800 ppm, Fig. (**12a**). In 18%Cr-8%Ni-Mo cast steel (Fig. **12b**), solubility of nitrogen reaches even up to 1000 ppm. In corrosion-resistant cast steels, nitrogen acts as an alloying element and is a replacement of expensive nickel [5, 24].

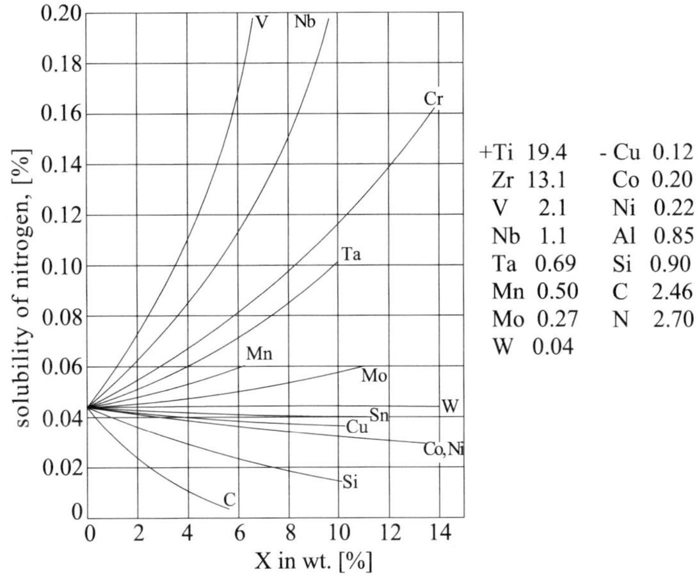

Fig. (11). The solubility of nitrogen in binary Fe – X element at 1600°C [12].

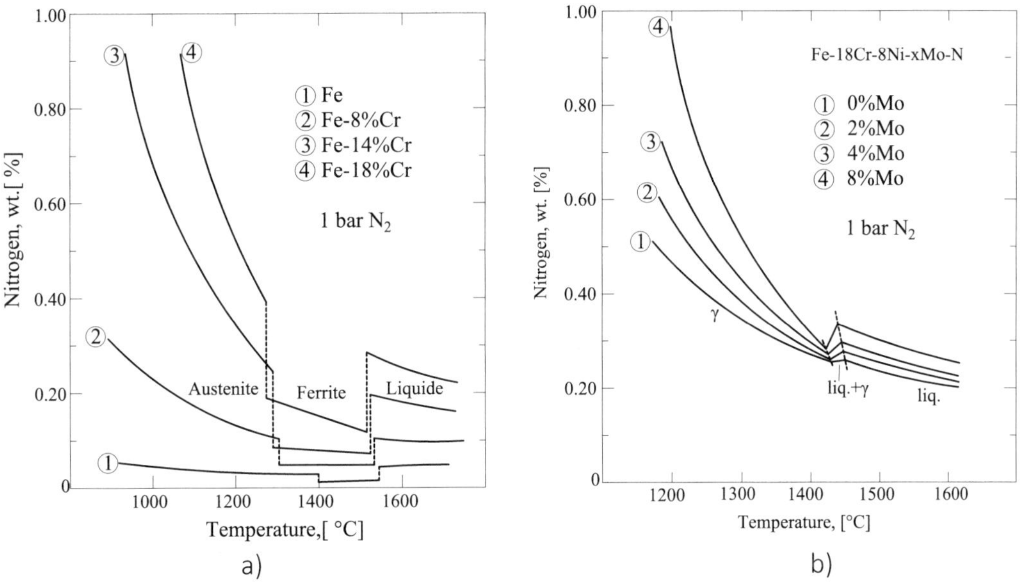

Fig. (12). The effect of chromium (**a**) and molybdenum (**b**) on the nitrogen solubility in high-alloyed cast steel [25, 26].

Nitrogen, as a highly stabilising austenite, increases the UTS and YS strength of the corrosion-resistant cast steel, and also its corrosion resistance.

Carbon oxide {CO}

{CO} and argon are gases that are practically insoluble in molten steel. In gas bubble - molten steel (with dissolved nitrogen or hydrogen in it) contact, the reaction occurs as a result of very low partial pressure of, *e.g.* nitrogen, inside the bubble. Thus, this reaction has similar effect to a vacuum. The results of this process depend on the time of flowing out gas bubbles, bath temperature, interfacial tension and the chemical composition of molten steel.

On this principle, {CO} formed in the process of oxidation of carbon leads to the removal of both nitrogen and hydrogen (it also accelerates the outflow of non-metallic inclusions into the slag). The intensity of the elimination reaction of gases from the bath depends on the course of oxidation. Since nucleation of {CO} bubbles takes place on the furnace hearth (in the pores of the ceramic lining – Fig. **13**), the deeper the bath, the better the degassing effect can be.

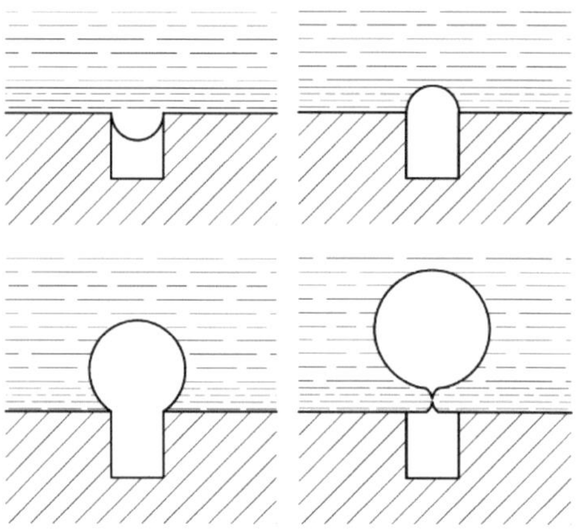

Fig. (13). Formation of bubble in the pore of a lining.

However, the possibility of separation of the bubble depends on the atmospheric pressure p_{at}, density of the liquid steel ρ_m and slag ρ_s, metal height h_m and slag height h_s over the bubble and surface tension of the liquid steel σ_m according to the equation:

$$p_{\{CO\}} \geq p_{at} + \rho_m \cdot h_m + \rho_s \cdot h_s + 1.02 \cdot 10^{-6}\, \sigma_m / r \quad \textbf{(15)}$$

where: r – bubble radius,[m]

σ_m – surface tension, [mN/m]
ρ_m and ρ_s – [kg/m^3]
h_m and h_s – [m]

An intensive oxidation of the bath brings more effective results in the case of nitrogen removal than hydrogen removal.

5. INCLUSIONS IN STEEL CASTINGS

Obtaining good quality of steel for castings is associated with the least amount of gases dissolved in it and precipitated non-metallic inclusions or other impurities (Pb, As, Sb, Bi, Sn). The first two parameters are the number of formed inclusions, but also the most favourable form of their occurrence.

There are a lot of classifications of inclusions. One of them is the division into inclusions:

- exogenous - arising from external causes (as components of the furnace lining, casting ladle, moulding sand, steel slag or protective slag in the ladle or as particles, leached from the gating system during filling the moulds).

This group also covers inclusions resulting from the oxidation of the metal stream and the formation of oxides (and nitrides or silicates). According to [27 - 29], in these conditions, there are inclusions with a diameter over 100µm formed, Fig. (**14**). In most cases, these are macroscopic inclusions visible even on the surface of the casting (as particles of moulding sand, mould or the core of the protective coating, ceramic components, which have not had time to flow out from the molten steel as differences in steel density and inclusions);

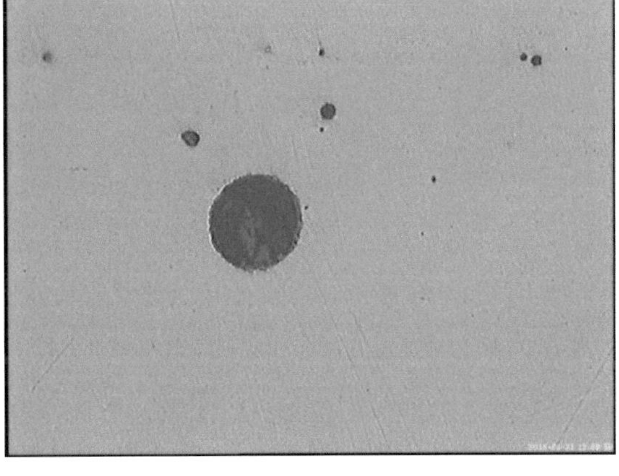

Fig. (14). Macro-inclusion (silicate) in carbon cast steel, Magn.250X.

-endogenous - formed in molten metal, in the processes of:

oxidation of silicon, manganese, aluminium and chromium (and also Ti, Ca, V, Nb, and other elements), deoxidation of steel in the furnace or casting ladle, modification of steel after deoxidation.

reaction of dissolved gases (nitrogen, oxygen) with elements dissolved in molten steel (Si, Mn, Cr, Ti, Al, Mg, Ce and Fe matrix); according to K. Leitner and K. Ploeckinger [5], more than 80% of the inclusions in steel is of endogenous origin; according to [20], as many as 83% of the inclusions are due to secondary oxidation of molten steel, 14% - from the metal - mould reaction, 2% - from slag and only 1% - from the steel deoxidation process;

reactions occurring during the solidification of steel.

Distinction between these two types of inclusions can be made on the basis of their size: exogenous inclusions are large, randomly distributed [30], with a diameter of more than 100 μm [28]), and stand out from the rest of inclusions with their shape (*e.g.* sharpness of shapes). While endogenous inclusions are small, numerous and evenly distributed in the metal matrix. Very often, the inclusions have a complex structure, *e.g.* oxy-sulphides nucleate at silicates, Fig. (**15**).

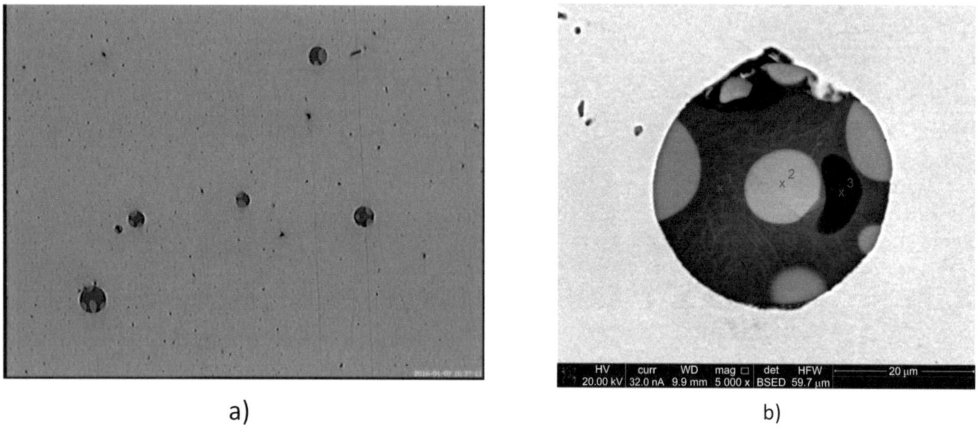

Fig. (15). Type I inclusions in carbon cast steel (point 1 - $MnO{\cdot}SiO_2$; 2 - $MnO{\cdot}MnS$; 3 – $SiO_2{\cdot}MnO$); a- Magn. 400X; b-Magn.5000X.

The formation mechanism of exogenous inclusions is based on a chemical reaction and mechanical erosion [7] rather than on a chemical reaction and a change in the chemical composition of the inclusion at the time of its outflow from the bath [8, 9]. Leaching of tapping spout, ladle, plug and fitting during pouring steel is an example of this type of mechanism of inclusion formation. Here, steelmakers use an example of enlarging the diameter of inclusions after pouring steel into the ladle, which has a capacity of over 100 tons. Founders do not have that option because the moulds should be filled already after a maximum of 7 - 10 minutes from the steel pouring. Most often, steel in the casting ladle has a capacity of 7 - 10 tons, and the bath loses temperature at a rate of about 6-8^{0}C/min. In the casting ladle with a capacity of *e.g.* 160 tons, the temperature drop equals about 1^{0}C/min.

In the second division, there are **primary and secondary** inclusions distinguished. This division is based on the time and place of their nucleation:

- **primary inclusions** are formed as products of deoxidation at temperatures above the liquidus temperature. Their type depends on the applied deoxidiser, and the amount is proportional to the total content of oxygen and nitrogen in molten steel. At a constant temperature above the liquidus temperature, there are both low-melting silicates, oxy-sulphides and nitrides formed, Fig. (**15**). The latter are always in the form of polyhedral, suggesting that their melting temperature is above the temperature of liquid steel, in which they nucleated.

- **secondary inclusions** are formed during the solidification of steel in the range of liquidus - solidus. The amount of these inclusions depends on the concentration of oxygen remaining after deoxidation and on the concentration of the element used for deoxidation.

Another division distinguishes: single oxides (SiO_2, MnO, Al_2O_3, Cr_2O_3, TiO_2 *etc.*),

sulphides: MnS, FeS, CaS, CeS,
nitrides: TiN, AlN, VN, ZrN, CrN,
silicates: FeO SiO_2, MnO SiO_2,
aluminosilicates: Al_2O_3 SiO_2,

Another division of inclusions can be found in steels and cast steel. It is very important for founders of cast steel because, in contrast to steel, morphology of inclusions in castings does not change after the solidification of steel. However, in the processes of forging, rolling and continuous casting, non-metallic inclusions are crushed, shifted and moved in the direction of plastic work. The conditions are not as favourable in order to control the morphology of inclusions in steel castings. Therefore, the inclusions, formed in the process of deoxidation and solidification of steel, have the final shape and characterise the subsequent properties of cast steel.

In cast steel, there are three types of non-metallic inclusions:

***type I** - spherical (oval) precipitations arranged more or less evenly throughout the alloy matrix, Fig. (**15**). This type of inclusions is formed in the molten steel, containing more than 100 ppm of oxygen [31], according to [32] - more than 200 ppm of oxygen. These inclusions are formed during the monotectic reaction of liquid rich in sulphur and oxygen. These are heterogeneous inclusions (Fig. **15b**), evenly distributed in the metal matrix (with high concentrations of oxygen) and also often - at primary grain boundaries (Fig. **16a**).

These types of inclusions are present in cast steel, even at a lower oxygen content (higher than 80 - 100 ppm) but after addition *e.g.* rare earth metals. High oxygen content is connected with low aluminium content. It is usually caused by oxygen and sulphur segregation up to grain boundaries and interdendritic areas during solidification. The slower the process of cooling and solidification, the more likely the occurrence of inclusions of the I type. Both sulphides and oxides nucleate in molten steel at the same time, forming two-phase inclusion, Fig. (**15**). In the crystallisation process (at further lowering the temperature), from the sulphur and oxygen-rich bath, first sulphide, and then - manganese silicate or aluminosilicate is released.

Complex sulphides (Fe,Mn)S; (Mn,Cr)S; (Mn,Ti)S or (Mn,V)S can occur in alloy cast steel [8]. The presence of these inclusions can be found in steel castings after a misconducted process of deoxidation or during tapping, pouring steel or filling casting moulds, and if the stream of metal was subject to strong turbulent movements (stream of metal on the tapping spout is not uniform, too high stream of molten steel into the ladle, no slag in the ladle, keeping the molten steel for a long time without covering it with exothermic powder, secondary oxidation of molten steel in the mould).

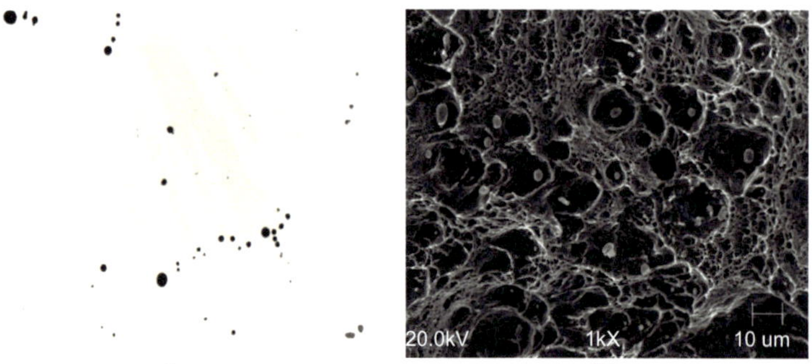

a) b)

Fig. (16). Type I MnS inclusions in carbon cast steel:b– SEM after impact toughness test, Magn. 1000X, a– optical microscope without etching, Magn.630X

There are data [30] indicating that the cause of inclusion formation during steel tapping, leaching materials from tapping spout, ladle and fittings of the pouring system, is weakening of bounding strength of ceramics at near-pouring temperature of liquid steel.

***Type II** during solidification of steel, concentration of elements with low partition coefficient k_0 increases in the interdendritic areas and at the grain boundaries, which promotes the formation of inclusions in the eutectic form, (Fig. **17a**).

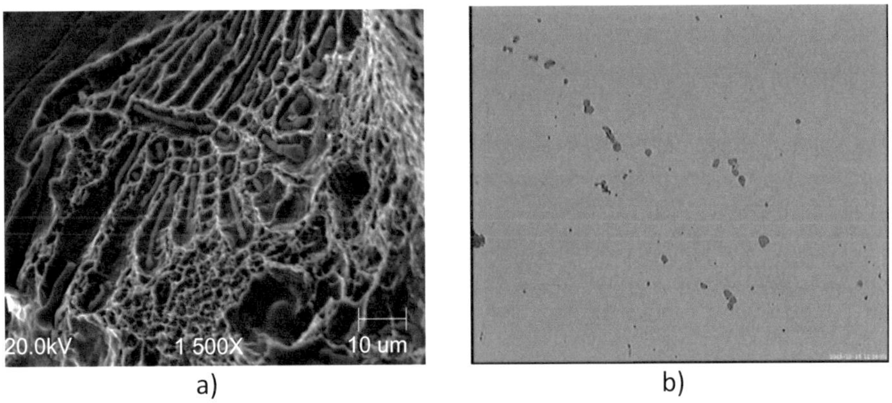

Fig. (17). Type II inclusions in low carbon steel castings: a – SEM after impact toughness test, Magn.1500X, a – SEM after impact toughness test, Magn.1500X, b – optical microscope, without etching, Magn.100X.

Scanning microstructure of eutectic precipitates of II type indicate Fig. (**17a**) that they form extensive colonies of single sulphides in the form of oval rods. On the brittle fracture surfaces, they form a network of extensive eutectics, particularly harmful for plastic properties of cast steel.

Inclusions of the II type are the result of element segregation during the solidification of steel. Elements whose partition coefficient k_0 is small, segregate the most. Assuming that

$k_0 = C_s / C_l$ where: C_s and C_l are the content of the element in solid

and liquid phases at solidification front.

$k_0^C = 0.13$; $k_0^O = 0.10$; $k_0^S = 0.05$; $k_0^{Al} = 0.92$;

$k_0^{Mn} = 0.84$;$k_0^{Si} = 0.66$; $k_0^{Cr} = 0.95$; $k_0^{Ti} = 0.05$ [12].

The lower the value of the k_0 coefficient, the more the element segregates.

Sulphides of the II type are formed under solidification conditions, but at too low aluminium content and higher oxygen content. As a result, the liquid steel has high solubility of sulphur and inclusions nucleate in the last drops of the liquid before its solidification (at primary grain boundaries in the interdendritic areas) [8]. Oxide Al_2O_3 formed after deoxidation of Al is usually surrounded by a sulphide MnS, but it is a conglomerate of two different phases, (Fig. **15**).

***Type III** – during solidification (decreasing of temperature), the segregation of carbon, silicon, manganese and other elements results in the formation of irregular, angular inclusions. From Fig. (**18**) can be seen their angular form with planar planes. This morphology indicates that they nucleate already in the liquid (above the liquidus temperature). They are pure manganese sulphides. Their distribution in the cast steel matrix is accidental, unlike in the case of inclusions of the II type - at the grain boundaries. Inclusions of III type occur most often when, during deoxidation of steel with aluminium, too much of it (more than 0.03 - 0.04%) is introduced. Under these conditions, the oxygen content is low, and at a lower sulphur content (after desulphurisation of steel in the refining period), sulphides nucleate at a higher temperature, earlier than sulphides of the II type. Their characteristic feature is also that they do not contain other phases. Despite deoxidation of steel with aluminium, sulphides of the III type do not contain Al, though nuclei Al_2O_3 form heterogeneous complexes with sulphides [8].

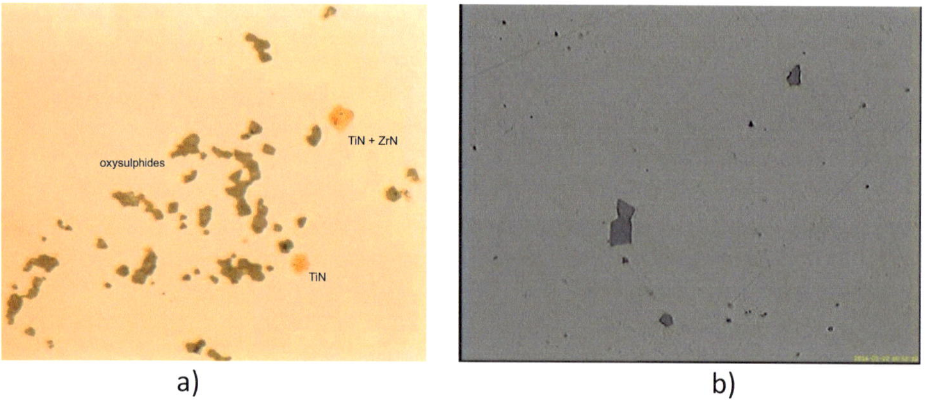

Fig. (18). Third type of non-metallic inclusions in cast steel; a – oxides and oxy – sulphides, Magn.1000X; b – oxides, oxy-sulphides and nitrides (yellow), Magn.1000X

5.1. Conditions for the Formation of Sulphides

The mechanism of the formation of all three types of inclusions is given by C. Sims and C.H.Briggs [22, 24]. It is based on the assumption of an increase in the solubility of sulphur in molten steel, if the solubility of oxygen decreases. This compound was presented in the form of a dependence shown in Fig. (**19**). An increase in the concentration of aluminium after secondary deoxidation of steel in the ladle, indicates ensuring inclusions of the third type in castings. This is important for an engineer, who should provide the required plastic properties of cast steel. However, it should be noted here that an excessive concentration of 0.04% of Al may lead to brittleness, if the nitrogen content (usually over 100 - 120 ppm) is critical in the matrix.

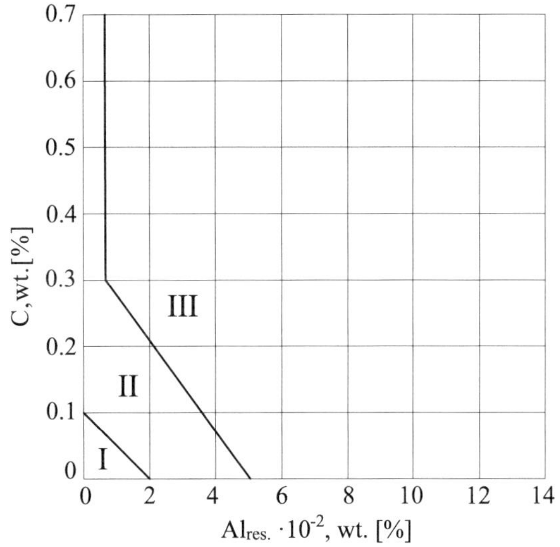

Fig. (19). Conditions for formation of three types of inclusions.

For typical grades of carbon cast steel with the content of 0.2 - 0.3%C, changing sulphides of II into the III type therefore requires the addition of aluminium. Only for final contents of Al above 0.02-0.04% in cast steel, inclusions of the third type are observed. This is confirmed by data of W. Dahl and H. Hengstenberg [33], and others [34, 35]. Formation of inclusions of the third type is obtained even after the introduction of other deoxidisers:

Fe-SiAl after the introduction of 0.5 kG/t ($Al_{res.}$= 0.03%),

Fe-CaSi 4.0 kG/t ($Al_{res.}$= 0.007%),

Fe-CaSiAl 3.0 kG/t ($Al_{res.}$= 0.01%),

Fe-MnSiAl up to 5.5 kG/t only type II.

The above-given values should be treated with caution because during solidification of castings, there are two factors that hinder the process of deoxidation: segregation of elements and the cooling rate in the mould. Therefore, steel for thick-walled castings (g > 70mm) should be deoxidised in two stages

(Al+Fe-CaSi), and the type of moulding sand applied must correlate to the deoxidiser applied in practice. However, the best recommendation is to use oxygen sensors and create dependence of the amount of oxygen deoxidiser introduced from the oxygen activity in steel prior to pouring.

Because sulphides comprise about 80 - 90% of all inclusions in steel [17, 32], their morphology significantly affects the properties of cast steel. The parameters, which characterise the morphology of inclusions, are: chemical composition of cast steel, type of deoxidiser applied, final degree of deoxidation ($\%Al_{res.}$, $\%O_{sol.}$, %C, Si and Mn) and also solidification conditions (cooling rate, number of nuclei, degree of undercooling). In the Fe-MnS-MnO system (Fig. **20**), two monotectics (Fe - MnO and Fe - MnS) have a common line C-D. At low amounts of oxygen in the system (point x) during cooling, iron solidifies first and the remaining liquid is enriched with sulphur and oxygen. During further cooling, at the point O, there is a monotectic reaction $L_1 \rightarrow Fe_m + L_2$, indicating the formation of the oxide-sulphide phase L_2. During further cooling, this phase changes the chemical composition along the line p`- p``- p to the point F, where the liquid disintegrates according to the reaction $L_2 \rightarrow Fe + MnS + MnO$, giving triple eutectics. Under the conditions of high initial concentrations of oxygen, the solidification of first crystals of iron is associated with the enrichment of liquid with oxygen, and as a further consequence, it leads to the separation of the liquid phase of oxides and sulphides with the chemical composition of point B. These oxides contain more oxygen than the oxides from point O.

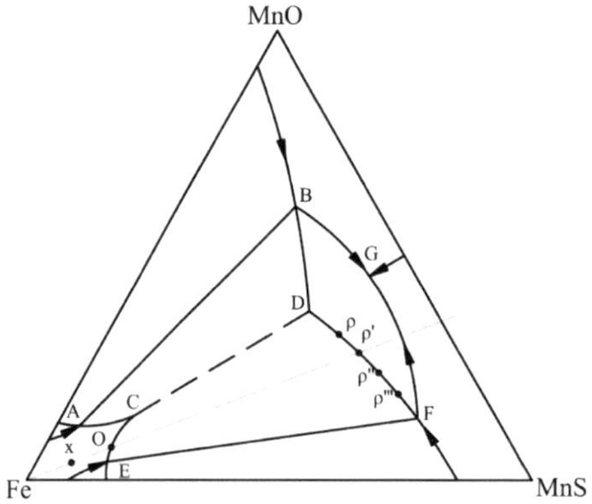

Fig. (20). Phase diagram Fe-MnS-MnO [32].

At a temperature of about 1200°C, the solubility of oxygen and sulphur is lower, and a disintegration of monotectics occurs, which leads to a separation of the oxygen phase in the spherical form. These are oxides of I type, which surround the sulphide phase (oxysulfides).

During deoxidation of steel, the introduced deoxidisers react with oxygen, enriching molten steel with sulphur. This process occurs in the interdendritic areas, and in the last phase of solidification, it leads to the formation of sulphides in the eutectic form, (Fig. **17a**).

In the presence of strong deoxidisers (*e.g.* Al), oxygen is practically bound and the liquid steel is strongly enriched with sulphur, which creates conditions for conversion of sulphides of II to III type.

Under practical conditions, the solidification of steel at the time of non-equilibrium solidification should be considered. Up to solidus temperature T_s^0 (Fig. **21**) steel with composition C_0 has 100% solid phase in the equilibrium conditions. Under these conditions, the first crystals have a chemical composition equalling C_s^1, and the last crystals - C_s^2. However, under actual conditions, both in the molten steel and crystals, diffusion does not occur completely, whereby the element dissolved originally in the liquid iron remains partly in the liquid. The increase in its concentration lowers the solidus temperature along the line 1` to 3`. It should be emphasised that the last portions of the molten steel can solidify even below the peritectic temperature (point 3`). In the end, the part of molten steel that solidifies last (interdendritic areas, connections of three grains, and in the macro scale - hot spots and casting - riser connections) is enriched in elements with a low coefficient of distribution k_0. These are: sulphur, oxygen, nitrogen, carbon and manganese. It is well illustrated by Fig. (**15**) and practical example of precipitates of oxides and oxy-sulphides by Fig. (**16**).

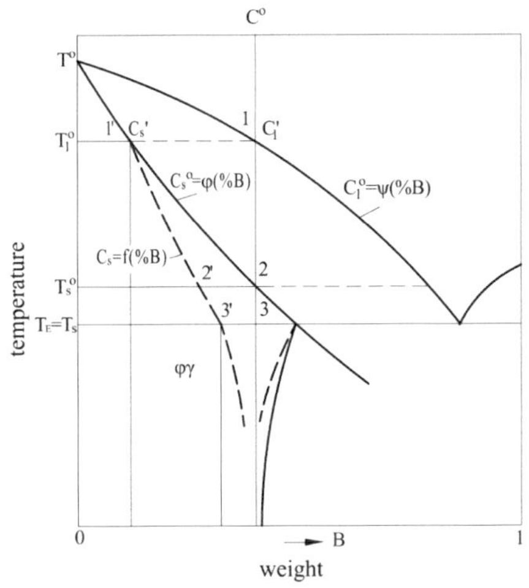

Fig. (21). Segregation of elements with low k_o in real conditions

5.2. Modification of Sulphides

The morphology of sulphides in various grades of cast steel is of fundamental importance in keeping the castings in a variety of operating conditions, *e.g.* at high or low temperatures and corrosion resistance. Thus, modern technologies of steel smelting ensure obtaining sulphur content less than 0.01% (0.005% and below), at the oxygen content less than 20 - 30 ppm, (VOD, VAD, AOD or LF processes). The listed processes of secondary metallurgy are the basis technologies that ensure the required quality of steel for castings.

The characteristics of sulphides in steel are based on two binary systems Fe-S and Mn-S, (Fig. **22**).

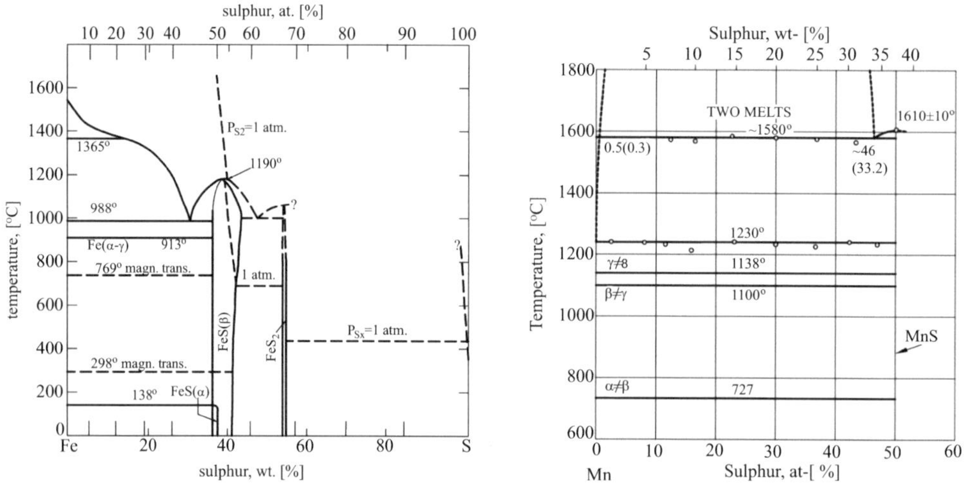

a) b)

Fig. (22). The system iron-sulphur [16] -a; and manganese-sulphur [15]-b.

Although there are other types of sulphides (FeS_2, $Fe_{1-x}S$), FeS sulphide is the most important in steel and cast steel. Its melting temperature equals 1190°C and density equals 4.77 g/cm^3 [8]. Precipitates of FeS may contain additions of up to 20%Cr [36], up to 7% Mn [8], titanium and vanadium [37]. Solubility of carbon and oxygen in precipitates of FeS is small. On the other hand, two-phase precipitates of FeS with FeO next to precipitates of 2FeO with SiO_2 are observed very often.

Virtually, there is no solubility of sulphur in iron in the solid state, although there is sulphur solubility in liquid iron, Fig. (**22**). During solidification, sulphur nucleates as solid, primary phase $Fe_{1-x}S$ (T_m = 1190°C), or eutectics (T_m= 988°C).

At high oxygen concentrations, there are eutectics FeO - FeS with even lower melting temperature T_m= 940^{0}C formed, Table **1**. Due to the low melting temperature, these forms of sulphide precipitates can be observed along the primary grain boundaries, (Fig. **16a**).

Table 1. Physical properties of sulphides and oxy-sulphides of interest in steelmaking [32].

Type of inclusion	**Melting temp. [^{0}C]**	**Density [g/cm^3]**	**Chem. composition, wt.%** **Mn S Fe Ce Ti**	**Ref.**
MnS	1610	3.99	63.1 36.9	[8, 38]
FeS	1190	4.77		[8]
CeS	2450	5.88		[8, 10]
TiS				
ZrS_2	1550			
$Mn_{35}Fe_{65}S$			22.0 36.5 41.5 (different amounts of Mn, Fe in solid solution)	
$Mn_{96}Ti_{03}S$			61.1 37.2 - - 1.7	
CaS	2500 2525 2400	2.18		[39] [38] [32]

Therefore, they affect the properties of steel (called hot shortness), causing brittleness and cracking. The effect of sulphides in cast steel is much more unfavourable, since the primary precipitations of sulphides are not crushed and modified, as during the steel plastic work.

The presence of manganese in steel leads to binding of sulphur in MnS (T_m= 1610^{0}C) at the expense of FeS. Manganese sulphide has a smaller value of free energy than the iron sulphide Fig. (**23**). This conversion occurs in steels in which the ratio of Mn: S is higher than 4.

According to Fig. (**22**), in the system Mn-S, there are three modifications of sulphide MnS (δ, γ, β) and the area of two immiscible sulphides: MnS, and MnS_2 (in the range of 0.3 to 33.5wt% of S). The solubility of oxygen in MnS is negligible, but the precipitations form conglomerates with oxides FeO and silicates. In alloy steels, *e.g.* 18%Cr - 9% Ni sulphides MnS contain certain amounts of chromium (4-12%) and calcium [8]). Application of Fe-CaSi to deoxidation or complex deoxidisers of Al-Ca20, Al-Ca30-Si55, Al-Ca14-Si41, Fe-Ca16-Si48-Ce16 or Al-Ca6-Si23-Ce26 type [35], also leads to a change in the sulphide morphology. These changes are caused by calcium and cerium. Low activity of sulphur after the introduction of calcium, cerium and lanthanum to the molten steel leads to the formation of spherical oxy-sulphides, whose chemical composition is totally different from the sulphides of type I, (Fig. **24**). In the case of steel deoxidation with content of 0.2%C-1.4% Mn by rare earth metals (mischmetal), oval manganese sulphides of complex chemical composition were obtained [29]. Like the III-type inclusions, they are evenly distributed throughout the matrix of cast steel, but have a spherical shape. Also, due to the oval shape of these inclusions, they are classified as type Ib according to Z. Bużek [34], or as the type IV according to C. E. Sims and C.H. Briggs [24].

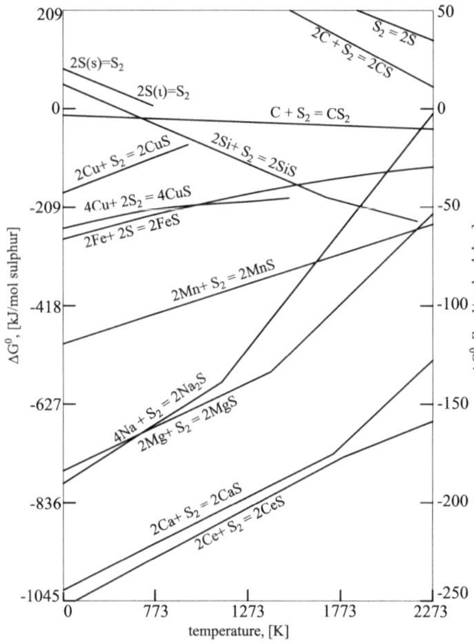

Fig. (23). Formation free energy of sulphides [5, 27].

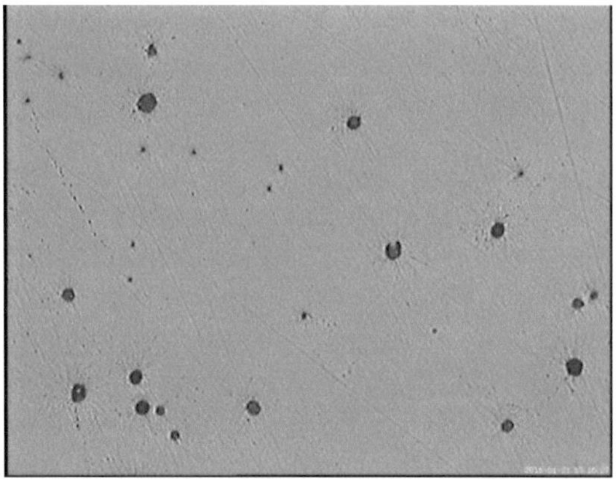

Fig. (24). Morphology of sulphides after deoxidation of low-alloyed cast steel by Ce-additions, Magn.200X [29].

Rare earth metals (Ce, La, Pr, Nd, Y) have a high affinity both for oxygen and sulphur, (Fig. **26**). Therefore, they are used in metallurgical processes as desulphurisers and deoxidisers. They reduce the activity of oxygen and sulphur, but also modify the type of sulphides, leading to a substantial increase in the plastic parameters of cast steel, particularly impact toughness in minus temperature. Due to the high melting temperature (CeS= 2450^{0}C; LaS= 2200^{0}C; Ce_2O_2S= 1950^{0}C), they are formed in the molten steel in the form of polyhedral. The structure of cerium and lanthanum sulphides is closely related to manganese sulphides MnS, hence the suggestions that they form solid solutions of the type (MnS· CeS) [8].

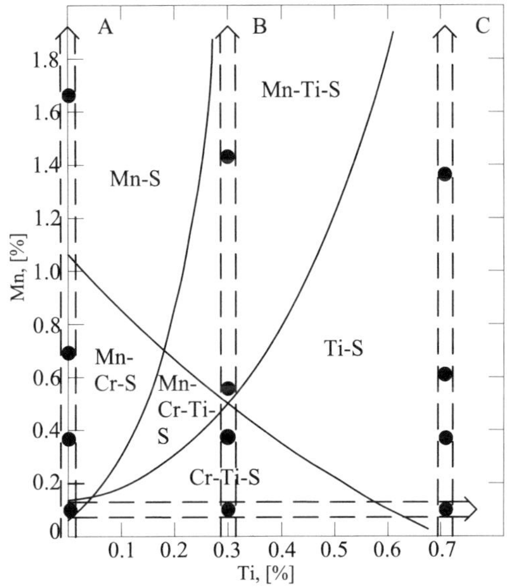

Fig. (25). Effect of titanium and manganese on the formation of sulphides [40].

Sulphides of Ti and Zr

The chemical affinity of elements for sulphur has the following tendency: Zr, Ti, Mn, Nb, V, Al, Mo, W, Fe, Ni, Co and Si (in descending order) [24]. This order, accepted in steel practice, is partly consistent with the data given in Fig. (**23**).

It follows that calcium, magnesium and cerium react with sulphur stronger than manganese. However, while decreasing the temperature (solidification), manganese increases the activity of sulphur in the molten steel [32].

Elements with the highest affinity for sulphur are used in the foundry industry to modify steel in the ladle. Although their additions are small (below 0.1%), and the final concentrations in the metal matrix from 0.01 to 0.06%, their presence has a significant influence on the properties of cast steel. Sulphides of these metals are formed generally at increased sulphur content in the final chemical analysis.

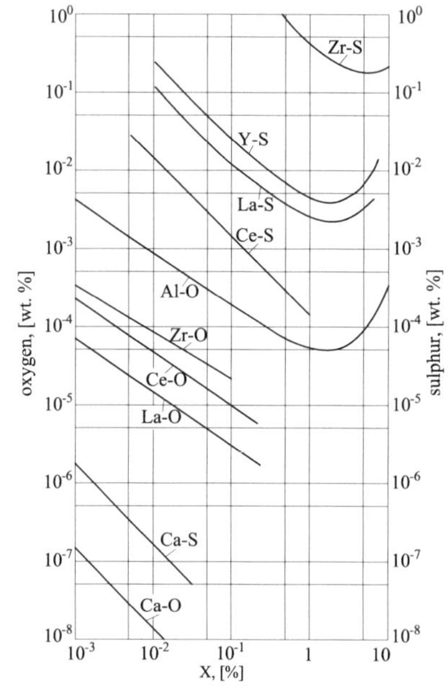

Fig. (26). The free energy formation of oxides and sulphides in the steel [42].

According to F. Jellinek [40], there are four types of sulphides of titanium: Ti_6S; TiS_{1-z} where z = 0.2-0.3; Ti_{1-x} where x = 0.11-0; and phase τ (tau), rich in Ti and S. Sulphides encountered in metallurgical studies are mainly TiS.

The type of sulphides in steel depends primarily on its chemical composition. In steel 0.1%C-18%Cr-2%Mo, changes in sulphide morphology are dependent on the content of titanium and manganese, (Fig. **25**). At the low content of titanium and manganese, sulphides of chromium can be formed, and at higher manganese content, only sulphides of manganese. With the increase in titanium (above 0.25%) and manganese (above 0.6%) content, there are complex sulphides of MnS·TiS formed. At high contents of titanium (over 0.3%), there are only titanium sulphides formed. In cast steel, there are no studies of this type. It should also be noted that, in cast steel, there is not such high titanium content (over 0.3%).

Sulphides of zirconium found in steel and steel castings are mainly ZrS, although according to [40], in the Zr-S system, there are four intermetallic compounds. Sulphides of zirconium have a similar structure to the sulphides TiS and MnS. With the increase in zirconium additions in steel (up to 0.2%), MnS sulphides disappear at the expense of pure sulphides ZrS [41].

Influence of Calcium

As one of the very commonly used precipitation deoxidisers, calcium is introduced into the molten steel in the ladle, in the form of Fe-CaSi or wire feeding of Fe-CaSi powder to the ladle, (calcium treatment of liquid steel). The product of the reaction: (MnS) + [Ca] = (CaS) + [Mn] that takes place in parallel to steel deoxidation by Al additions, are stable sulphides of calcium. After simultaneous deoxidation of steel with Al and alloys containing Ca, following inclusions have been identified (Table **2**; in wt.%). In practice, there are more complex types of inclusions in steel deoxidised with ferro-alloys containing calcium.

Table 2. The possible inclusions after deoxidation of steel by aluminium [30].

Phase	CaO	Al_2O_3	Mn	S	Ca	T_m [^{0}C]	Density [g/cm^3]
$CaO{\cdot}6Al_2O_3$	8	92				~1850	3.38
$CaO{\cdot}2Al_2O_3$	22	78				~1750	2.91
$CaO{\cdot}Al_2O_3$	35	65				1605	2.98
$12CaO{\cdot}7Al_2O_3$	48	52				1455	2.83
$3CaO{\cdot}Al_2O_3$	62	38				1535	3.04
MnS CaS			63	37 45	55	1610 >2000	

These are: oxy-sulphides containing Al, Ca, S, O compounds $CaO{\cdot}Al_2O_3$ as core with a ring of CaS around it $CaO{\cdot}Al_2O_3$ or pure CaS.

In accordance with the affinity of elements for oxygen (E-R diagram, Chapter 2) and free energy of formation of oxides and sulphides, (Fig. **26**), CaO is a more stable oxide than Al_2O_3, hence calcium easily forms calcium aluminates in steel.

Calcium solubility in liquid iron is small, and depends mainly on the carbon, silicon and nickel contents. In low- and medium-carbon steels (up to 0.4%C), calcium solubility equals about 0.03 - 0.04%. Thus, after deoxidation with Al, the following reaction occurs:

$$3[Ca] + (Al_2O_3) = 3(CaO) + 2[Al] \quad \textbf{(16)}$$

and with high sulphur content: [S] + (CaO) + 2/3[Al] = (CaS) + 1/3 (Al_2O_3).

lgK= 48 088/T – 17.687 [30]

Introduction of calcium into the bath facilitates the control of non-metallic inclusions. The inclusions Al_2O_3 formed at the beginning of deoxidation are modified by calcium, and lead to the formation of liquid calcium aluminates. The chemical composition of these aluminates depends on the content of oxygen, aluminium and other components in the liquid steel. Excessive calcium reacts with sulphur, giving CaS inclusions.

Final concentrations of calcium in cast steel lead to a change in the morphology of inclusions: their oval shape is extremely beneficial for the final properties of cast steel (not only carbon, but also corrosion-resistant cast steel). On the basis of metallurgical studies of carbon cast steel (0.2%C - 0.015%S), smelted in arc furnaces, J. Senberger with co-authors showed, (Fig. **27**), that at the content of 40ppm of calcium, inclusions of II type can be completely eliminated. In order

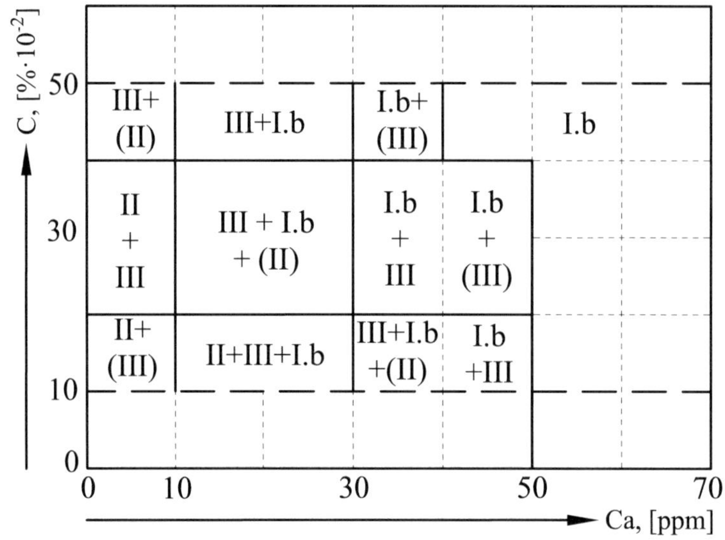

Fig. (27). The effect of deoxidation by Ca-alloys on the type of inclusions [43].

to obtain good impact toughness, the required calcium content should equal 0.045%. However, it is dependent on the sulphur content, in accordance with the following equation:

$$KCV_T = b_o + b_1 \cdot \ln Ca/S$$

where: b_o and b_1 are constants, depending on the temperature and chemical composition of cast steel,

Ca and S - concentrations of calcium and sulphur.

The advantages of application of complex deoxidisers containing calcium are:

- obtaining spherical non-metallic inclusions in cast steel,
- significant improvement of plastic parameters of cast steel (especially impact toughness and reduction of area),
- reduction in the total amount of inclusions in steel owing to lowering their melting temperature and increasing the rate of their outflow from the liquid steel,
- improving machinability of castings.

The Effect of Rare Earth (REs) Treatment

According to Fig. (**26**) the rear earth metals (available in practice as Mischmetall of approx. 50%Ce + 25%La + other REs) have higher affinity for oxygen than aluminium. In the case of using them for deoxidation or desulphurisation of steel, they modify the character of inclusions, especially sulphides. First, oxides (RE_2O_3; La_2O_3) are formed on oxy-sulphides and RE sulphides nucleate.

In the entire temperature range (up to 2000^{0}C), CeS is the most stable sulphide. However, since the compound Ce_2O_2S is more stable than the sulphide CeS, in improperly deoxidised steel (containing a certain amount of oxygen), after the addition of rare earths alloys, oxy-sulphides and impure RE sulphides will be formed. These processes are dependent on the concentrations of oxygen and cerium in liquid steel. At low concentrations of these elements, during solidification of steel, oxy-sulphides RE with a small diameter, flowing slowly from steel, are formed. Control of these precipitations is to maintain the dependence: $[\%\ RE]\cdot[S] < 4\cdot10^{-4}$ [30].

Therefore, steel should be first deoxidised by aluminium or Fe-CaSi, and then the RE-alloys should be introduced. Unlike calcium, rare earth metals have good solubility in molten steel. However, their density is even more related to the density of steel than of other oxides and sulphides, whereby they flow out slowly from the liquid steel.

5.3. The Influence of Non-Metallic Inclusions on the Properties of Cast Steels

In casting practice, any content of non-metallic inclusions in the casting is harmful. Their presence primarily deteriorates the plastic parameters of cast steel (reduction of area, elongation and impact toughness) as well as corrosion resistance, fatigue strength and behaviour of casting at high temperatures (in heat-resistant conditions). The references on beneficial effect of inclusions on improvement of machinability of steel (and steel castings) are not comparable to the previously mentioned inclusion effects.

The increase in the number of inclusions, defined as contamination factor I_{nm} on ductility of carbon cast steel, was presented by Ju. A. Szulte [17] (Fig. **28**).

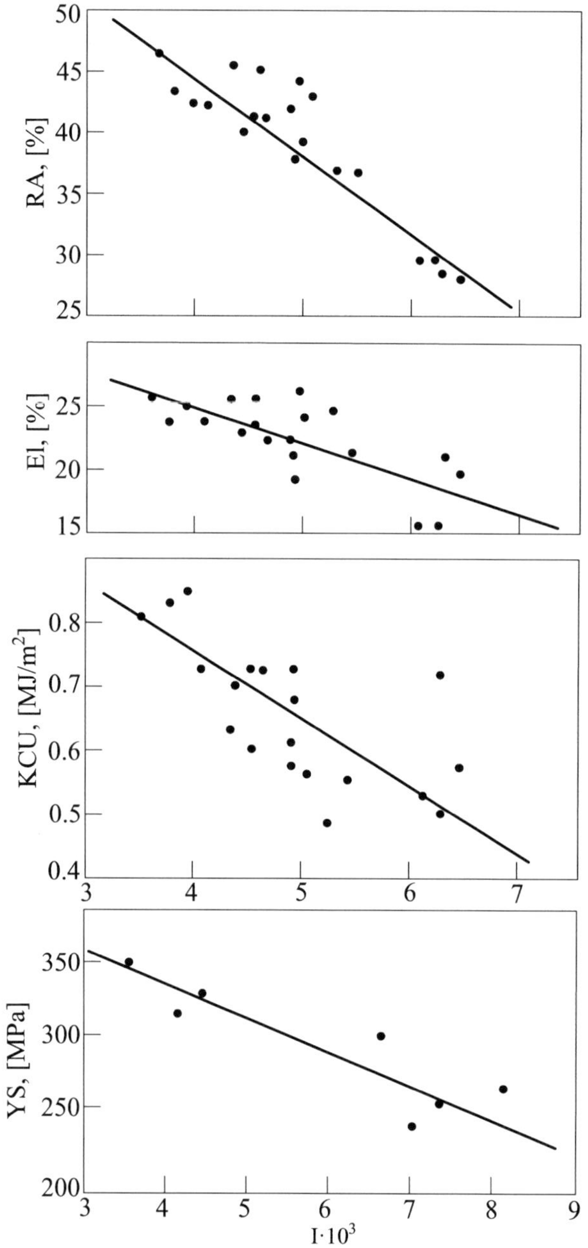

Fig. (28). Effect of inclusion content on the mechanical properties of cast steel [41].

$$I_{nm} = b\Sigma a_i \cdot m_j / l \quad \textbf{(17)}$$

where: b – scale numbering,
a – average inclusion diameter,
m - number of inclusions in each class,
l - the length of inclusions

The data presented in Fig. (**28**) confirm the most adverse effects of inclusions. The most adverse effect of inclusions refers to the reduction of an area and impact toughness of cast steel. These effects are intensified even more at sub-zero temperatures. The data presented in Fig. (**28**) for cast steel, but also studies [44] for steel, indicate that the inclusions also affect strength parameters. The yield strength of cast steel decreases with the number of inclusions. In these dependencies, the volume fraction of inclusions is more important than the size of the inclusions [8].

Such effect of inclusions is an important factor of the need to increase the quality of steel during smelting, as well as elimination of the secondary oxidation of steel during its pouring and filling the mould.

The effect of non-metallic inclusions on steel (and cast steel) properties is based on the principle that each inclusion is a defect of discontinuity of the metal matrix. Hence, the critical value of inclusions (its diameter) determines the conditions of cracking the material.

The other parameter, which is dependent on the morphology of inclusions, is **impact toughness.** Resistance to sudden increase in stress decreases with an increasing amount of inclusions. This dependence is well-documented in the literature for steel [44] and cast steel [17, 35, 41]. As an example, (Fig. **29**) specifies here the results of studies on impact toughness of low-manganese cast steel.

Cast steel of chemical composition; 0.2%C - 1.2-1.4% Mn was double-deoxidised (Al + Fe-CaSi) and argon treated in ladles with a capacity of 8.5 tons. After modification of steel during tapping in the ladle Fe-Ti and Fe-V, changes in the impact toughness were obtained, depending on the modifier used and thickness of the walls. By little additions of both titanium and vanadium, the impact toughness values at sub-zero temperatures can be significantly improved. Here, the effect of vanadium is, however, more beneficial than that of titanium.

It is important that in the castings with 50mm wall thickness, the impact toughness values at minus 40°C fulfil the requirements in industrial conditions. In the wall thickness up to 25 mm, modification with vanadium provides impact toughness of 35 J/cm^2 even at the temperature of minus 60°C. However, at even lower temperatures or in castings with thicker walls, the chemical composition of cast steel must be modified by introducing, *e.g*. nickel, or by lowering the carbon content in cast steel.

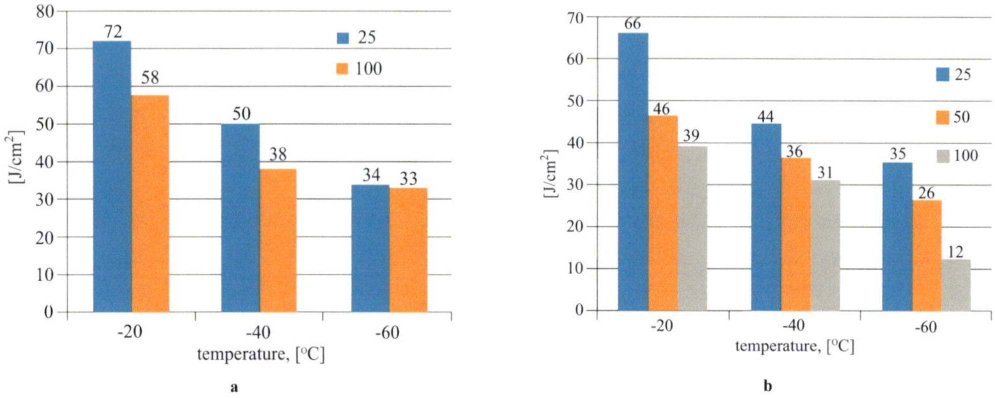

Fig. (29). The effect of temperature on the impact toughness of low-manganese cast steel by modification with vanadium (b) and titanium (a) [29].

At "-40^{0}C", the impact toughness is also closely dependent on the volume fraction of inclusions V_v and the final quantity of modifier, (Fig. **30**). However, it should be emphasised here that the reduction of the quantity of inclusions is affected not only by deoxidation and modification processes, but also by simultaneous argon treatment of steel in the ladle and application of appropriate slags.

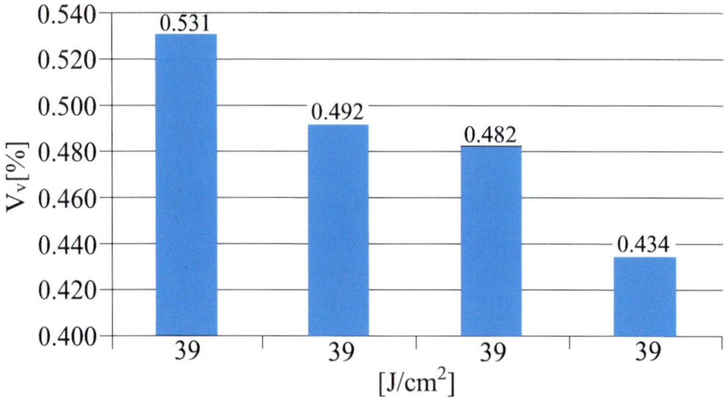

Fig. (30). The influence of volume fraction of non-metallic inclusions and V-additions on the impact toughness of low-alloyed manganese cast steel [29].

The very similar relations obtained in carbon cast steels [13, 22, 40] in low alloyed cast steels [45, 46] and also in corrosion resistant 18%Cr – 11%Ni cast steels [47].

CONFLICT OF INTEREST

The author (editor) declares no conflict of interest, financial or otherwise.

REFERENCES

[1] A. Ghosh, *Secondary Steelmaking Principles and Application.,* C.R. Press, Ed., Taylor and Francis Group, 2016.

[2] A. Ghosh, "Materials and Metallurgical Thermodynamics", In: *Easter Economy Ed.* Prentice-Hall.

[3] E.T. Turkdogan, *Fundamentals of Steelmaking.* The Institute of Materials, The University Press: Cambridge, 1996.

[4] Steelmaking Data Sourcebook, *The Japan Society for the Promotion of Science.* Gordon and Breach Science Publishers, 1988.

[5] K. Leitner, and K. Ploeckinger, *Die Edelstahlerzeugung, 2.* Springer Verlag: Wien, N.Y., 1965.

[6] A. McLean, and I.D. Somerville, *Proc. Int. Symp. on Modern Developments in Steelmaking,* 1981 p. 739 Jamshedpur.

[7] C.E. Sims, and W.D. Forgang, *Electric Furnace Steelmaking* vol. II. USA, 1963.

[8] R. Kiessling, and N. Lange, *Non-Metallic Inclusions in Steel.* Part I-IV, The Metals Soc.: London, 1978.

[9] L.S. Darken, and R.W. Gurry, "The System iron-oxygen II. Equilibrium and thermodynamics of liquid oxide and other phases", *J. Chem. Soc. A,* vol. 68, pp. 798-816, 1946. [http://dx.doi.org/10.1021/ja01209a030]

[10] P.J. Spencer, and O. Kubaschewski, "A thermodynamic assessment of the iron-phosphorus system", *Arch.Eis,* vol. 49, no. 5, p. 225, 1978.

[11] M. Hansen, *Constitution of binary alloys.* 2nd ed. McGraw-Hill with Elliot: N.Y-Toronto-London, 1958.

[12] H. Larson, and J. Chipman, *Iron and steel division-oxygen activity in iron oxide slags, The American Inst.of Mining.* Metallurgical and Petroleum Engrs, 1954.

[13] F.P. Edneral, *Electrometallurgy of Steel and Ferro-alloys.* vol. Vol. I. MIR Publishers: Moscow, 1979.

[14] K.G. Trubin, and G.N. Oiks, *Steelmaking. Open-hearth and combined processes, Izd.Mir.* Moskwa, 1951.

[15] C.L. Coudurier, D.W. Hopkins, and L. Wilkomirsky, *Fundamentals of Metallurgical Processes.* Pergamon Press: Oxford, 1978.

[16] H. Schenck, and H. Wünsch, "Über die Gleichgewichtslöslichkeit des Wasserstoffs im flüssigen reinen Nickel und Eisen und die Beeinflussung im Eisen durch Sauerstoff", *Arch. Eis,* vol. 32, pp. 779-790, 1961.

[17] Ju. A. Szulte, Elektrometalurgija Stalnowo Litja, Metalurgija, Moskwa, 1970.

[18] H. Kaneko, T. Nishizawa, and K. Tamaki, "Decomposition of Austenite in Austenitic Stainless Steels", *J.Japan Inst.Met,* vol. 27, pp. 299-304, 1963. [http://dx.doi.org/10.2320/jinstmet1952.27.7_299]

[19] H. Schenck, and H. Wünsch, "Über die Gleichgewichtslöslichkeit des Wasserstoffs im flüssigen reinen Nickel und Eisen und die Beeinflussung im Eisen durch Sauerstoff", *Arch.Eis,* vol. 32, pp. 779-790,

1961.

[20] Y. Fukai, *The Metal-Hydrogen System – Basic Bulk Properties.* Springer Verlag: Berlin, 2005.

[21] M. Weinstein, and J.F. Elliott, "The Solubility of Hydrogen in Molten Iron-Silicon Alloys", *Trans. AIME,* vol. 227, pp. 285-286, 1963.

[22] J. van Eeghem, and P. Altmann, "A.de Sy, Desoxydation, Morphology of Sulphide Inclusions and Properties of Cast Steel", *30th Intern. Foundry Congress,* 1963 Prague.

[23] A. Sieverts, G. Zapf, H. Moritz, and Löslichkeit von Wasserstoff, "Deuterium und Stickstoff im Eisen", *Z. Physic. Chem.,* vol. A 138, pp. 19-37, 1938.

[24] C.E. Sims, and C.H. Briggs, "A Primer on Deoxidation", *J. Met.,* vol. 11, pp. 815-822, 1959.

[25] M.O. Speidel, and P.J. Uggowitzer, "Cast high nitrogen steels", In: *60eme Congress Mondial de la Fonderie, CIATF* vol. 3. la Hague, 1993.

[26] R. Kirchheim, and A. Pundt, *Hydrogen in Metals,* D.E. Laughlin, K. Homo, Eds., vol. III. 5th ed.. Elsvier, 2014.

[27] G. Wrangler, J.J. de Barbadillo, "Sulfide Inclusions in Steel", In: *Proc. Intern. Symposium,* Port Chester, N.Y., 1974.

[28] A. McLean, and I.Q. Somerville, *Proc. Int. Symp. on Modern Developments in Steelmaking,* Jamshedpur, 1981.

[29] Z. Sierant, *"Technology of heavy castings for minus temperature",* PhD Dissertation, Univ.of Sc.a.Technology, Cracow, Poland, 2015. in Polish

[30] A. Ghosh, *Secondary Steelmaking Principles and Application.,* C.R. Press, Ed., Taylor and Francis Group, 2016.

[31] G.O. Rhodes, and J.J. Conway, "High-nitrogen austenitic stainless steels with high strength and corrosion resistance", *JOM,* vol. 48, no. 4, pp. 28-31, 1996. [http://dx.doi.org/10.1007/BF03222915]

[32] W. Luniow, and W. Awierin, *Sera i Fosfor w Stali, Metalurgija.* Moskwa, 1988. (in Russian)

[33] W. Dahl, and H. Hengstenberg, "Entstehungsbedingungen der verschiedenen Sulfideinschli essenformen", In: *Stahl u. Eisen* vol. 13. , 1966, no. 86, pp. 782-795.

[34] Z. Buzek, and V. Schindlerova, "The effect of additions on the solubility of sulphur in iron", In: *Hutnicke Listy* vol. 7. , 1969, pp. 493-497.

[35] J. Głownia, *"The effect of comlex deoxidizers on the basis of Al-Mn on the sulphide inclusions and mechanical properties of carbon cast steels",* PhD, AGH-Univ. of Sci. and Technology, Cracow, Poland, 1971. (in Polish)

[36] J. Philibert, "L`analyse quantitative en micro-analyse par sonde electronique", *Publication IRSID, serie B,* no. 51, 1965.

[37] R. Vogel, and G-W. Kasten, "Das Zustandsschaubild Eisen-Schwefel-Titan", In: *Arch. Eis.,* 1948, no. 19, pp. 65-71.

[38] K.A. Gschneider Jr, Sulphide Inclusions in Steel, ASM, Materials/Metalworking, Technology Series, Metals Park, Ohio, no.6, 1975.

[39] R.C. Weast, *Handbook of Chemistry and Physics* The Chemical Rubber Co.: Cleveland, Ohio, 1969.

[40] F. Jellinek, "Sulfides of the transition metals of groups IV,V and V", *Ark. Kemi,* vol. 20, pp. 447-480, 1963.

[41] L.J. Kozlow, W.M. Kolokolcew, E.B. Ten, K.N. Wdowin, L.B. Dolgopolowa, and A.A. Filipienkow, *Proizwodstwo Stalnych Otliwok, MISIS.* Moskwa, 2005.

[42] T. Ototani, *Calcium Clean Steel.* Springer Verlag: Berlin, N.Y., 1986.

[http://dx.doi.org/10.1007/978-3-642-82752-5]

[43] J. Senberger, Z. Buzek, A. Zadera, K. Stransky, and W. Kafka, *Metalurgie Oceli na Odlitky, VUTIUM.* Czech Rep: Brno, 2008. (in Czech)

[44] P.F. Wieser, Ed., *Steel Castings Handbook* 20. 6th ed.. Steel Founders` Soc., 1980.

[45] D. Wittekopf, "Qualitätsentwicklung der Stahlgusswerkstoffe im Ergebnis sekundärmetallurgischer schmelzebehandlung", *Giessereitechnik,* vol. 35, no. 4, p. 111, 1989.

[46] T. Wachelko, and M. Soinski, "Impact toughness of L15G cast steel in minus 40^0 C", *51 Intern. Foundry Congress,* pp. 1-9 Lisboa.

[47] Z. Szklarska-Sialowska, "Influence of sulphide inclusions on the pitting corrosion of steels, Intern.Conf.on Localized Corrosion, 6-10 Dec.,1971, Williamsburg, Va, USA. Also in", *Corrosion,* vol. 28, pp. 388-396, 1972.

CHAPTER 5

Deoxidation

Abstract: Deoxidation of steel intended for castings is particularly important. Because rimmed steel, or even half-killed steel, cannot be used for castings, the oxygen content must be as low as possible. Low oxygen content is also associated with non-metallic inclusions nascent during solidification. Their morphology (number, location and size) decides on the mechanical properties in service conditions (*e.g. sub*-zero temperatures and in heat resistance conditions). Elimination of oxygen from steel must be combined with the removal of inclusions and their impact on the quality of cast steel.

Keywords: Effect of deoxidation, Morphology of non-metallic inclusions, Oxygen content, Type of deoxidation.

1. INTRODUCTION

Solubility of oxygen in liquid iron equals 0.16% at the iron's melting temperature (1535°C). During solidification, the value is rapidly decreased due to the reduction in oxygen solubility along with the temperature (particularly at the liquidus temperature), resulting in the formation of gaseous ({CO}) and liquid (SiO_2, MnO, TiO_2 *etc.*) inclusions. Reactions during steel deoxidation occur in the refining period, when initially, after the oxidation period, the oxygen concentration in the bath is high, and at 1600°C, it equals 0.23% [1, 2].

Due to the conditions of desulphurisation and deoxidation, and also the necessity to ensure a final oxygen content below 50-60 ppm (0.05 wt.%) in steel castings during the refining period and during the taping of steel into the ladle, steel deoxidation must be performed. At higher oxygen contents in steel during its solidification, bubbles of {CO} are released, which are the cause of microporosity and reduction in the plastic parameters of cast steel. Therefore, not half-killed or rimmed steels, but only **killed** (deoxidised) steels are used for castings. This is the fundamental technological principle in the process of making steel castings.

The total amount of oxygen dissolved in liquid steel is greater than in the equilibrium with carbon content. So, actual oxygen content (excess oxygen above the equilibrium with carbon) forms oxides with other elements that are present in liquid steel (Al, Si, Mn, S, Fe and Cr, Ti, V, Nb for example). These oxides

Jan Głownia

remain within liquid steel as inclusions and are detrimental on the mechanical properties (mainly plastic parameters: elongation, reduction of area and impact toughness). The deoxidation process then covers the binding of oxygen in proper oxides, and their removal from the bath to the slag before the solidification of steel.

These elements (deoxidisers), which have a higher affinity for oxygen than iron, are used for binding oxygen in oxides. In the Ellingham - Richardson diagram (E-R diagram in Chapter 2), there are elements whose standard free energy ΔG^o is lower than ΔG^o_{Fe} of the iron oxidation reaction (below the red line in the E-R diagram). Whereas these elements, whose ΔG^o is higher than $\Delta G^o_{Fe,}$ do not oxidise in the process of melting and oxidation, and remain in steel before tapping in such concentration as in the charge. This means that, if the alloyed scrap is in the charge, some elements into the charge, they can be recovered by reducing the cost of smelting into the charge. These include: Ni, Co and Cu.

2. TYPES OF DEOXIDATION

There are three basic ways of deoxidation: extractive/diffusion deoxidation, precipitation/final deoxidation and vacuum deoxidation.

Diffusion Deoxidation

The basis of this deoxidation method is the equilibrium of oxygen between liquid steel and slag:

[FeO] = (FeO) with the partition constant of $L_{FeO} = a_{(FeO)} / a_{[O]}$ = const.

The course of deoxidation depends mainly on the reduction of activity $a_{(FeO)}$ by reduction reactions with additions: C, FeSi, Fe-Ca-Si, Al, CaC_2. In the initial period of refining, the so-called deoxidising mixture, with the composition of: $CaO+Fe\text{-}Si+C+CaF_2$ in an amount of 4: 1: 1: 1, is introduced onto the slag. As a result, the following reactions take place in the slag:

$$2(FeO) + Si = (SiO_2) + 2[Fe] \quad \Delta G^o = -299\,960 + 98.13 \cdot T \text{ [J/mol] [8b]} \quad \textbf{(1)}$$

$$(FeO) + C = \{CO\} + [Fe] \quad \textbf{(2)}$$

In the latter refining period, CaC_2 is formed from limestone and carbon around the electric arc (in higher temperature). The calcium carbide has a stronger reduction effect and leads to lower concentrations of (FeO) in slag.

$$(CaC_2) + 3(FeO) = (CaO) + \{CO\} + 3[Fe] \quad \textbf{(3)}$$

The course of diffusion deoxidation depends on the diffusion rate of (FeO) at the metal - slag boundary. This is a slow process, requiring the largest possible interfacial surface. To accelerate the course of these reactions in an arc furnace, where the movement of the bath is limited within the range of the electric arc, induction coils causing additional mixing of molten steel are mounted into the furnace walls, or foamed slags are applied.

Another method is performing the initial deoxidation of the bath before the introduction of refining slag. Reducing the oxygen activity in the bath $a_{[O]}$ by adding aluminium also accelerates the process of deoxidation because, according to the partition coefficient $L_{(FeO)}$ = const., this affects the equilibrium (reduction of $a_{[O]}$ leads in parallel to reduction of $a_{(FeO)}$). The final content of (FeO) in slag should be lower than 1%, (FeO + MnO <3%).

The formation of the reaction product in slag instead of molten steel is the beneficial effect of diffusion deoxidation. A deoxidation time lasting from 40 to 60 minutes is the disadvantage. In practice, the process of diffusion deoxidation is not sufficient for obtaining fully killed steel; therefore, additional precipitation deoxidation must be performed before tapping of steel to the ladle, or in the ladle.

Precipitation Deoxidation of Steel

This deoxidation method consists in introducing a deoxidiser into the molten steel. This treatment is performed:

- after slag-off the oxidising slag (initial deoxidation),
- prior to tapping metal (introduced to the molten steel in EAF)
- in the ladle, during tapping of steel (deoxidiser is introduced onto the bottom of the ladle and then tapping of steel is performed).

The chemical reaction of deoxidiser with liquid steel is performed according to the equation:

$$[O] + [Me] = [MeO] \tag{4}$$

$$\lg K = a_{[MeO]} / a_{[O]} \cdot a_{[Me]} = -\Delta G^{o}/4.575 \cdot T \tag{5}$$

The reaction product is formed in the bath, and then it flows into the slag. The rate of outflow depends on the type of the formed oxide, its size and melting temperature, its interfacial tension and movement of the bath.

Deoxidisers in steel are elements, whose value of free energy is lower than ΔG^{o} of iron. In the Ellingham-Richardson diagram, these are all elements, whose values ΔG^{o} are below the line [Fe] + [O] = (FeO) (marked in red). Deoxidation capability of a given element is the higher, the more negative the ΔG^o value is. However, the cheapest and easily accessible ones are applied in steel practice.

The most common deoxidisers are:

aluminium - introduced into liquid steel as an alloy with a purity of >98%;

silicon - used as Fe-Si alloy, containing from 50 to 90%Si, with various content of carbon (up to 0.15% or up to 0.5%) and the lowest possible content of sulphur and phosphorus (up to 0.05% each); and as: Fe-Ca-Si alloy with the content of 20-30%Ca - 60-65%Si - 1 to 2%Al;Fe-Ca-Si-Al (15-20%Ca – 25-60%Si – 10-30%Al); Fe-Ca-Si-Mn (15-25%Ca – 40-55%Si- 10-20%Mn);

manganese - used as Fe-Mn alloy containing more than 75% Mn - 15%Si

In addition to these deoxidisers (ferro-alloys used for complementing chemical composition before tapping), in practice, there is a number of other multi-component alloys, differing both in the content of Si, Mn, and the content of C, P, S. In steel practice, multi-component alloys containing calcium (7 - 30%), in combination with additions of Si, Al, Mn, Ti, are also available [1, 3, 4]. The oval shape of inclusions and their low melting temperature are beneficial effects of using deoxidisers containing calcium.

3. DEOXIDATION WITH ALUMINIUM AND OTHER ELEMENTS

One of the strongest and cheapest deoxidisers, which is commonly used in steel smelting is aluminium. Deoxidation with aluminium proceeds according to the reaction [2]:

$$2[Al] + 3[O] = [Al_2O_3] \uparrow \quad \textbf{(6)}$$

$$\lg K_{Al\text{-}O} = a_{[Al_2O_3]} / a_{[Al]}^2 \cdot a_{[O]}^3 = 62\,870/T - 19.53\ [2] \quad \textbf{(7)}$$

$$\Delta G^o = (-287\,560 + 89.32 \cdot T) \cdot 4.184\ [J/mol]\ [2] \quad \textbf{(8)}$$

Deducing the formula for the activity of oxygen:

$a_{[O]}^3 = a_{[Al_2O_3]} / \lg K_{Al\text{-}O} \cdot a_{[Al]}^2$

from the equation for constant K, the definition of the element **capability of deoxidation** can be introduced as follows: oxygen activity that is in equilibrium with the activity of aluminium, at the given temperature.

For the pure system Fe - Al - O, D.C. Hilty and W.Crafts [3] presented an apparent equilibrium constant in the form of the equation:

$$lgK' = lg[\%Al]^2 \cdot [\%O]^3 = -58\,600/T + 22.75 \quad \textbf{(9)}$$

which in Fig. (**1**) indicates a significant influence of aluminium on the final concentration of oxygen. At the content of 0.2% Al, it is possible to lower the oxygen content down to 0.002%. The increase in temperature leads to a lower capability of deoxidation with aluminium.

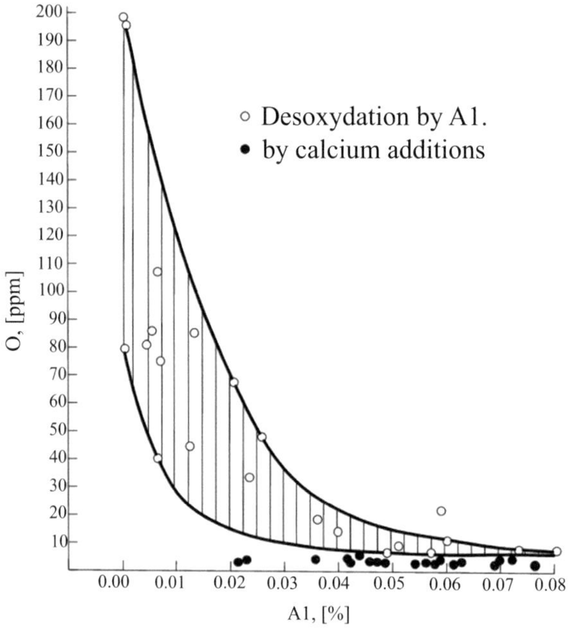

Fig. (1). Effect of Al – and calcium additions on oxygen contents in pure iron [5].

Deoxidation with Silicon

The reaction between silicon and oxygen dissolved in liquid iron is presented in the equation [2, 3]:

$$[Si] + 2[O] = [SiO_2]\uparrow \quad \textbf{(10)}$$

$$lgK_{Si\text{-}O} = lg\ a_{(SiO_2)}) / a_{[Si]} \cdot a_{[O]}{}^2 = 30\,350/T - 11.64\ [2] \quad \textbf{(11)}$$

$$\Delta G^o = (-138\,720 + 53.27 \cdot T) \cdot 4.184\ [J/mol]\ [2] \quad \textbf{(11a)}$$

$$lg\ K_{Si} = 29\,700 / T - 11.24\ [6, 7] \quad \textbf{(12)}$$

Fig. (**2**) indicates the equilibrium between dissolved oxygen and silicon. Already, at contents above 0.2% Si in a pure Fe-Si-O system, the oxygen concentration decreases rapidly, and at a temperature of 1600°C, it equals about 0.012%. This dependence refers to the conditions when the metal bath is under pure SiO_2 slag. Under these conditions, the product of [%Si] $[\%O]^2 = 2.8x\ 10^{-5}$. In the presence of slag containing 90%CaO + 10%CaF_2, the value of the reaction product equals $5.8x10^{-6}$, which means that oxidation of silicon under basic slags is more efficient than under acid slags [2, 6].

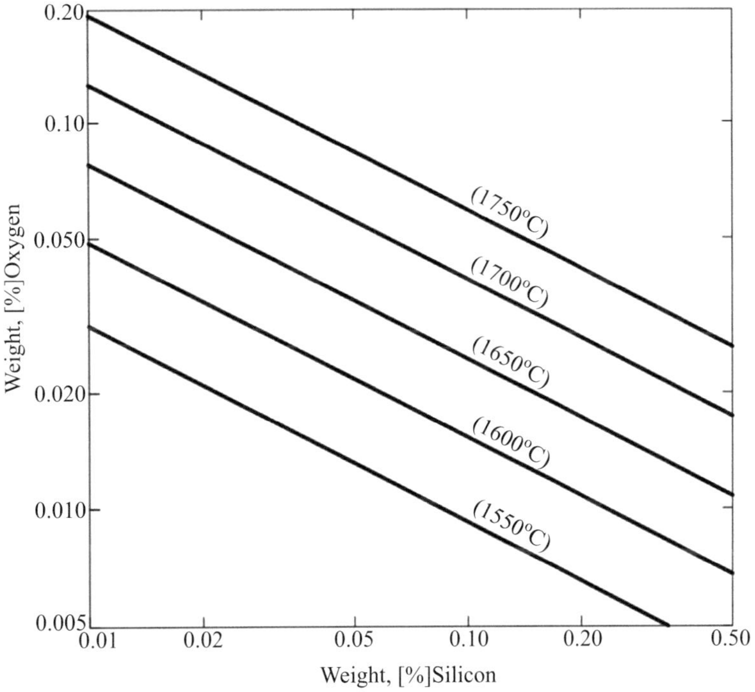

Fig. (2). Equilibrium between oxygen and silicon in pure Fe-O-Si system at SiO_2 slags [2].

A reduction the temperature leads to lower oxygen content. This effect is beneficial, when the temperature declines from the moment of tapping until the steel is poured from the ladle. However, the nascent inclusions of silica have a low tendency to flow out of the bath (they have high viscosity).

Deoxidation with Manganese

Manganese forms ideal solutions with liquid iron, so for the reaction:

$$[Mn] + [O] = MnO_l \qquad \textbf{(13)}$$

the value of free energy of solution of oxygen in liquid iron is [2]:

$$\Delta G^o = (-67610 + 29.6 \cdot T) \cdot 4.184 \text{ [J/mol]} \quad \textbf{(13a)}$$

and the reaction constant K_{Mn}

$$\lg K_{Mn} = \lg (a_{MnO})_l / [a_{Mn}] \cdot [a_o] = 14\,780/T - 6.5 \quad \textbf{(14)}$$

Under conditions when a pure oxide MnO is formed, manganese is a deoxidiser weaker than silicon (compare equations 11a and 13a). In practice, the oxide MnO is bound with oxide FeO, forming complexes that change the chemical composition, along with decreasing the temperature and changing the concentration of manganese in liquid steel. These conditions are illustrated in Fig. **(4)** (chapter 2). There is effective deoxidation of the bath only at a content of more than 0.25% Mn.

Deoxidation by Titanium

Titanium has a high affinity for oxygen (E-R diagram in chapter 2), and hence, a high capability of deoxidation. Due to its price, it is very rarely used for deoxidation. However, in corrosion-resistant cast steel, it is introduced as a carbide stabiliser (instead chromium carbides), and at a content of 0.03%C in cast steel, it can equal 0.15% Ti. With such titanium content, the elements in charge can react with oxygen during the melting period.

For reaction:

$$[Ti] + 2[O] = TiO_2(s) \quad \textbf{(15)}$$

$\Delta G^o = (-150\,358 + 54.56 \cdot T) \cdot 4.184$ [J/mol] [2]

There is data indicating that there is a product other than the oxide TiO_2 in the deoxidation reaction with titanium. Thus, for example, according to [8], products of deoxidation with titanium form large liquid inclusions of coagulated titanium oxides, with different chemical composition [3]. According to [7], these are oxides TiO, TiO_2, Ti_2O_3 and $FeO \cdot TiO_2$. Research by J.Chipman [9] indicates that these titanium oxides occur depending on both the concentrations of titanium and oxygen.

Deoxidation by Rare Earths REs

This group of alloys contains most of all Mischmetal, consisting of 14 elements, mainly cerium 50%; lanthanum 25%, neodymium, praseodymium and others. All these elements have a high affinity for oxygen, nitrogen and sulphur. Due to their price, they are very rarely used in practice for deoxidation. The deoxidation process is presented as a reaction of cerium:

$$[Ce] + 2[O] = CeO_{2\,(s)} \quad \textbf{(16)}$$

$\Delta G^o = 852\ 887 - 250 \cdot T$ [J/mol] [10, 10a, 11]

$$2[Ce] + 3[O] = Ce_2O_{3\,(s)} \quad \textbf{(17)}$$

$\Delta G^o = 1\ 428\ 766 - 362 \cdot T$ [J/mol] [10, 10a, 11]

All cerium oxides (also Ce_2O_3) are in the solid form in molten steel, since their melting temperature is very high (T_m of CeO_2=2600°C; T_m of Ce_2O_3=1690°C [3];). The scope of occurrence of oxides and cerium sulphides is presented in Fig. (**3**). Cerium oxides have density similar to the density of liquid steel (5-6 g/cm^3 [5]); therefore, they flow out of steel slowly.

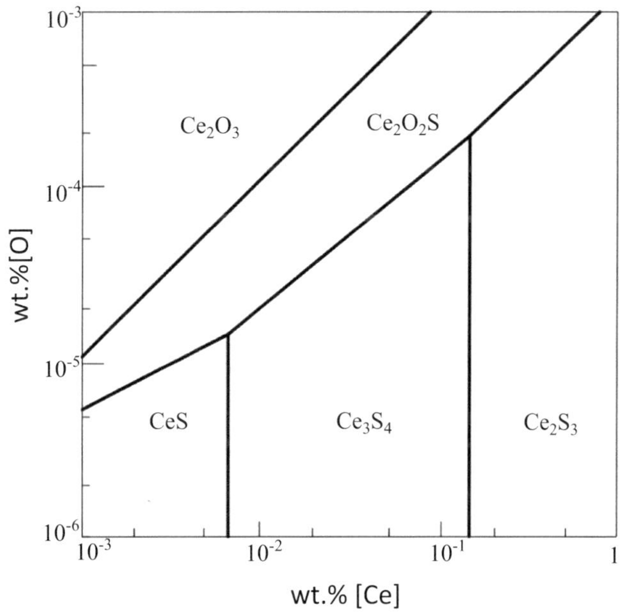

Fig. (3). Phase stability diagram for Ce-O-S system [12].

Deoxidation by Complex Deoxidisers

The deoxidation capability of elements is significantly improved during the oxidation of steel by adding double or multi-component alloys. For example, deoxidation of steel with the alloy Al + Mn allows to obtain a definitely lower oxygen content in steel, (Fig. **4**). Similar effects can be achieved by deoxidation of steel with alloy Al+Mn+Si [3, 8]. When introducing 0.05% of Al + Mn alloy into steel, the oxygen concentration can be reduced twice. During the taping of steel into the ladle, at intensive mixing of liquid steel, there is coagulation of inclusions, their faster outflow and a lower degree of contamination of steel with inclusions are obtained [1]. This is essential in order to decrease the amount of secondary non-metallic inclusions, formed during the solidification of steel.

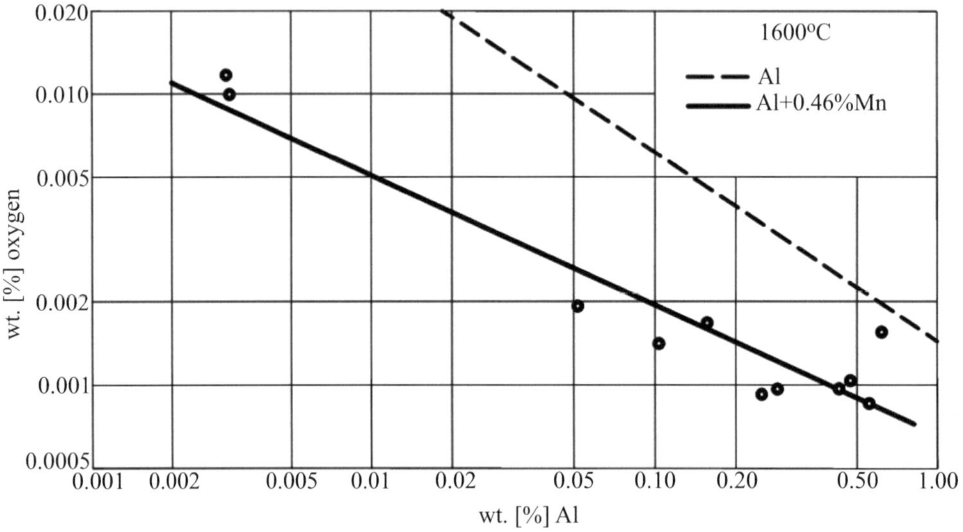

Fig. (4). Comparison of the deoxidation degree of alloys by Al with silicon and manganese [8].

Deoxidation by Ca-Additions

Calcium has a higher affinity for oxygen than other metals (E-R diagram in chapter 2). Additionally, both calcium and other metals of this group (Mg, Ba, Sr) strongly react in molten steel with sulphur, forming stable sulphides. According to the data [13], calcium and magnesium first react with oxygen and then with sulphur (Fig. **26** in chapter 4). Therefore, if these elements are used for desulphurisation, the first bath should be deoxidised with additions of aluminium or with another deoxidiser. Calcium and magnesium are not used in the pure form as deoxidisers because their evaporation temperature is low:

$T_{vap}^{Mg} = 1378°C$,

$T_{vap}^{Ca} = 1693°C$,

and depends on the temperature according to: $\lg p^{Ca} = - 8\,040/T + 4.55$ [1, 14].

The calcium solubility in liquid iron is low: at 1600°C, it equals 0.02 - 0.03% (at the pressure of 0.1 MPa [15]), and according to [16] – 0.024%. Carbon, aluminium, silicon and nickel increase the calcium solubility in liquid iron.

Possibility of the reaction of calcium with: O, S, Sn, Pb, As, P, N is important for steel practice. These detrimental admixtures have a large effect on the plasticity of steel and cast steel, even at very low concentrations (less than 0.001%).

Calcium is introduced into steel as a deoxidiser in the form of alloys: most often Fe-Ca-Si (with a calcium content of about 30%Ca and 60%Si); Fe – Ca(20-28%) - Si (40-55%) - Mg (10-30%); and Al (10-50%) - Ca (10-30%) - Si (25 - 55%) [1]. Liquid alumina-silicates, coagulating slower and flowing out of liquid steel, as well as products of deoxidation reaction with silicon, are the reaction product of deoxidation with these alloys.

The introduction of an deoxidiser containing calcium into steel prior to being deoxidised with aluminium has a favourable effect on the final concentration of oxygen in steel. Already existing inclusions Al_2O_3 react with forming inclusions CaO (dissolved calcium has a higher affinity for oxygen than aluminium) and form conglomerates solid Al_2O_3 - liquid alumina-silicate CaO SiO_2. This way, there is faster outflow of non-metallic inclusions. The effects of this action were presented by B.Tivelius and T.Sohlgren [17]. The combined effect of both of these deoxidisers is characterised by the fact that at a content of 0.02%Al, regardless of the amount of introduced Fe-Ca-Si, the oxygen content in the bath remains constant and is below 10ppm [17]. Application of multi-component alloys, or total additions of Al + Si + Ca for steel deoxidation, leads to improve in the steel quality (reduction in the number of non-metallic inclusions) and a lowering of the oxygen content in steel [17].

4. REMOVAL OF THE NON-METALLIC INCLUSIONS FROM THE MOLTEN STEEL

In the first moments after the introduction of deoxidisers into steel, the oxygen content is not subjected to quantitative change. A large amount of oxygen dissolved in liquid steel passes into the products of the deoxidation reaction (inclusions). The overall amount of oxygen, total oxygen (dissolved oxygen and bound in inclusions) can be reduced by removing the inclusions. Due to differences in the density of the inclusion ($\rho_{incl} \approx 2.5$ - 5.5 g/cm^3) and density of molten steel ($\rho_l \approx 7.0$-7.3 g/cm^3), inclusions flow out of molten steel in a natural manner, but the rate of outflow depends on a lot of factors (Table **1**):

- time
- type of inclusion (its chemical composition, size, melting temperature, interfacial tension inclusion-liquid steel σ_{inter} or viscosity η)

Table 1. Densities of different inclusions [3, 14].

Inclusion	ρ_{incl}	$\rho_l - \rho_{incl}$
SiO_2	2.3	4.8
$FeO \cdot SiO_2$	3.0	4.1
$MnO \cdot SiO_2$	3.6	3.5
Al_2O_3	3.9	3.2
MnS	4.0	3.1
$2FeO \cdot SiO_2$	4.3	2.8
FeS	4.6	2.5
MnO	5.5	1.6
FeO	5.9	1.2

- movement of the bath (argon stirring, induction mixing, movement of metal during tapping into the ladle).

The principle: "inclusion from the molten steel must be removed until the time when the molten steel reaches liquidus temperature", is the basis for the elimination of inclusions from molten steel. In later stages of solidification, between the liquidus and solidus temperatures, when precipitations of dendrites have already been formed, their removal is very difficult, (Fig. **5**).

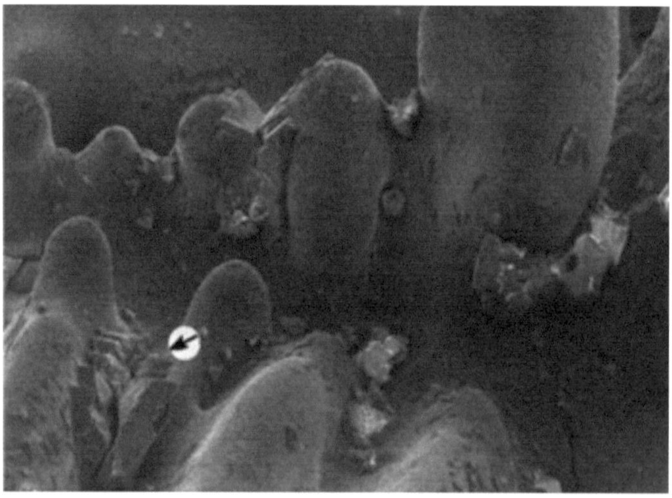

Fig. (5). Non-metallic inclusions on the molten surface of the dendrites in cast steel,Magn.3 500X

Assessment of the inclusion outflow rate V_{incl} was based initially on the Stokes' law:

$$V_{incl} = 2 \cdot g \cdot r^2 (\rho_l - \rho_{incl}) / 9 \cdot \eta \qquad \textbf{(18)}$$

where: g – force of gravity,[m/s^2]
r – radius of the inclusion,[m]
η – viscosity,[kg/m·s]
ρ_l – density of molten steel,[kg/m^3]
ρ_{incl} – density of inclusion.

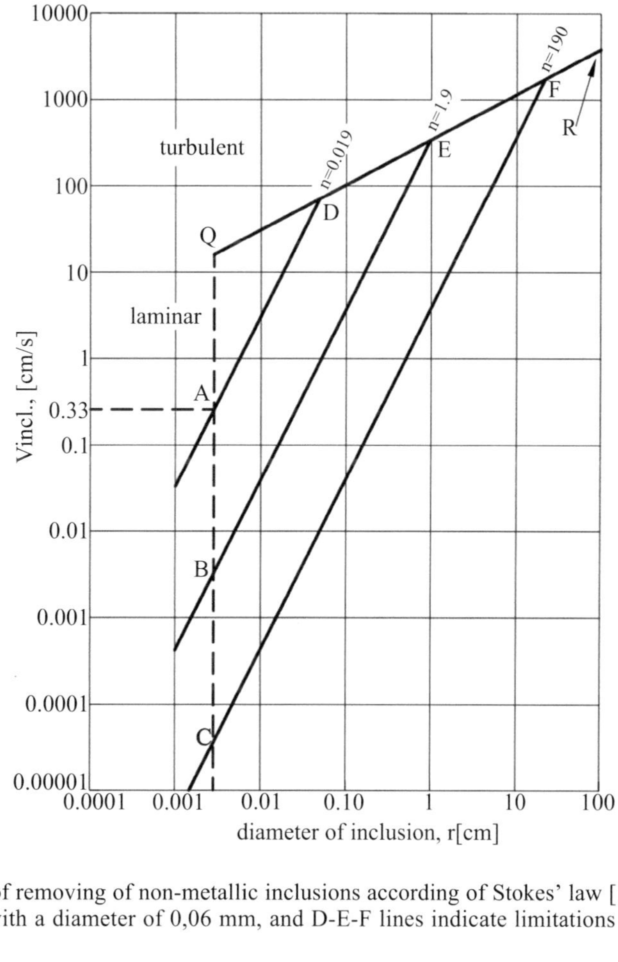

Fig. (6). The velocity of removing of non-metallic inclusions according of Stokes' law [1, 18]. A-B-C lines give V for inclusions with a diameter of 0,06 mm, and D-E-F lines indicate limitations of the Stoke`s law application.

According to this equation, under normal conditions (η and ρ_{incl}), the time of outflow is mainly dependent on r as well as the distance between the place where the inclusion was formed and the metal-slag surface. Inclusions with small diameters (*e.g.* 0.1 mm) have a very long outflow time [2] and cannot be removed from the bath during the time that the liquid steel is kept in the casting ladle (with a capacity of 15 tons -7 minutes) until the moment of pouring it into moulds. Mechanism of removing inclusions from liquid steel should also take into account the type, form and size of the deoxidation product. For example, liquid products of a deoxidation reaction (SiO_2, CaO SiO_2) may coagulate during the outflow, and solid inclusions (Al_2O_3) - are unlikely to increase their diameters.

Practical studies [19] indicate that the concentration of oxygen dissolved in molten steel decreases very rapidly (Fig. **7**), when compared to the total oxygen concentration. These differences disappear only when pouring the steel. In contrast, differences in the number of inclusions remain very divergent, indicating the presence of inclusions when pouring the steel and when its solidification is finished.

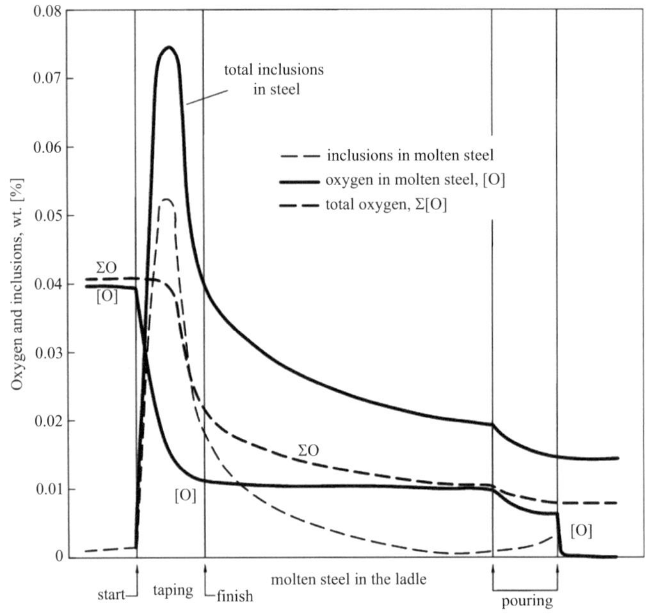

Fig. (7). Effect of time on the oxygen and non-metallic inclusions in molten steel [19].

New remarks to the Stokes` law brought the studies of E. Ploeckinger and M. Wahlster [19]. Deoxidation with pure elements indicates (Fig. **8**) that both large (diameter over 75 μm) and very small (diameter below 50 μm) inclusions flow out of liquid steel within two minutes. Thus, Stokes` law is of a limited scope. These data also indicate that each deoxidiser leads to the formation of a characteristic size of inclusions. The largest ones are formed after deoxidation with titanium and cerium, while the smallest ones after deoxidation with zirconium, silicon and aluminium. Titanium, cerium, aluminium, Fe-Ca-Si and Fe-Ca-Al form liquid reaction products, and zirconium - fine and solid inclusions

ZrO_2 [19].

Comparison of the results of studies by E. Ploeckinger and M. Wahlster (Fig. **7** and 8) indicates that the rate of inclusion outflow is not consistent with their size. Fine inclusions ZrO_2 flow out faster than the inclusions of TiO_2 with significantly larger diameter, (Fig. **8**). These data also indicate that the inclusions of similar size (SiO_2 and Al_2O_3) lead to completely different values of total oxygen content over time, Fig. **(7)**.

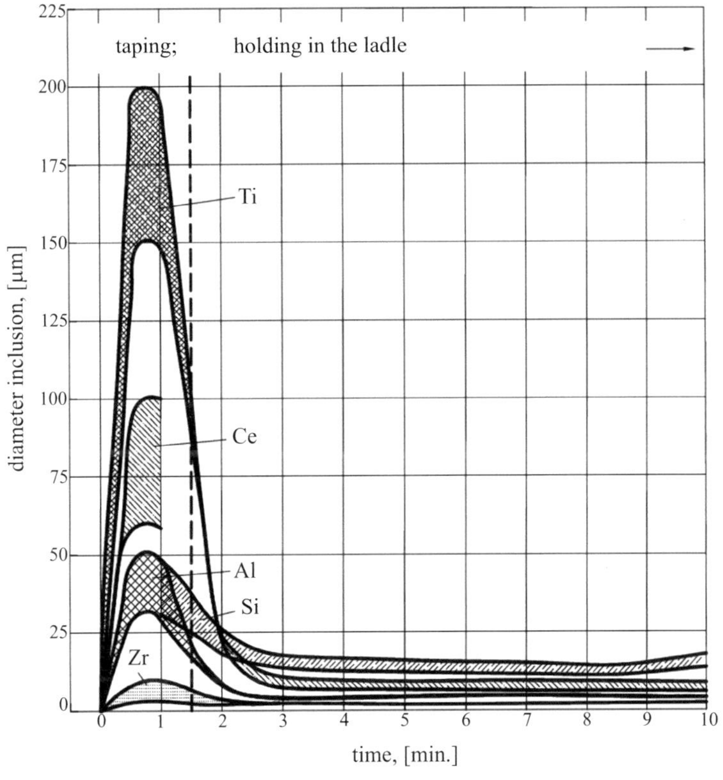

Fig. (8). Diameters of inclusions after deoxidation by different elements [19].

These differences are largest during the taping of steel into the ladle, when the turbulent movement of liquid steel creates conditions for the coagulation of inclusions. According to [3, 19], physical and chemical properties of inclusions (interphase tension of inclusion - liquid steel) affect the rate of inclusion outflow. In the second stage, when the steel is kept in the ladle, the bath is quite and the oxygen removal rate decreases. Another period, in which there is a significant change in the oxygen content, is the process of pouring steel into moulds. It is connected with the process of lowering the temperature of the liquid steel quickly and the formation of dendrites (solidification of steel). A characteristic feature of this period is the rapid decrease in oxygen solubility (chapter 4. Gases). Very often, nascent secondary inclusions (in the liquidus - solidus range) are not able to flow out of liquid steel, (Fig. **5**). Therefore, in order to achieve the least amount of inclusions, the largest amount of oxygen should be removed in the first period of deoxidation (above the liquidus temperature).

In conclusion, it should be emphasised that the preferred conditions for the removal of inclusions from molten steel are related to:

- multi-component deoxidizers that form liquid products during deoxidation,
- creation of conditions for coagulation of inclusions (*e.g.* argon stirring of molten steel in the ladle),
- introduction of these deoxidisers, whose reaction products are characterised by high interfacial tension.

5. DEOXIDATION IN VACUUM CONDITIONS

This method of deoxidation is very beneficial for the reduction of the amount of inclusions in steel because the reaction product is in a gaseous form and can be easily removed from molten steel. The deoxidation mechanism is based on the following reaction of carbon with oxygen:

$$[C] + [O] = \{CO\} \qquad \textbf{(19)}$$

$K_{C\text{-}O} = p_{\{CO\}} / [a_C] \cdot [a_O] = p_{\{CO\}} / [\%C] \cdot [\%O] \cdot f_C \cdot f_O$

$\Delta G^o = -\ 22\ 200 - 38.34 \cdot T\ [J]$

$\lg K_{C\text{-}O} = 1\ 160/T + 2.003$ [3, 20, 25]

By reducing the partial pressure of {CO} over the bath, the course of the reaction is shifted to the right. Therefore, the lower partial pressure of {CO} over the bath, the lower concentration of oxygen and carbon in the molten steel, Fig. (**8**) in chapter 2. Because the effect of temperature on the equilibrium constant is small, the value of the reaction product is calculated by:

$[\%C] \cdot [\%O] = 2.5 \cdot 10^{-3}\ p_{\{CO\}}$

as the so-called Vacher - Hamilton constant (assuming that, at low activity values, they may be replaced with percentages).

Theoretically, the oxygen concentration can be lowered to 0.002% at a pressure of 0.1 atm. and a carbon content of 0.1%. However, such values are difficult achieve in practice because the pressure inside the bubble is higher than the pressure over the bath (vacuum in the furnace chamber).

Nucleation of {CO} bubbles on the walls of the crucible also depends on the existing pressure. The lower the vacuum pressure, the smaller the radius of the cavity in which the gas bubble nucleates.

During long-term keeping of steel in a vacuum, a reaction between the carbon dissolved in liquid steel and refractory of crucible may also occur. The reduction of oxygen from the crucible material (Al_2O_3, MgO) causes an increase in the concentration of carbon, which in total, may lead to an increase of the oxygen content in the bath. In the case of crucibles with MgO, the following reaction occurs:

$$[Fe] + MgO_s = [FeO] + \{Mg\}.$$

Magnesium evaporates as a product of this reaction and the bath is enriched with oxygen. In the case of using corundum or zirconia crucibles, an increase in the content of aluminium or zirconium is observed in molten steel [3].

High vacuum and high temperatures accelerate the reduction process of crucible materials and increase the practical differences between theoretical and practical values of the product [%C] [%O]. The example of holding molten steel in vacuum is show on the Fig. (**9**). After 10 minutes melting in vacuum the oxygen content increased very several times.

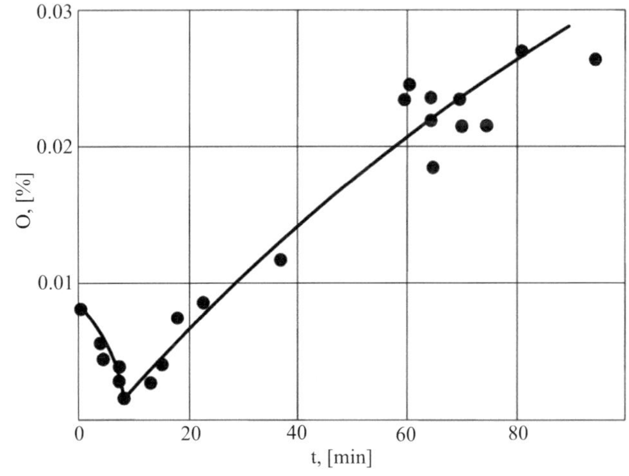

Fig. (9). The changes of oxygen content during vacuum melting of Cr18-Ni9 steel in MgO crucible (at pressure 0.02mmHg) [21].

The beneficial effects of deoxidation in vacuum can be summarised as follows:

- reaction products do not form slag. {CO} bubbles flow out of the bath and are removed through the vacuum into the atmosphere;
- in the case of smelting corrosion-resistant steels (*e.g.* 18%Cr-11% Ni-2% Mo) in a vacuum, simultaneous deoxidation reduces the carbon concentration in steel, which increases its resistance to intercrystalline corrosion;
- the gaseous product of the deoxidation reaction in a vacuum accelerates the outflow of non-metallic inclusions. The collision of {CO} bubbles with existing inclusions in the bath gives them additional acceleration;
- detrimental admixtures present in the liquid steel can be removed from the bath. These elements, which evaporate at a pressure higher than their partial pressure, can be removed from the bath. These are: manganese, copper, tin and zinc. According to the studies [18], in double alloys Fe - X, already at the pressure of 10^{-3} Tor, significant amounts of Pb and Cu can be removed after few minutes.

6. EFFECT OF DEOXIDISERS ON THE MECHANICAL PROPERTIES OF CAST STEELS

Mechanical properties of cast steel are most often characterised by such technological processes as: method of smelting (*e.g.* EAF, VAD, LF) or heat treatment (normalisation, quenching and tempering, annealing). The final deoxidation process in the ladle is also important, especially for plastic parameters: impact toughness, elongation and reduction of area. In the event of impact toughness, an effect of deoxidation appears, in particular at sub-zero temperatures. The direct causes of changes in the properties are: morphology of inclusions (number, size and distribution), segregation of elements along the grain boundaries, solidification conditions (dendrites, globular grains, porosity, shrinkage), grain size and temperature.

The impact of deoxidation is revealed not only by reducing the oxygen concentration, but most often by changing the morphology of non-metallic inclusions. The beneficial effects of deoxidation are induced in order to determine the optimal amount and the type of deoxidisers introduced into the liquid steel. Examples of such studies are included in Table **2**.

Table 2. Effect of Al, Ca, Ce additions on the mechanical properties of cast steel (0.25%C-1.2%Mn-0.6%Si) after normalisation and tempering [22].

Al[%]	Ce[%]	Fe-Ca-Si [%]	Oxygen [ppm]	Nitrogen [ppm]	YS [MPa]	UTS [MPa]	El. [%]	RA [%]	KCU [J/cm^2]	HB
0.10	-	-	40	100	377	630	20.7	29.7	52	178
0.10	-	0.15	30	80	385	667	23.2	44.0	44	187

(Table 2) contd.....

Al[%]	Ce[%]	Fe-Ca-Si [%]	Oxygen [ppm]	Nitrogen [ppm]	YS [MPa]	UTS [MPa]	El. [%]	RA [%]	KCU [J/cm²]	HB
0.10	0.15	-	25	90	325	610	22.2	43.5	84	183
0.10	0.10	0.10	25	80	360	645	24.5	51.0	75	179
0.15	0.15	0.15	25	70	375	642	24.0	44.2	92	183

The best plasticity and strength were achieved during combined deoxidation of cast steel with additions of Al + Fe-CaSi + Ce. Triple deoxidation of carbon cast steel also leads to low concentrations of oxygen and nitrogen.

Beneficial results of strength parameters in cast steel containing 0.3%C- 0.9% Mn- 0.6%Si were obtained in [23], after deoxidation with additions 0.1%Al + 0.4% Fe-Ca-Si.

The results of the deoxidation of carbon cast steel with complex deoxidisers on the alloy matrix Al-Mn are also interesting [24]. In cast steel containing 0.25%C- 0, 8% Mn-0.4%Si, the highest values of strength and plastic parameters were obtained after deoxidation with alloy Al - 12%Mn - 20%Ca, (Table **3**).

Table 3. Mechanical properties of carbon cast steel after deoxidation by Al – Mn additions [24].

Deoxidiser	UTS[MPa]	YS[MPa]	El.[%]	RA [%]	KCU [J/cm²]
Al – Mn14	440-480	320 - 340	30 - 34	60 - 70	140 - 154
Al - Mn30	620 - 660	310 - 370	17 - 18	33 - 35	47 - 71
Al - Mn60	436 - 457	227 - 254	19 - 21	32 - 37	47 - 58
Al - Mn8-Mg15	495 - 504	270 - 285	19.5 - 20.5	30 - 31	48 - 58
Al- Mn11- Ca8	500 - 520	340 - 350	19 - 21	31 - 35	52 - 60
Al- Mn12-Ca20	528 - 557	330 - 360	25	43	65 - 82
Al- Mn12-Ca20	515 - 525	305 - 310	29 - 30	60 - 60.5	108 - 128

These studies were carried out in a basic induction furnace with a capacity of 30 kg. The deoxidisers were introduced with a lance onto the bottom of the crucible, at the amount of 0.15%. The best results are obtained by oxidation with the complex deoxidiser Al-Mn-Ca. The very high reduction of area and impact toughness should also be emphasised. It is characteristic that the yield strength correlates with the plastic parameters.

All of these studies confirm the effectiveness of steel deoxidation with multi-component deoxidisers, especially those containing calcium.

CONFLICT OF INTEREST

The author (editor) declares no conflict of interest, financial or otherwise.

REFERENCES

[1] K. Leitner, and K. Ploeckinger, *Die Edelstahlerzeugung.* Springer-Verlag: Wien, N.Y., 1965.

[2] C. Bodsworth, *Physical Chemistry of Iron and Steel Manufacture.* Longmans, 1973.

[3] D.C. Hilty, and W. Crafts, "The solubility of oxygen in liquid iron containing aluminium", *J. Metals Trans.*, no. 2, pp. 414-424, 1950.

[4] M.O. Speidel, and P.J. Uggowitzer, "Cast high nitrogen steels", *60th Inter. Foundry Congress, CIATF,* vol. 3, 1993 Hague.

[5] A. Ghosh, *Secondary Steelmaking: Principles and Application.* CRC Press, 2016.

[6] T.J. Baker, "Sulfide Inclusions in Steel", In: *Proc. Intern. Symp., Port Chester, ASM, Metals Park, Ohio*, 1974.

[7] C.L. Coudurier, D.W. Hopkins, and L. Wilkomirsky, *Fundamentals of Metallurgical Processes.* Pergamon Press: Oxford, 1978.

[8] C.E. Sims, and W.D. Forgang, *Electric Furnace Steelmaking* vol. II. USA, 1963.

[9] J. Chipman, "The deoxidation equilibrium of titanium in liquid steel", In: *Trans. 218 AIME*, 1960, pp. 767-768.

[10] E. Ploeckinger, M. Wahlster, and P.E. Waudby, "Untersuchungen über die Bildung und Abscheidung von Desoxydationsprodukten", *Stahl Eisen,* vol. 80, vol. 2, pp. 659pp. 74-669-98, 1978196019781978.P.E. (a). Waudby, "Rare earth additions of steel", *International Metals Reviews,* vol. 80, vol. 2, pp. 659pp. 74-669-98, 1978196019781978.

[11] F.D. Richardson, *Physical Chemistry of Melts in Metallurgy.* vol. 1-2. Academic Press INC: London, 1974.

[12] Z. Buzek, and V. Schindlerova, "Vliv nekterych prvku na rozpustnost siry v zeleze, Hutnicke Listy,vol.7,pp.493-497,1969 (in Czech)",

[13] T. Ototani, *Calcium Clean Steel.* Springer Verlag: Berlin, N.Y., 1986. [http://dx.doi.org/10.1007/978-3-642-82752-5]

[14] G. Wrangler, J.J. de Barbadillo, "Sulfide Inclusions in Steel", In: *Proc. Intern. Symposium*, Port Chester, N.Y., 1974.

[15] M. Wahlster, A. Choudhurry, and L.E. Rohde, "Die metallurgischen und werkstoffkundkichen Auswirkungen der desoxydation mit Calcium, Aluminium und Silicium auf die Eigenschaften eines feinkornbaustahles", *Stahl Eisen,* vol. 90, pp. 1013-1019, 1970.

[16] C.E. Sims, and C.H. Briggs, *Electric Furnace Proceedings, AIME,* 1959pp. 104-122

[17] B. Tivelius, and T. Sohlgren, "The influence of different de-sulfurizers on the properties of solid material during TU ladle treatment", *Ironmak. Steelmak.*, vol. 6, no. 11, pp. 38-46, 1979.

[18] F. Hartmann, "Die Bewegung von Schlackenteilchen im flüssigen Stahl", *Stahl Eisen,* vol. 65, pp. 29-36, 1945.

[19] E. Ploeckinger, and M. Wahlster, "Untersuchungen über die Bildung und Abscheidung von Desoxydationsprodukten", *Stahl Eisen,* vol. 80, pp. 659-669, 1960.

[20] A. McLean, and I.Q. Somerville, *Proc. Int. Symp. on Modern Developments in Steelmaking,* Jamshedpur, 1981.

[21] F.P. Edneral, *Electrometallurgy of Steel and Ferro-alloys.* vol. Vol. 1. Mir Publ.: Moscow, 1979.

[22] W.W. Łuniow, and W.W. Awierin, *Sera i Fosfor w Stali, Metałurgija.* Moskwa, 1988. (in Russian)

[23] J. Van Eeghem, and A. DeSy, "Side effects of deoxidation of steel", *J. Steel Castings Res.,* vol. 33, p. 1, 1964.

[24] J. Glownia, *"The Effect of complex deoxidizers (Al-Mn) on the sulphides inclusions and mechanical properties of carbon cast steels"*, PhD., Cracow, Univ.of Sc.a.Technology, 1971 Gordon and Breach Science Publishers, 1988.

[25] Steelmaking Data Sourcebook, *The Japan Society for the Promotion of Science.* Gordon and Breach Science Publishers, 1988.

CHAPTER 6

Secondary Metallurgy for Small Ladles

Abstract: Improvement in the quality of cast steel is based on the application of secondary metallurgy methods. The differences in the techniques used partly result from the capacities of melting furnaces and ladles. In small casting ladles, the temperature of the molten steel falls much faster, which requires attention to timescales with respect to argon blowing or standing in vacuum, or else additional heating of the metal in the ladle. For these reasons, typical techniques for foundries have been presented, such as argon stirring, AOD, VOD, VD, LF and DETEM. For each technique, the benefits and effects of its use in practice have been given.

Keywords: Argon blowing, Decarburisation, Degassing, Desulphurisation, Vacuum technologies.

1. INTRODUCTION

The need to increase the quality of cast steel has led to the introduction of new techniques, smelting processes or furnace constructions in the metallurgical industry. This is not due to the continuous growth of steel production in the world, but mainly due to the material requirements of various cast steel grades in certain sectors of industry. The importance of the following parameters should be emphasised: impact toughness at low temperatures, the operating temperature of heat-resistant cast steel, creep resistance, and near net shape casting. Under all these casting conditions, the quality of the material is the basis of successful marketing. The quality of cast steel is characterised by the amount of non-metallic inclusions in the cast steel matrix, the content of gases (oxygen, nitrogen and hydrogen), sulphur and phosphorus, as well as the content of harmful elements, such as Pb, Sn, Zn, Sb, Bi and As. These parameters also include: the reproducibility of properties, the application of micro-additives (V, Ti, Zr or B) for improving the properties, the homogeneity of the matrix and the surface quality of the casting surface. To eliminate the risks of failing to meet the required parameters, many smelting process technologies have been developed and implemented. However, many of these are for large capacity units (150 - 350 tons), being unsuitable in casting processes, due to the impossibility of performing such a large number of foundry moulds and the difficulty of cleaning such a large

Jan Głownia

number of castings. However, the use of some known metallurgical processes in foundries has been implemented by their application in small casting ladles (with a capacity of 10 to 30, and sometimes 50 tons).

A significant improvement in the quality of steel for castings is obtained by the use of [1]:

- deoxidation process with argon blowing of the molten steel,
- more durable ceramic materials for lining of tapping spouts, ladles and gate systems, and as protective coatings for moulds,
- modifications of non-metallic inclusions by injection of calcium into the melt, double deoxidation and the addition of rare-earth metals,
- introduction of vacuum during the melting and pouring of steel.

The modern steel smelting process has been changed from traditional processes. It is based on two principles:

- so-called primary steelmaking, which is employed for speedy scrap melting in a furnace and transfers all others processes to another melting unit,
- secondary metallurgy in a special unit for further refining, alloying and quality control.

The latter process has received wide acceptance and is currently adopted in steel foundries. Historically, the 20th century brought a number of new techno-logical solutions:

1. application of synthetic slag in the Perrin process, 1933,
2. introduction of Vacuum Degassing (VD) in 1950 - 1960,
3. argon stirring of liquid steel in the ladle,
4. development and implementation of the processes: Vacuum - Oxygen Decarburisation (VOD) and Argon - Oxygen Decarburisation (AOD),
5. injection of calcium silicide with synthetic slags treatment, 1960 - 1970,
6. introduction of continuous casting of steel instead of ingot castings, after 1960,
7. due to the lowering temperature of steel when pouring it from the tundish and pouring steel from the furnace into a ladle, new technologies were developed: ladle furnace (LF), Vacuum Arc Degasser (VAD), ASEA- SKF, and DETEM

At present, there are several design and technological procedures that are a combination of these processes, and lead to the removal of carbon from the bath, desulphurisation, dephosphorisation and microalloying of steel.

2. CRITERIA FOR THE SELECTION OF SECONDARY METALLURGY TECHNOLOGIES

Each of the secondary metallurgy processes has its advantages and limitations. According to [2], they can be divided into the following criteria:

I. Metallurgical

- temperature control, the aim of which is to prevent heat loss and too great a temperature drop after tapping the metal and before the introduction of the ladle into the device (*e.g.* VD or VOD), keeping the liquid steel inside the ladle for a longer period of time;

- removal of gases, mainly hydrogen and nitrogen, as well as oxygen and carbon by the reaction $[C] + [O] = \{CO\}$;

- decarburisation of steel, *e.g.* in corrosion-resistant steels;

- desulphurisation; successful removal of sulphur must be ensured by having a basic lining of the crucible, a low activity of oxygen in the bath, high temperature and intensive stirring of the bath;

- control of the chemical composition of molten steel: this is possible only to a small extent, because large additions of alloying elements cause temperature losses;

II. Technological

- freeboard ensures adequate space above the bath for safely conducting argon treatment, decarburisation or manipulating reaction slag;

- metal: a large volume of freeboard reduces the weight of steel and increases the height of the ladle; the following heights of freeboard are recommended for individual processes:

LF	300 mm
AP/IP	300 – 500 mm
VD	500 mm
VAD	500 mm
VOD	1000 mm

- method of stirring the liquid steel: there are three possibilities, induction stirring, blowing with argon through the bottom of the ladle and injection of argon with a lance;

- methods of heating the bath:

1. by oxidation with Al and the supply of gaseous oxygen; when adding 1kg Al/ton of steel, the temperature can be increased by 25°C [2];

2. by heating of the bath with an electric arc: electrodes work either beneath the sealed cover of the ladle, or additionally at the applied vacuum;

- methods of slagging-off from the ladle.

Criteria for the selection of one of the secondary metallurgy technologies are also dependent on the existing equipment in the foundry and the devices possessed. They can be presented in the following order: energy demand, melting furnaces possessed and their capacity, the transport system in the casting shop, free space to use, and the construction of ladles in relation to the intended secondary metallurgy device.

3. THE CHOICE OF SECONDARY METALLURGY PROCESS

In previous practice, there has been a choice of one of more of the secondary metallurgy processes from among the many solutions that have been implemented (over 50 [3]). In cast steel foundries, a lot of these processes (such as DH, RH and ESR) are not applied, as these are typically metallurgical technologies with a large capacity of 100 to 400 tons. In the vast majority of foundries, secondary metallurgy processes are associated with the smelting of steel in electric arc furnaces (EAF) or in induction furnaces. The following solutions are used in such conditions:

AP (Argon Pouring, argon stirring), VD (Vacuum Degassing),
IP (Injection Process), VOD (Vacuum Oxygen Decarburisation),
TN (Thüssen Niederrhein), VAD (Vacuum Arc Degassing),
AOD (Argon Oxygen Decarburisation), ASEA - SKF
LF (Ladle Furnace) DETEM

The main advantages and possible effects offered by each of these technologies are presented in Table **1**.

Table 1. Technical possibilities of individual processes [3 - 6].

Process	Condition of operations	Effects
AP	Argon stirring + active slag	Desulphurisation, decreasing of non-metallic inclusions, homogenisation of the molten steel.
IP, TN	Injection of CaO, CaSi deoxidisers	Deoxidation, desulphurisation, decreasing and modification of inclusions.

(Table 1) contd.....

Process	Condition of operations	Effects
AOD provided in convector or the ladle	Oxygen blowing to decrease C and Si; argon blowing through a purge plug (in the ladle).	Decarburisation (first step) by the mixture O_2 +Ar, reduction of Cr and Mn from the slag (second step), desulphurisation with new slag (FeSi+CaO+FeCaSi - third step), refining for increasing temperature and alloying (fourth step).
VOD duplex process (EAF + VOD unit)	Melting of steel in EAF, than pouring it to the ladle and installing the ladle into vacuum chamber.	Decarburisation of molten steel by blowing oxygen in vacuum. Stirring of the bath by argon through purge plug installed in the bottom of the ladle. In the process, desulphurisation, degassing and alloying are also possible.
VD ladle with molten steel is placed in vacuum chamber	Under pressure 10-140 Pa and argon blowing – hydrogen, nitrogen and oxygen are decreased.	This process eliminates up to 80% of hydrogen, 50% nitrogen and 60% oxygen.
VAD ladle with molten steel is placed in vacuum chamber	Heating by electric arc under vacuum conditions, active slags are used.	At vacuum pressure approx. 100-130 Pa (first step) degassing occurs (decrease of H,N,O but also temperature). After that temperature and the pressure are increased (second step – deoxidation, alloying). In the third step the pressure is decreased for better degassing and desulphurisation.
ASEA - SKF	This process consists of a ladle furnace, mobile induction coil stand (bath stirring), a vacuum cover fitted with carbon electrodes (heating).	A pressure <100 Pa can give: deoxidation of about 80%, decreasing the sulphur approx. 60-70% and hydrogen approx. 60%.
DETEM: ideal solution for small heats (5-20 tons)	One ladle for tapping, vacuum degassing, oxygen blowing and pouring. Small freeboard, vacuum cover and particular refractory materials.	A temperature 70°C above liquidus, after 20-25 minutes of degassing of 316L and argon stirring gave: <2 ppm of H, approx. 0.005% S and improvement in non-metallic inclusions [14].
VIM melting and pouring in vacuum conditions (castings or ingots)	Prepared charge is heated and melted under vacuum. Deoxidation, modification and alloying also occur under vacuum.	In foundries with capacity up to 40-50 kg, VIM is used for high alloyed castings (mostly 18Cr-11N--Mo, 25Cr-20Ni or Ti alloys). At a pressure of 10^{-1}-10^{-2}Pa, the final content of oxygen is lower than 10ppm, hydrogen <1-2ppm and nitrogen <30ppm.
EMS electromagnetic stirring	The main application is in continuous casting process. An induction coil is installed around the nozzle (first method); it makes an irrotational force and retards the flow); or a rotating EM stirrer is set up around the nozzle (second method).	Higher mixing energy provides better homogenization of the bath than argon stirring. The freeboard in the vessel is not required. The EMS induces force in the liquid steel and not in the slag (slag-free tapping and teeming). Mixing energy is 10-100 times as large as in the ASEA-SKF process with argon stirring [12].

4. PRINCIPLES OF THE PROCESSES AND LAYOUT

4.1. AP process

- blowing of a steel bath with argon brings a number of benefits.

- enhances the homogeneity of the chemical composition and temperature.
- enhances the steel's purity level by improving the flotation of non-metallic inclusions into the slag and the partial removal of gases.

The mechanism of inclusion removal is based on the formation of a thin film at the bubble - liquid steel boundary, which adheres by the suction of the inclusion onto the bubble [7]. Film formation is dependent on the interfacial tension between the inclusion – bubble – molten steel.

- facilitates the reaction course at the metal - slag boundary.
- reduces the tendency to form bigger skulls.

It also has the disadvantage that it reduces the temperature of the steel bath, which hinders the ability to obtain the optimum temperature of pouring steel from the ladle.

In practice, there are two different procedures for purging inert gas, *i.e:*

- by means of an injection lance.
- by means of a purge plug.

In the first method, argon blowing lances are introduced into the ladle from the top, through the slag, (Fig. **1a**). At the end of the lance, there is a graphite head through which argon is blown. The location of the lance has a significant impact on the volume of liquid steel through which argon is flowing, (Fig. **1a**).

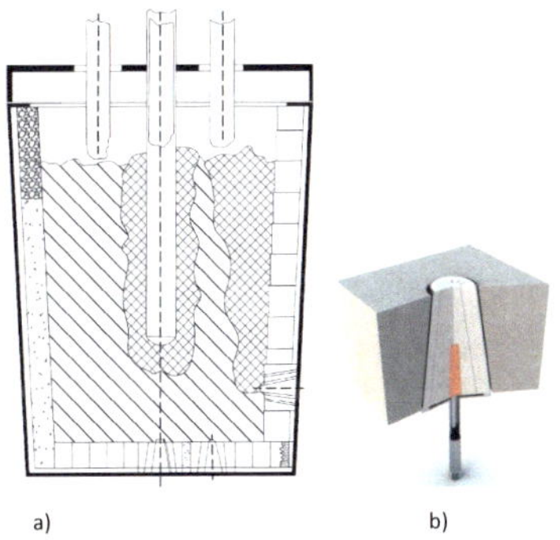

Fig. (1). Argon stirring of molten steel in the ladle: 1- by lance (a), 2- by purge plug (b) [8].

The largest effect of argon on liquid steel is given by the purge plug method presented in Fig. (**2**). In most cases, the purge plug is installed into the bottom (Fig. **1a**), in an area 1/2 - 1/3 of the ladle radius. Such a solution must also meet safety requirements. The purging plug is placed in a ceramic block equipped with a breakthrough safety system.

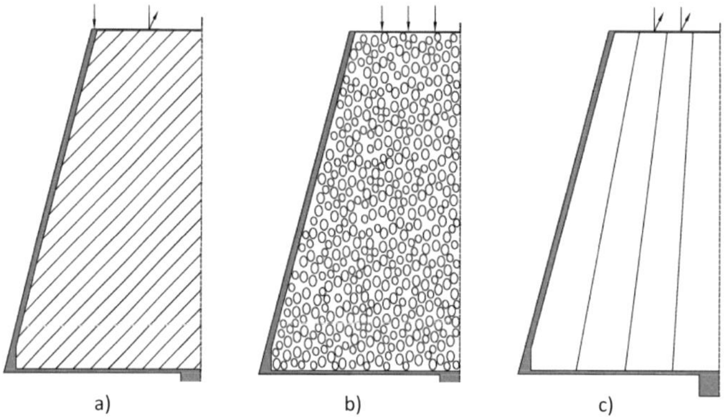

Fig. (2). The possible configurations of argon flow through purging plugs [6, 9].

The appropriate plug is made of magnesia materials, magnesia chromite, chrome - corundum or high-alumina corundum with high purity [7]. The purge plugs are pressed and burnt. There are three basic purge plug designs: one with directional porosity-c, and two with non-directional porosity – a, b, (Fig. **2**).

The initial blowing design consists in bringing argon through the joint between the ceramic cone and the steel sleeve, (Fig. **2a**). Wear attacks the joint and thus affects this area. The degree of safety of this design is not very high [8].

Introduction of porous ceramics changes the location of the liquid steel attack. In this case, the porous material is subject to wear. Argon flows through the grain gaps or by burning out material. The liquid steel infiltrates into the pores and consumption of the ceramic material occurs due to erosion. In the third case, thin directed pores (Fig. **2c**) (diameter below 1 mm), increase the flow of argon. The ceramic material between the pores is subject to wear, since it is exposed to chemical action (reactions between liquid steel and ceramic material), thermal action (temperature shocks) and mechanical action (wear by erosion and infiltration). The broad application of such a design is related to the easier flow of argon and the possibility of using resistant ceramic material on the purge plug. This design can withstand 20 heats, while purge plugs made of non-directional material can survive only 10 heats [6]. Comparison of the properties of purge plugs' ceramic materials can be found in these two papers [6, 7].

The application of argon blowing into the molten steel in the ladle leads to the homogenisation of the chemical composition and the temperature, as well as an improvement in the steel's purity. However, it does not give any clear changes in the content of hydrogen and nitrogen. Reducing the amount and size of inclusions after blowing the steel with argon is an easy and inexpensive procedure, usable in all foundries. In contrast, reduction in the gas content (hydrogen and nitrogen) in molten steel by argon blowing requires very large amounts of argon (2 to 3 m^3/t to obtain 3 ppm of hydrogen) [3]. Such large quantities of argon lead to large drops in the steel's temperature. In small casting ladles, with a capacity of up to 7 tons, blowing of argon for 10 minutes lowers the temperature of the steel even by 150°C [3]. This creates difficult conditions for pouring steel into moulds.

More advantageous conditions of argon treatment are obtained by the combined introduction of synthetic slags. Not only an improvement in the steel's purity, but also its desulphurisation, are then possible. An example of the changes in the content of oxygen and nitrogen during the smelting of Hadfield steel and during its argon treatment in the ladle is shown in Fig. (**3**).

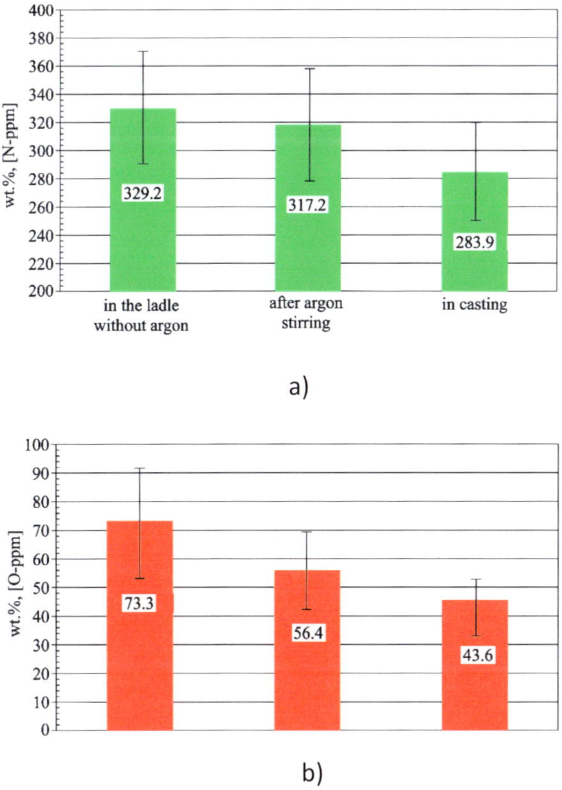

Fig. (3). Effect of argon stirring in the ladle on the oxygen (b) and nitrogen (a) contents [8].

Another increase in the concentration of nitrogen during the refining of steel justifies the usefulness of its subsequent argon treatment in a ladle. There is then a visible reduction in the oxygen content, and a slight reduction in the nitrogen content, (Fig. **3a**). The use of synthetic slags during argon injection or double deoxidation of steel in the ladle, with additions of Fe-CaSi and Fe-Ti, brings beneficial effects in the form of degassing of the molten steel and a reduction in non-metallic inclusions [8].

4.2. IP – Injection Processes

There are several technological designs for this process. Some of them involve the blowing of powder into the molten steel in a stream of argon. In SL (Scandinavian Lancers) technology [10], Fe-CaSi, and in TN (Thüssen Niederrhein) technology, synthetic slag (on a CaO base), are blown into the molten steel [11]. The main objective of the IP is deoxidation, desulphurisation and the modification of non-metallic inclusions, Fig. (**4**).

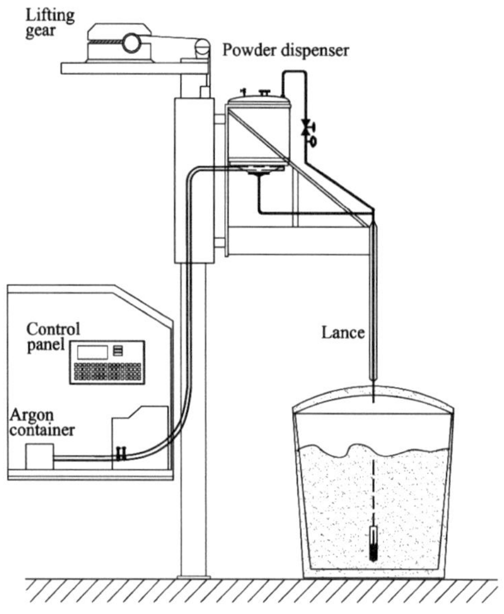

Fig. (4). Principles of powder blowing in SL and TN processes [13].

Introduction of pure calcium into the molten steel leads to its strong evaporation because, at a temperature of 1600°C, it generates a pressure of 1.81 atm [12]. As a result, the residence time of calcium in steel is very short. In contrast, injection of the alloy Fe-CaSi (Ca≈30% - Si≈69% - Al≈10%) gives an evaporation pressure of Ca below 1 atm. When introducing Fe-CaSi at a depth of 1-1.5 m, there can be a reaction of calcium with the dissolved sulphur and oxygen in the liquid steel.

The productivity and efficiency of the injection process depends on the feed rate (which is related to the design and dimensions of the lance outlet), the refractory lining materials (burned dolomite, magnesia, chrome magnesia) and the desulphurisation agents used. For a successful injection process, factors such as the storage and handling of powders are also important, since they can react with air or moisture. The diameter of the particle size should be below 0.1 mm, and the lance for argon injection should be multi-hole (T-shape; multiport).

The ladle lining is a very important parameter for the whole process. The practice of powder injection should be based on the following principles:

- dolomite lining,
- steel is pre-deoxidised with aluminium,
- introduce synthetic slag while tapping,
- inject CaSi powder at 0.1- 0.3 kg/t for 8 to 10 minutes at an argon flow rate of 0.001 - 0.003 Nm^3/ min.

Newer designs for introducing CaSi powders are based on the so-called wire feeding. Powders of metals or alloys are introduced *via* a cored wire (with a diameter of 3 - 18 mm and a sheet metal thickness of 0.2 - 0.6 mm) wound on coils (in sizes from 100 to 3000 kg). Injection is carried out at either a ladle station or ladle furnace (LF) by using specially designed machines. Cored wires are introduced into molten steel by pneumatic or hydraulic methods. Injection stations have been designed with the capability of feeding in many different materials (Al, Fe-CaSi, C, Fe-Ti, Fe-Mn, Fe-Si and others). The primary objectives of this technology are: deoxidation of steel (by Al, Fe-CaSi or Ca-Al [14]), desulphurisation, completing the chemical composition adjustment of alloy steel, the modification of non-metallic inclusions and an improvement in the castability of the molten steel.

Wire Applications

- aluminium feeding: the main advantage is reduced Al consumption. The recovery of Al by using this method is 80 – 90% [15, 16]. Another advantage is that nitrogen pick-up can be reduced.
- calcium treatment: is used for deoxidation, desulphurisation and inclusion modification in steel. The use of calcium is successful with Al-killed steels, in which it forms liquid calcium aluminates. These inclusions agglomerate and float to the slag during argon blowing. The recovery of calcium in cored wire processes is higher than with powder injection [17].

In steels deoxidised with Al, calcium recovery reaches 23% [14]. This ratio increases in steels with higher carbon or silicon contents and in those with lower sulphur contents (less than 40 ppm). To reduce the vaporisation of calcium, it is important to regulate the injection rate to 0.1 kg of Ca/ton per minute.

Calcium treatment enhances the formation of globular inclusions and reducing the tendency of MnS to accumulate at grain boundaries. The effect of the injection of CaSi and CaAl-wire feeding is shown in Fig. (**5**). By modifying non-metallic inclusions, the reduction of area becomes more than twice as great [18].

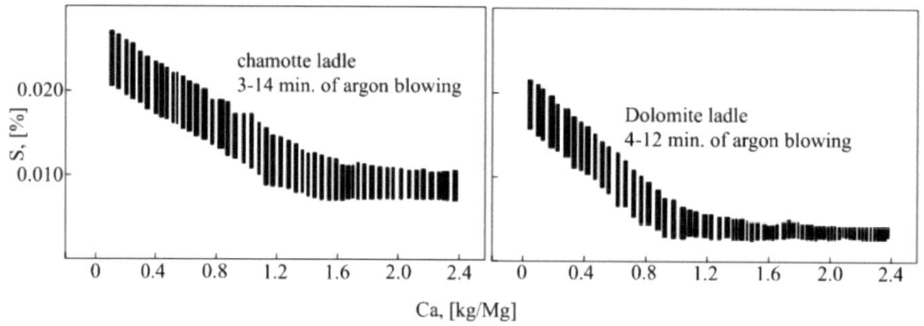

Fig. (5). Effect of CaSi injection and CaAl-wire feeding on the reduction of area in structural steels [13].

- alloying: application of powdered alloying elements inside the cored wire allows for their introduction directly into the bath, without contact with the slag. This way, oxidation is avoided. This method of introducing alloy additions increases the yield to 90 - 98% for such elements as: Fe-B, Fe-Ti, Fe-Nb, Fe-Mn and Fe-Si, and up to 95 - 100% for C [15].

The technique of introducing alloy additions into the liquid steel consists of previous deoxidation during tapping into the ladle, blowing with argon, then checking the chemical composition and temperature, and only then introducing the alloy additions with cored wire. It should be emphasised here that some of the elements have a melting point below the liquidus temperature of steel (Al, Fe-Mn, Fe-Si, Fe-SiMn, Fe-CaSi), and some have higher (Fe-V, Fe-Nb, Fe-Mo). Dissolution of the latter can be accelerated by longer argon treatment and greater fragmentation.

5. SECONDARY METALLURGICAL UNITS WITHOUT VACUUM TREATMENT

This group of secondary metallurgy processes covers a smaller quantity of procedures, although one of them (LF) is widely used in the foundry industry. The following processes should be mentioned here:

- ladle furnace (LF),
- ladle treatment station (LTS),
- desulphurisation station (MHD),
- chemical heating station (CHS).

LF Process: during the process of degassing or desulphurisation, there is a reduction in the temperature of steel. The temperature losses have to be compensated either by superheating in the melting furnace prior to tapping, or by using a ladle furnace (Fig. **6**).

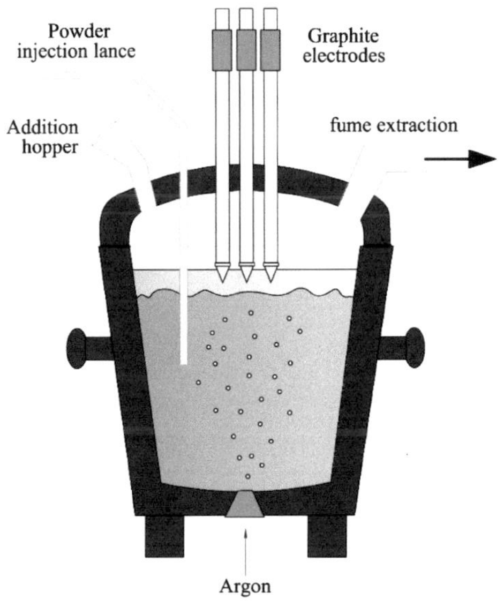

Fig. (6). Schematic drawing of LF process.

However, overheating of steel in the furnace reduces the melting performance, and increases its gassing, deoxidation and desulphurisation, alloying and slag management. A ladle furnace usually works together with an EAF, where the charge is melted, dephosphorised and decarbonised, and the whole refining process (desulphurisation, diffusion deoxidation and basic alloying) is shifted to the LF. Thereby, the efficiency of melting units is improved and the control of the steel quality is performed in a specialised station. An LF unit is typically constructed to enable argon purging from the bottom, electric arc heating through the top cover, sampling and temperature measurement.

LTS Process: applied among large units for melting and continuous deoxidation of steel with Al.

- homogenisation of steel, adjustment of chemical composition by wire feeding and alloy additions.
- homogenisation of liquid steel by bottom purging.
- desulphurisation by addition of synthetic slag or by powder blowing.

It is not applicable in cast steel foundries due to the large capacity of liquid steel.

MHD Process: applied mainly for the desulphurisation of steel in large furnace units (*e.g.* 350 tons). Pig iron from the blast furnace contains 0.04 - 0.12% of sulphur and requires desulphurisation to a level below 0.01% before blowing in an oxygen converter. In the MHD process, the liquid metal is desulphurised by a metal-slag reaction, and then its chemical composition and temperature are homogenised to the level required in subsequent technological processes.

CHS Process: introduction of aluminium and blowing of pure oxygen increase the temperature of liquid steel. The addition of 1kg Al/t and the introduction of 0.6 m^3 of oxygen increase the temperature by 25°C [2]. The temperature increase is due to the exothermic reaction of aluminium dissolved by oxygen blowing. The highest temperature increase is given by [4]:

0.1%Al +34°C

0.1%Si + 33°C

0.1%Cr + 14°C

0.1%C + 11°C

0.1%Mn + 7°C

The addition of Al is more preferable, because silicon results in the formation of SiO_2 inclusions, which react with the basic lining of the furnace and reduce the basicity of the slag.

The Al_2O_3 inclusions resulting from the process should be modified with additions of CaO at a level of 1.8 kg per 1kg Al [19]. This process is also applicable to large units with a capacity of several hundred tons (for converter processes, *e.g.* VOD and VODC).

AOD Converter Process: the principle of the process consists of blowing oxygen and argon into liquid steel containing more than 18% chromium. As a result of the reaction

$$[C] + [O] = \{CO\} \quad (1)$$

and the reaction constant: $K_{[CO]} = p_{[CO]}/ a_{[C]} \cdot a_{[O]}$

inside the converter, there is a reduction in the partial pressure $p_{[CO]}$ and, as a further consequence, a reduction in the activity (and hence concentration) of both oxygen and carbon. In this process, a carbon content below 0.03% can be obtained with a chromium content of about 18%.

An AOD converter works either with an EAF or with an oxygen converter (LD). After tapping steel into the ladle, the temperature is measured, slag taken off and the chemical composition is controlled. The molten metal is then poured from the ladle into a converter. The converter lining has a significant impact on the process of smelting of corrosion resistant steels in the AOD converter. Magnesia - chromite bricks (>50% MgO) are used as the protective working lining. The durability of the lining depends not only on the type of the ceramic materials applied, but also on the technology used for smelting and the conditions prevailing in the steelworks (foundry).

The process includes periods of oxidation and reduction. In the initial stage of oxidation, pure oxygen or a mixture of oxygen and argon in a ratio of 3: 1 are injected. Only argon is blown during the reducing period. Bubbles of {CO} and pure argon lead to the removal of hydrogen and nitrogen from molten steel, because the partial pressure of hydrogen and nitrogen inside these bubbles is virtually zero, allowing atoms of hydrogen and nitrogen to diffuse into the bubbles. The outflow time of the bubbles, their size and the amount of injected argon determine the level of degassing in the bath.

The technology of steel smelting in the AOD process is shown below, (Fig. 7):

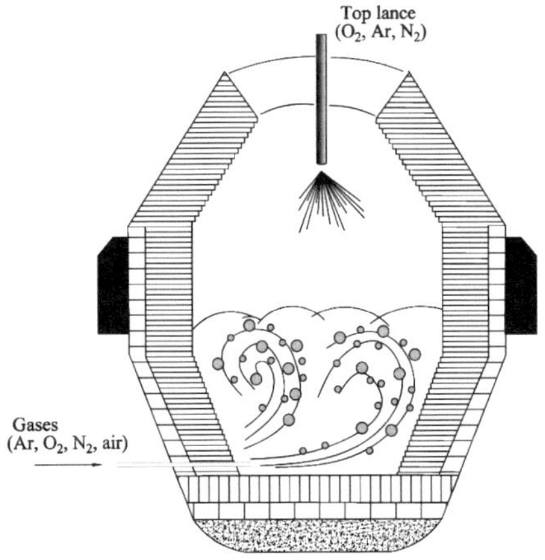

Fig. (7). Schematic layout of AOD process.

1. charge, *i.e.* liquid steel from an EAF furnace, is poured from a ladle to a converter. The initial temperature of steel in the converter should equal 1520-1530°C. The chemical composition of the steel should be close to the required one, with the exception of carbon, sulphur and silicon. In fact, the levels of these elements guide the correct course of the process (as they are removed from the steel). Typically, the selected chemical composition contains up to 1.5% carbon, up to 0.5% Si and 0.05 - 0.15% S.

The process consists of the following periods: decarburisation, reduction, desulphurisation and refining.

Decarburisation (oxidation) is initiated by the blowing of a mixture of oxygen and argon, starting with an oxygen-rich mixture and ending with a mixture with a greater participation of argon. The composition of this mixture is dependent on the ratio of carbon concentration to nitrogen, Table **2**. The primary objective of this period is the maximum removal of carbon, with the lowest possible chromium oxidation.

Table 2. Recommended mixtures of oxygen to argon.

Ratio O_2/N	**5:1**	**3:1**	**2:1**	**1:1**	**1:2**	**1:8**
%C	Till 1700°C	0.55	0.35	0.20	0.04	0.02

The oxidation period is carried out not only with a variable ratio of O_2/Ar, but also with a variable rate of delivery. In converters with a capacity of 8 tons, initially 0.4 m^3/t·min of oxygen is provided and the final rate is 0.24 m^3/t·min. The appropriate values for argon were 0.17 - 0.48 m^3/t·min [20]. The consumption of the mixture equals 1 Nm^3/t·min at a pressure of 1 - 1.2 atm.

Reduction: after obtaining the required carbon content, there is a reduction of the chromium oxides and manganese in the slag. Owing to additions of Fe-Si and CaO, more than 95% chromium and 93% manganese are recovered [20]. During this part of the process, argon is injected intensively for 5 - 10 minutes, the temperature is measured and a sample is collected for designation of the chemical composition.

Desulphurisation: while not interrupting the injection of the mixture, 2/3 of the slag is taken off, and new portions of slag (CaO + CaF_2) are added for the desulphurisation of the steel. Again, additions of Fe-Si, Fe-CaSi and CaO are introduced with continuous argon treatment of the steel.

Refining: this period is intended for the control of chemical composition, with continuous argon treatment of steel. Ferro-alloys are introduced into the bath, in order to complete the missing elements. Temperature control determines the moment of tapping. In the case of a too high a temperature, steel scrap with a similar chemical composition is introduced into the bath.

The total smelting time is about 90 minutes, while the oxidation period is 50 minutes and the time of argon treatment is 15 minutes.

Advantages: The AOD process allows a composition of up to 0.01%C, below 0.01%S and less than 6 ppm of hydrogen to be obtained in corrosion resistant steels [21]. A significant reduction in non-metallic inclusions, up to 70% less than in molten steel in an EAF, is very beneficial [20]. Improvement in technology provides new outlets and a new field of purchasers.

Limitations: Capital costs of a new metallurgical unit, additional costs of production, additional space required in the foundry and more staff needed.

6. VACUUM PROCESSES

The introduction of a vacuum to the smelting processes brings a number of benefits:

1. allows for the removal of hydrogen, nitrogen and oxygen.
2. with application of oxidation with gaseous oxygen, decarburisation is also possible.
3. enables the lowering of impurities through evaporation processes.

The principle of removing oxygen is given in chapter 5, Deoxidation. With lower pressures, the deoxidising capacity of carbon is higher, so that less oxygen remains in equilibrium with carbon. However, the deoxidising capacity of carbon remains unchanged at pressures from 10^{-3} to 10^{-5} mmHg. It is related to the kinetics of the reaction $[C] + [O] = \{CO\}$. The partial pressure of $\{CO\}$ concerns only the vacuum and the surface of the bath; inside, the pressure of $\{CO\}$ must be higher. Deoxidation of steel only by the reaction of carbon and oxygen is incomplete, especially in low-carbon steels. It is then necessary to introduce strong deoxidisers (with a high deoxidising capacity). Removal of oxygen from liquid iron is possible if the vacuum pressure is lower than $2.25 \cdot 10^{-5}$ mm Hg [22].

Removal of nitrogen and hydrogen from liquid steel is in accordance with the Henry's law. This is multi-stage, and one of the most important steps of the process is the removal of gas molecules from the interface. In addition to the vacuum pressure, the final nitrogen content is affected by the presence of active elements in the bath. Oxygen has the largest impact; therefore, before the process of removing nitrogen and hydrogen, the bath should be deoxidised with additions of species such as aluminium.

Evaporation of impurities

The possibility of removing detrimental admixtures from steel is an extremely beneficial advantage of degassing in a vacuum. This applies to such elements as bismuth, antimony, arsenic, lead, zinc, tin or copper. One of the most important factors determining the evaporation rate is the vapour pressure of an element above molten steel at a given temperature. These values are shown in Table **3**.

Table 3. Vapour Pressures of some elements in molten iron at 1600°C [22, 23].

Element	Vapour pressure, [Torr] Pure Fe	in 1% solution	Element	Vapour pressure [Torr] Pure Fe	in 1% solution
Zn	$4 \cdot 10^4$	-	Mn	20	0.25
Bi	10^4	-	Sn	1.9	$4.7 \cdot 10^{-3}$
Mg	1150	-	Cu	0.8	$9 \cdot 10^{-3}$
Sb	1000	-	Si	0.2	$3.5 \cdot 10^{-3}$
Pb	400	0.88	Cr	0.19	$1.7 \cdot 10^{-3}$
			Fe	0.1	

These data indicate that when smelting alloy steels, the melting loss of manganese (70%) and chromium (5-7%) should be taken into account.

The following possible processes can be applied in cast steel foundries:

1. VD – Vacuum Degassing,
2. VOD – Vacuum Oxygen Decarburisation,
3. VAD – Vacuum Arc Degassing,
4. VIM – Vacuum Induction Melting (a group of units for melting and casting in vacuum) [21].

VD Process

This is one of the simplest and oldest treatments of secondary metallurgy. It consists in the introduction of the ladle with liquid steel into a vacuum chamber and subjecting it to the effect of a vacuum, (Fig. **8**). There are also methods that involve the direct sealing of the ladle by installing a vacuum cover. In foundries where there is no space for the use of a converter, the use of a proper vacuum chamber solves many technological problems.

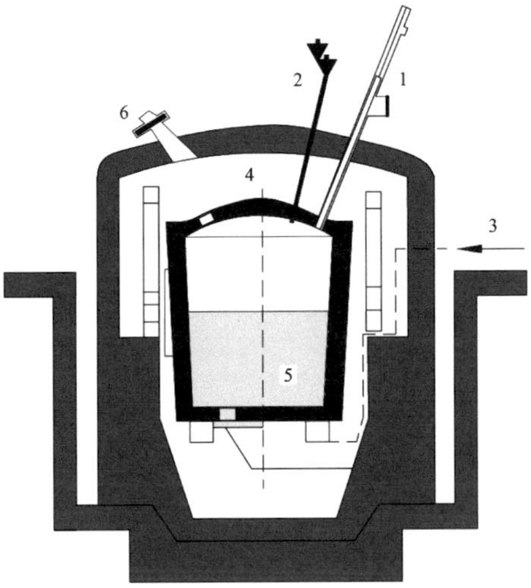

Fig. (8). Schematic illustration of a VD unit, [3]. 1-oxygen lance; 2- alloying unit; 3-argon; 4- vacuum chamber; 5- molten steel; 6- window.

Technology of the VD Process

The heat is tapped without a slag into a ladle with a basic lining. In such conditions, it is easier to perform deoxidation and add ferro-alloys. The ladle with liquid metal is transferred to the vacuum unit and lowered into the tank.

After connecting the argon, it is necessary to check if the gas flows smoothly through the porous plug. The tank is then closed and the vacuum pumps are started. Operation of the vacuum is monitored by the observation of the foaming of the slag and the boiling of the steel bath (through the window, 6). At a pressure of less than 300 Pa, the argon flow is increased gradually for effective degassing. The degassing time depends on the necessary time for the removal of nitrogen and hydrogen.

Other stages include taking a sample for analysing chemical composition, deoxidation and adding ferro-alloys for the final adjustment of the composition.

Advantages: The VD process allows for steel with a hydrogen content of 1 ppm and less than 35 ppm of nitrogen, at a concentration of sulphur of less than 40 ppm, to be obtained [13]. The efficiency of this process is increased by using a ladle with a capacity of over 100 tons. The process has good points in increasing the temperature of molten steel, decreasing the quantity of inclusions and the homogenization of the chemical composition of the bath.

Limitations: Very high installation costs, requirements of new position, very high quality staff required.

The VOD Process

The construction of the device for the VOD process (Fig. **9**) allows for oxidation with gaseous oxygen in vacuum, with simultaneous injection of argon, to be performed. The charge to the process is liquid steel melted in an EAF, with a chemical composition similar to the VD process (0.3 - 0.6% C, below 0.3% Si and a high concentration of chromium (≈ 18%) or Cr + Ni (18% Cr - 11% Ni or 20% Cr - 24% Ni). Oxygen is introduced with a lance from the top, and argon through a porous plug from the bottom. Injection of these gases is performed at a pressure of 10^{-2}atm.

A VOD device comprises of: a casting ladle (2), a vacuum chamber (4,1) with a lid (9), pumping plant (10), installations for oxygen (6) and argon (7) blowing, a vacuum feeder for dosage of ferro-alloys (5) and control and measurement equipment (7,8), (Fig. **9**).

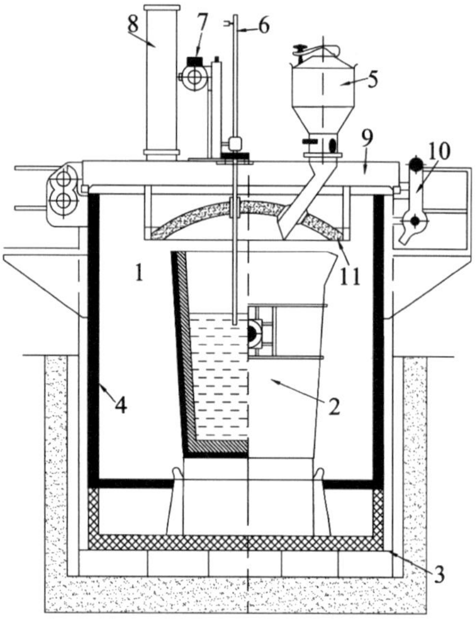

Fig. (9). Schematic illustration of VOD unit [20].

The ladle has a high quality magnesia-chromite lining, because it is subjected to high temperature changes, the reaction of the slag with the steel under vacuum, intensive stirring in the bath during oxidation and argon treatment, and large thermal stress. It fulfils two functions: the transport of liquid steel from the moment of tapping to the vacuum chamber, and it also ensures the conditions for oxidation. Therefore, its height must take into account the so-called freeboard.

The vacuum chamber has a roof on which a movable oxygen lance (water-cooled, sealed with a cover), tanks for ferro-alloys, sight glasses and equipment for sampling and temperature measurements are built. The lance has a ceramic tip, allowing for its immersion in the liquid steel during the oxidation period.

Control and measurement devices enable continuous measurement and control of the pressure in the vacuum chamber, the pressure and the amount of oxygen and argon injected, the temperature, the chemical composition of the bath and the amount of ferro-alloys and slag introduced.

The course of the VOD process: The first stage is smelting chrome or chrome-nickel steel in an electric arc furnace (EAF), and the second one is decarburisation of the steel in a vacuum chamber. Steel smelting in an EAF is typically performed by reversible melting using oxygen blowing. The final composition of steel corresponds to the desired grade of chrome or chrome-nickel steel provided that the carbon content is in the range of 0.3 - 0.5%, and silicon < 0.3%. Flow rate 20-30 $m^3/(h \cdot t)$ [18], or 4l/t min [24]). The main periods of the VOD process are:

1 – decreasing the pressure to 8.000 Pa and pulling down the oxygen lance.
2 -stopping the injection of oxygen, and with a vacuum of 1 Torr only argon is injected. Under these conditions, decarburisation occurs by the reaction [C] + [O] → {CO}, up to a content of 0.05% carbon.
3 - the vacuum is released, the temperature (1650-1700°C) measured and samples are taken for chemical composition analysis. The bath is deoxidised (Fe-Si, Al or Fe-CaSi) and the slag ($CaO + CaF_2$) is put in. Desulphurisation is started.
4 - vacuum pumps are re-activated and argon is injected. In this period, there is a reduction of Cr_2O_3, desulphurisation of the bath and dissolution of ferroalloys. Argon treatment of liquid steel facilitates the outflow of inclusions and levels the chemical composition of steel.
5 - pumps are turned off, the temperature and the chemical composition of the steel are measured. If the temperature is too high (more than 1650°C), alloy scrap is added. At a temperature of 1550 - 1570°C, the molten steel is tapped to the ladle.

The introduction of the process in a steel work or foundry offers the following **advantages:**

- increase in the quality of the steel and the productivity of the furnace.
- utilisation of cheaper materials (scrap), with high carbon and chromium contents.
- decrease of production costs.
- a better yield of chromium (97-98%).
- allows for steel to be obtained that contains less than 0.01% C, below 0.01% S

and the following gas levels: hydrogen ≈ 2 ppm, oxygen 40 - 120 ppm and nitrogen 150 - 250 ppm (in 18% Cr - 11% Ni or 20% Cr- 25% Ni – Mo steels) [20].

Limitations: Application of small VOD converters (with a capacity of 10-15 tons) in a foundry requires heating of metal. Therefore, they can work together with a VAD unit. Such an arrangement is expensive, and it can be replaced by more comprehensive constructions (LF, DETEM). The foundry needs additional space for new equipment, more electric power and transport capacity.

VAD (Vacuum Arc Degassing) process is a favourable solution because:

- heating steel with an electric arc eliminates limitations of temperature (but the process is longer), ensuring appropriate refining of steel in a vacuum.
- it allows for the degassing of the bath, both by reaction of carbon with oxygen, and degassing with an inert gas (at the end of this process carbon content is <0.01%, sulphur < 0.01% and hydrogen 2-3 ppm.
- it leads to homogenisation of the chemical composition and the temperature by the argon stirring the molten steel in the ladle.

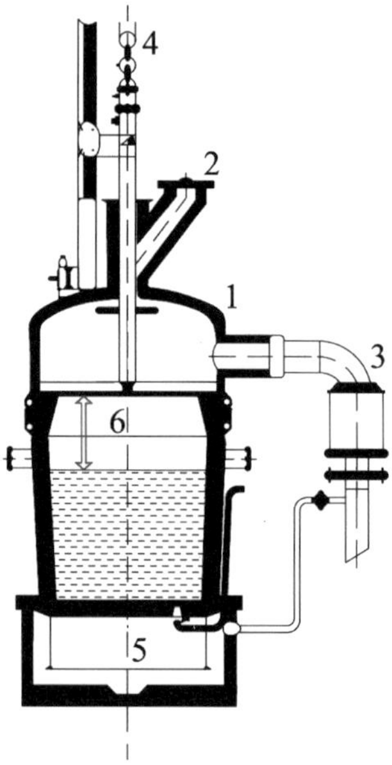

Fig. (10). Principle construction of DETEM unit [25]. 1- vacuum cover; 2- vacuum lock(ferro-alloys); 3- pump station; 4- oxygen lance;5- safety holder; 6- free-board.

The DETEM process: An original method for cast steel foundries is the DETEM construction, which creates the possibility of secondary metallurgical operations for capacities of 5 to 20 tons [1, 25]. It consists of a pouring ladle with an applied vacuum cover, which is sealed and connected to the vacuum, (Fig. **10**). Construction of the vacuum cover is of particular importance for the whole metallurgical process.

Through the use of direct sealing in the ceramic lining of the cover with the ladle lining, very little freeboard (100 mm) was obtained, which greatly improves decarburisation and argon stirring. After applying an injector, all stations occupy a small area in the shop (3x4 m) [1]. The small reaction space (freeboard) permits the use of small vacuum pumps.

The course of the metallurgical process:

1. tapping of steel from an EAF or from an induction furnace into the ladle, which has basic (dolomite) or neutral (80% Al_2O_3) lining. After filling the ladle, basic slag ($CaO + MgO + CaF_2$) is put in, the amount being 10 kg/t of liquid steel.
2. the transport of the ladle to the secondary metallurgy station, installing the vacuum cover and connecting the pumps, argon and oxygen lance; the vacuum operation time, oxidation and argon treatment are dependent on the grade of steel.
3. degassing under vacuum; the course of the process takes up to 20 minutes.

Results Obtained

* after 12 minutes a hydrogen content of <1 ppm (from 6 ppm) is achieved.

* at the same time, the nitrogen content is reduced from 160 to 50 ppm.

* temperature loss equals 3 to 5°C/min.

The DETEM process is applied in combination with other secondary metallurgy processes [1, 2]:

1. EAF – DETEM or Induction Furnace (IF) – DETEM
2. EAF - DETEM - VOD
3. IF – DETEM – VOD
4. EAF – DETEM – VD
5. EAF – LF – DETEM/VOD

The choice of one of the secondary metallurgy methods should be considered depending on the steel (cast steel) grades used, their required quality, and the investment and economic opportunities of the plant (foundry). However, it is important to implement new technologies in order to modernise the company and fulfil quality requirements. Finally, one should always take into account the following characteristics of secondary metallurgy:

- installation of one of the units expands production abilities, improves the quality of the produced goods and creates new competitive possibilities.
- application of small units with the use of a vacuum is particularly advantageous in foundries, since it eliminates gases (H, N and O), which are one of the major causes of defects in steel castings, (porosity, inclusions and cracks).
- vacuum converter processes shorten the processes of decarburisation, desulphurisation and the whole cycle of smelting, but they are expensive (new investment).
- good improvements in quality are brought about by the introduction of a combination of several processes, *e.g.*

IF – DETEM - VD

- implementation of a secondary metallurgy process shortens steel smelting by 5 - 22% [6].

7. VACUUM INDUCTION FURNACES

The process of melting steel in vacuum induction furnaces is widely used in the foundry industry. It is used for the production of castings made by investment casting (lost wax method or Shaw process) for the following industries: aviation, aerospace, power generation, medical (implants) and automotive.

It enables the production of steel and pure metals of very high quality. In the case of smelting pure metals with a high affinity for oxygen (Ti, Nb), or corrosion resistant steel, strong degassing of the bath (oxygen, hydrogen and nitrogen) and reduced formation of non-metallic oxide inclusions lead to a new generation of materials. In addition, the smelting process in high vacuum allows for very low concentrations of undesired volatile elements (Zn, Bi, Sb, Mn, Sn, Cu and Pb) to be obtained [26]. Other advantages of induction smelting in a vacuum are the ability to obtain a homogeneous chemical composition of the steel, rapid heating of the bath and the possibility of adapting the process to various grades of steel.

There are many designs of this process [18]. They can be divided into two groups of furnaces, enabling:

1. melting metals in a vacuum or protective atmosphere (Fig. **11**) and pouring the melt into the moulds (ceramic - investment casting or sand moulds) or metal (ingot moulds) under the air atmosphere.
2. melting and tapping of molten steel to different moulds under vacuum.

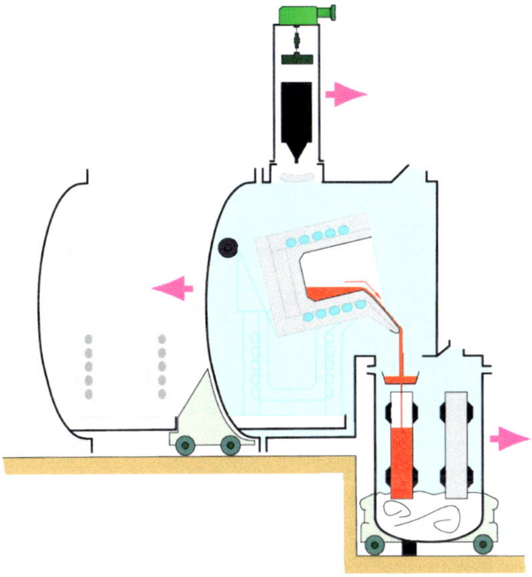

Fig. (11). Melting and casting in vacuum [24].

The process of smelting in a vacuum has its specific characteristics. They can be characterised as follows:

- the first portions of the charge fill half of the crucible and the rest is added during melting, so as not to form so-called bridges; if scrap from castings (gate system, risers, scrap castings) is used as a charge, it must be first cleaned (chargé items are shot-blasted);

- first, along with the charge, elements that do not oxidise during melting (nickel, cobalt, or molybdenum) or with a low affinity for oxygen (iron) are introduced; other elements must be introduced gradually during melting (chromium), or just before tapping (silicon, manganese, aluminium, titanium or vanadium);

- degassing already takes place in the initial vacuum during the heating of the charge (hydrogen). However, the removal of oxygen during the reaction with carbon takes place when the first portions of molten steel are obtained. At this moment, it is very important to avoid the boiling of the steel bath, and the vacuum should be carefully controlled. Introduction of argon is also beneficial at this point;

- the vacuum is increased after the melting of the charge and the degassing of the bath containing high-melting elements (Fe, Co, and Ni). Chrome is then also added. If the charge contains many impurities or volatile elements, changes in vacuum pressure should be performed gradually (after degassing);

- before the introduction of deoxidisers and easily oxidising elements, it is useful to increase the vacuum pressure by introducing argon. This reduces the dross from these elements;

- after dissolution of alloying elements, deoxidising the bath and temperature control, it is necessary to proceed quickly to tapping the steel.

A comparison of the results of the vacuum affects on the properties of carbon cast steel, smelted in an EAF (7-10 tons), VIM (1.5 ton), VARP, and EAF + AOD, as presented in [25]. The results concern desulphurisation, degree of purity, microstructure and mechanical properties, and the conclusions of this study can be presented as follows:

1. the lowest levels of sulphur are obtained with a combination of an EAF with an AOD process (0.002%), the oxygen converter and VARP converter (0.003%), and slightly higher (0.004%) with the EAF alone.
2. the lowest oxygen concentrations are obtained in the cast steel from an oxygen converter (31-45 ppm) from the process of EAF + AOD (65 ppm) and from a VARP converter (51-72 ppm). In the carbon cast steel smelted in an EAF, the oxygen content ranges from 38 - 49 ppm.
3. the concentrations of nitrogen are more diverse: the highest from the EAF (86 to 148 ppm), and the lowest from the converter processes (51 - 85 ppm).
4. tensile strength and yield strength are the highest in the cast steel smelted in the EAF (523 - 563 MPa), and in a vacuum converter - ≈500 MPa. The strength values demonstrate a relation with the degree of steel deoxidation: at a content of about 40 ppm of total oxygen, the tensile strength is at its highest (540-560 MPa), and at a content of 70 ppm, it falls below 500 MPa.
5. The secondary metallurgy processing of carbon steel (in the VOD converter) also brings an improvement in the plastic properties of carbon cast steel. In particular, this applies to the impact toughness; the transition temperature is shifted to more negative temperatures.

Melting and Casting in Vacuum

This process is applied to the casting of pure metals and high-alloy steels in the form of ingots or as investment castings. In the first method, there are possibilities of casting into one or more gravity dies (*e.g.* on a turntable), on a rectangular car or with a vacuum mould chamber, (Fig. **11**). These processes are used for casting ingots for wrought products, superalloys, high alloyed steels and bar stock for investment casting.

In the foundry industry, similar design solutions are applied. A mould, usually a Shaw shell mould, is placed in a vacuum chamber instead of gravity dies.

CONFLICT OF INTEREST

The author (editor) declares no conflict of interest, financial or otherwise.

REFERENCES

[1] F. Wagener, *Der Einsatz von Sekundär-metallurgischen Behandlungsverfahren für kleine Chargengrössen.* Technometal GmbH: Duisburg, 1989.

[2] K. Kmak, and F. Wagener, *Auswahlkriterien für den Einsatz von Sekundärmetallurgischen Anlagen.* Technometal GmbH: Duisburg, März, 1988.

[3] J. Senberger, Z. Buzek, A. Zadera, K. Stransky, and W. Kafka, *Metalurgie Oceli na Odlitky, VUTIUM.* Czech: Brno, 2008.

[4] G. Stolte, *Secondary Metallurgy.* Verlag Stahleisen GmbH: Düsseldorf, 2002.

[5] A. Ghosh, *Secondary Steelmaking: Principles and Applications.* CRC Press: Boca Raton, 2000. [http://dx.doi.org/10.1201/9781420042313]

[6] G. Paul, W. Klan, and H. Höffgen, "Betriebserfahrungen mit Gasspülsteinen mit gerichteter Porosität bei den Badischen Stahlwerken AG", *Radex-Rundschau,* vol. 1, pp. 288-302, 1987.

[7] W. Hammerer, G. Raidl, and H. Bartehl, "Bartehl, Gasspülsteine für Stahlpfannen", *Radex-Rundschau,* vol. 4, pp. 217-226, 1992.

[8] S. Sobula, *Effect of oxygen-blowing and argon stirring of Hadfield steel on the content of gases, PhD, Univ.of Sc.a.* Technology: Cracow, Poland, 2007.

[9] V. Caruso, "Erfahrungen mit dem Einsatz von CaSi-Fülldrähten in der Sekundärmetallurgie", *Stahl Eisen,* vol. 104, no. 6, pp. 299-304, 1984.

[10] G.P. Crawford, "Wire injection of metallurgical powders into molten metal", *Scaninject VI,* June 1992.p. 303 Lulea, Sweden *6th Intern.Conf. part II,* June 1992.,

[11] A. Herbert, G.K. Notman, D. Morris, S. Knowles, and C. Jemson, "Experiences with powder and wire injection at British Steel Corporation, Lackenby Works", *4th Intern.Conf. on Injection Metallurgy,* 1984, pp. 27-34 Lulea, Sweden.

[12] A. Ghosh, *Secondary Steelmaking: Principles and Application.* CRC Press: Boca Raton, 2016.

[13] H.P. Haastert, H. Maas, and H. Richter, "Anwendung des TN Verfahrens bei der Erzeugung von Stählen mit besonderen Zähigkeitseigenschaften", *Stahl Eisen,* vol. 100, no. 22, pp. 1298-1303, 1980.

[14] P. Critchley, E. Linder, H. Holtermann, and F. Wagener, Technometal GmbH; Duisburg, 1989.

[15] E. Martinez, "Anwendung des TN Verfahrens bei der Erzeugung von Stählen mit besonderen Zähigkeitseigenschaften", *Stahl u.Eisen,* vol. 100, no. 22, pp. 1298-1980, 1995.

[16] A. Anioła-Kusiak, A. Lux, K. Mamro, and M. Rzeszowski, *Metalurgia Argonowa Stali, Wyd.* Slask: Katowice, 1981. (in Polish)

[17] G. Stolte, *Secondary Metallurgy-Fundamentals Processes Application,* 1st. ed, Stahl und Eisen GmbH: Duesseldorf, 2002.

[18] F.P. Edneral, *Electrometallurgy of Steel a.* Ferro-alloys, Mir Publ.: Moscow, 1977.

[19] H. Knüppel, *Desoxydation und Vakuumbehandlung von Stahlschelzen, Band I, II.* Verlag Stahleisen, MBH: Düsseldorf, 1983.

[20] Secondary Metallurgy Solutions, Siemens, VAI, Metals Technologies.

[21] E. Hofmann, and W. Siefer, "Kennzeichnung signifikanter kriterien der Sekundär Metallurgie und ihre Übertragung auf den primärmetallurgische Bereich", *Giessereiforschung,* vol. 45, no. 16, pp. 42-51, 1993.

[22] G.M. Gill, E. Ineson, and G. Wesley, "The Behaviour of various elements in vacuum steelmaking", *J. Iron Steel Inst.,* vol. 191, pp. 172-175, 1959.

[23] D. Wittekopf, W. Nowak, and H. Riedl, "Zähigkeitseigenschaften der im VOD-Konverter sekundärmetallurgisch behandelten Stahlgussorten GS-C 25 und GS-25 CrMo 4", *Giessereiforschung,* vol. 42, no. 2, pp. 74-87, 1990.

[24] Vacuum Induction Melting and Casting, ALD Vacuum Technologies AG, Prospect "The Solution ALD"

[25] Ch. Wegler, W. Rubens, R. Seidel, G. Stolte, and W. Pulvermacher, "The Technology of "fast" removal of hydrogen in the ladle degasser", *Scaninject VI,* 1985 Linz, Austria *8th Inter.Conf.on Vacuum Technology,*

[26] L. Zemcik, "Specjalni slevarenske procesy", In: *Brno, Czech Rep.* 2nd ed. , 1987, p. 125. (in Czech).

CHAPTER 7

Cast Steels for Service Conditions

Abstract: Low-alloy cast steels are used in many branches of industry for various mechanical parts with very different wall thicknesses (40-400mm). The mechanical properties in such sections must be equal throughout. For these reasons, there are several grades of cast steels, which fulfil the requirements for different uses. This chapter characterises the parameters of low-alloy cast steels, intended for several service conditions.

Keywords: Cr-Ni-Mo, Effect of Cr –Ni, Low-alloy cast steel, Mechanical properties, Wall thickness.

1. INTRODUCTION

Cast steels of this group are the basis of all machine constructions due to their optimum strength properties compared to their plastic properties. The application areas of the castings made of these types of cast steel are primarily in the following industries: the machinery manufacturing industry; the automotive industry - heavy vehicles; the marine industry - ships, drilling platforms; extractive industries - excavators, bulldozers, mining crushers; and the energy industry - turbines. Castings designed for such working conditions have a wall thickness from 50 to 400 mm and are subject to strong dynamic loads. Very often, they work under varying climatic conditions, over large temperature differences, from + 20 to minus 60°C. Such items include wagon wheels, axle boxes and other parts of wagons and electric locomotives. Therefore, a high yield strength, a high value of YS/UTS, impact toughness and good weldability are required from these castings, as they are very often welded with rolled or forged components.

The use of carbon steel for castings of this type, with wall thicknesses above 30 mm, is limited, because without the introduction of alloying elements, a homogeneous structure and properties throughout the cross section of the casting walls cannot be obtained.

The limited application of carbon cast steel is also connected with the difficulty in obtaining a tensile strength above 650-700 MPa. Raising the carbon content or the

Jan Głownia

application of more complex heat treatments to obtain a UTS above 700 MPa can lead to elongation and too low an impact toughness (Fig. **1**). Thus, in these cases where a tensile strength of 700-1200 MPa is required in castings with wall thicknesses from 30 to 100 mm, alloying elements should be introduced into the cast steel. Therefore, cast steels containing silicon, manganese, chromium, nickel or molybdenum in particular are often found in use. The added elements' influence on the mechanical properties of cast steel should be considered comprehensively, because there are usually walls of various thicknesses in castings, and the properties and structure must be equal throughout the casting. However, due to unequal cooling rates in walls of different thickness or wall connections, each of them has a different structure, and often separation of carbides, sulphides or oxy-sulphides also occurs. Therefore, elements reducing the critical quenching rate are introduced into cast steel. These allow homogeneous structure and properties throughout the entire wall thickness of the casting to be achieved. These features are usually obtained after the introduction of not one, but several, alloying elements. The synergistic effect of elements such as Si, Cr, Mn, Mo and Ni means that cast steels, such as Mn-Si, Cr-Mo, Cr-Ni-Mo or Cr-Ni-Mo-V (Table **1**), are often used in practice.

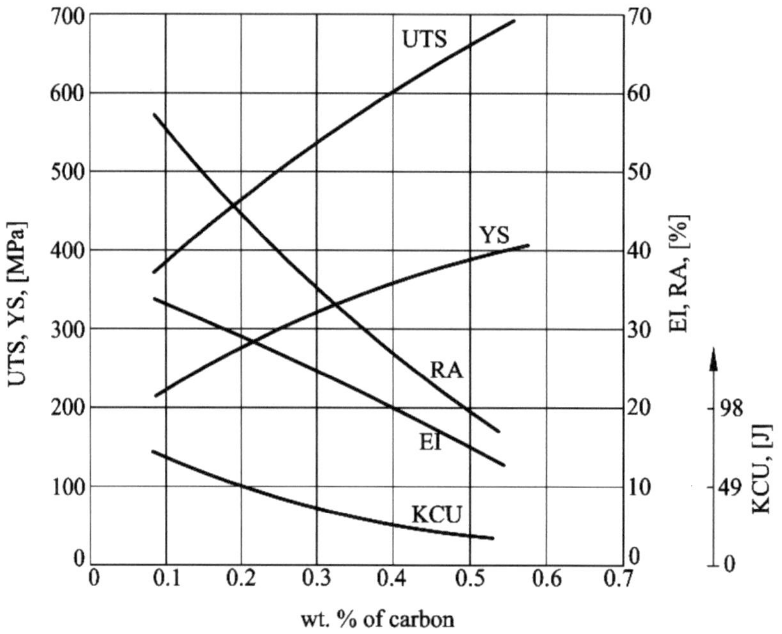

Fig. (1). Mechanical properties of carbon cast steel [1].

Table 1. Cast steel after quenching and tempering, intended for thick-walled castings.

ASTM	DIN	A27 60-30 62 –35 GS-Ck25		GS-30 Mn 5		GS-25 Cr Mo 4			GS–34 Cr Mo 4			GS–42Cr Mo 4			GS–34 Cr Mo 44			GS–30Cr Mo V 64			G–35 Cr Mo V 104			GS–40 Ni Cr Mo 656		
Quality class		I	II	I	II	I	II	III	I	II	III	I	II	III	I	II	III	I	II	III	I	II	III	I	II	III
Wall thickness, mm		Mechanical properties																								
up to 30	Re, MPa	245	275	345	440	390	510	635	510	635	765	540	655	835	510	635	765	540	685	835	540	685	835	540	685	885
	Rm, MPa	440-590	490-640	590-740	740-880	590--740	740-880	880-1030	740-880	880-1030	1030-1180	740-880	880-1030	1030-1180	740-880	880-1030	1030-1180	740-880	880-1030	1030 1180	740-880	880-1030	1030-1180	740-880	880-1030	1030-1180
	A5,%	22	20	16	12	16	12	9	12	9	7	12	9	6	12	9	7	13	10	7	13	10	7	13	11	8
	K, J	48	41	34	21	41	27	(14)	27	(14)	(7)	27	(14)	(7)	27	(14)	(7)	41	27	21	41	34	27	34	7	21
30-100	Re, MPa	215	–	255	–	345	440	–	440	540	–	490	590	–	470	570		490	590	685	540	635	785	540	685	835
	Rm, MPa	390-540	–	540-690	–	540-690	690-830	–	690-830	830-980	–	690-830	830-980	–	690-830	830-980	–	690-830	830-980	980-1130	740-880	880-1030	1030-1180	740-880	880-1030	1030-1180
	A5, %	22	–	15	–	16	12	–	12	9	–	11	8	–	12	9	–	13	10	16	13	10	7	12	10	7
	K, J	48	–	34	–	41	27	–	27	21	–	21	(14)	–	27	21	–	34	27	21	34	27	31	27	21	(14)

Chemical composition % weight		A27		GS-30 Mn 5		GS-25 Cr Mo 4		GS–34 Cr Mo 4		GS–42Cr Mo 4		GS–34 Cr Mo 44		GS–30Cr Mo V 64		G–35 Cr Mo V 104		GS–40 Ni Cr Mo 656	
Wall thickness, mm		up to 100	over 100	up to 100	over 100	up to 100	over 100	up to 100	over 100	up to 100	over 100	up to 100	over 100	up to 100	over 100	up to 100	over 100	up to 100	over 100
	C	0.20-0.28		0.27-0.34		0.22-0.29		0.30-0.37		0.38-0.45	0.38-0.45	0.30-0.37		0.27-0.34	0.27-0.34	0.32-0.39	0.32-0.39	0.37-0.44	0.37-0.44
	Si	0.30-0.50		0.30-0.50		0.30-0.50		0.30-0.50		0.30-0.60	0.30-0.60	0.30-0.5O		0.30-0.60	0.30-0.6	0.30-0.60	0.30-0.60	0.30-0.60	0.30-0.60
	Mn	0.50-0.80		1.20-1.50		0.50-0.80		0.50-0.80		0.60-1.0	0.60-1.00	0.50-0.80		0.60-1.00	0.60-1.00	0.60-1.00	0.60-1.00	0.60-1.0	0.60-1.00
	P	0.035		0.035		0.035		0.035		0.025	0.025	0.035		0.025	O. 025	0.025	0.025	0.025	0.025
	S	0.035		0.035		0.035		0.035		0.025	0.025	0.035		0.025	0.025	0.025	0.025	0.025	0.025
	Cr	–		–		0.80-1.20		0.80-1.20		0.80-1.20	0.80-1.20	0.80-1.20		1.30-1.70	1.30-1.70	2.20-2.70	2.20-2.70	1.10-1.50	1.10-1.50
	Mo	–		–		0.20-0.30		0.20-0.30		0.20-0.30	0.20-0.30	0.30-0.50		0.30-0.50	0.30-0.50	0.30-0.50	0.30-0.50	0.50-0.70	0.50-0.70
	Ni	–		–		–		–		–		–		–	–	–	–	1.30-1.70	1.30-1.70
	V	–		–		–		–		–		–		0.05-0.15	0.05-0.15	0.05-0.15	0.05-0.15	–	–

Mechanical properties depending on the cooling medium		A27		GS-30 Mn 5		GS-25 Cr Mo 4			GS–34 Cr Mo 4			GS–42Cr Mo 4			GS–34 Cr Mo 44			GS–30Cr Mo V 64			G–35 Cr Mo V 104			GS–40 Ni Cr Mo 656		
Wall thickness, mm		up to 150	over 150	up to 150	over 150	up to 150	150-250	250-400	up to 150	150-250	250-400	up to 150	15-250	250400	up to 150	150-250	250-400	up to 150	150-250	250-400	up to 150	150-250	250-400	up to 150	150-250	250-400
Air	Re, MPa											400	350	320				400	350	320	650	600	500	680	680	650
	Rm, MPa											700-850	650-800	650-800				700-850	650-800	650-800	800-950	750-900	700-850	850-1000	850-1000	850-1000
	A5, %											10	10	8				12	12	10	10	10	10	10	8	6
	K, J											12	(12)	9.6				20	(16)	(16)	20	(16)	(16)	20	(20)	(16)
Oil	Re, MPa											450	400	350				550	450	350	700	650	600	750	750	750
	Rm, MPa											700-850	650-800	650-800				750-900	700-850	650-800	850-1000	800-950	750-900	900-1100	850-1050	850-1050
	A5, %											15	10	8				15	12	10	10	10	10	8	6	6
	K, J											16	(12)	9.6				20	(16)	(16)	28	(24)	(20)	16	(16)	(9.6)
Recommended heat treatment												15h - 850°C/oil,						15 - 900°C/oil			15h - 900°C/oil			12h - 910°C/oil		
												15h - 640°C/ furnace						15h - 630°C/ furnace			15h - 645°C / furnace			12h - 600°C / furnace		
												4h - 880°C/ air						4h - 910°C/ air			4h – 910^C/ air			4h - 880°C/ air		
												4h - 600°C / water						4h - 620°C / furnace			4h - 625°C / furnace			4h - 640°C/ water		

The selection of a steel grade for thick-walled castings is also related to a change in the cooling rate in each of its parts, caused by the mass effect. Reduction in the cooling rate leads to changes in the matrix microstructure, including changes in the primary grain size and the distribution of non-metallic inclusions, along with an increase in the segregation of elements. These changes are usually intensified starting from the casting surface to the inside of the wall.

The effect of changing the cooling rate on the wall thickness is particularly important in castings, since control of the mechanical properties is performed on samples cast separately, of different thicknesses (cooling rate) in relation to some of the walls of the casting. In castings, this effect is particularly made visible by the segregation and the changes in the morphology of non-metallic inclusions in these parts, which solidify at the end of solidification (Fig. **2**).

The reasons for these changes result from the segregation of mainly sulphur and oxygen (formation of oxy-sulphides, chapter 4) and also the segregation of carbon and chromium, leading to the precipitation of carbides at prior grain boundaries (Fig. **2**).

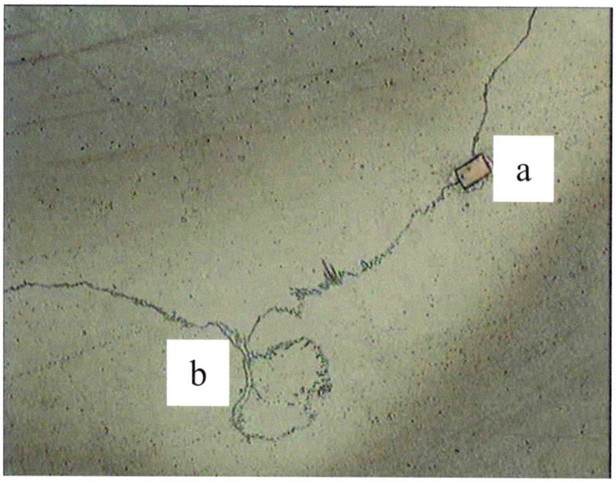

Fig. (2). Nitride (a - yellow) and carbides (b) on a prior grain boundary in Hadfield cast steel, Magn.400X.

Although the effect of wall thickness applies especially to castings weighing more than 2 to 5 tons (in castings with a wall thickness over 100 mm), it is difficult to produce similar components for devices that have a complex shape and heterogeneous wall thickness, and are made by plastic working. Therefore, production of machine castings with such a weight requires control of the influence of wall thickness on the homogeneity of the structure and its mechanical properties. Knowledge of metallurgical factors, such as the processes of deoxidation or modification, and technological factors (choice of moulding sand, its thermal conductivity coefficient, chills) is essential here [2, 3].

Homogenisation of the structure throughout the cross section of the casting wall is achieved by selection of the optimum chemical composition and appropriate heat treatment. These two parameters have the strongest influence on the hardenability of the material, which must increase along with the wall thickness of the casting. However, since cast steel has an as-cast structure coarser than forged steel, its hardenability is higher [4]. However, after normalising or other heat treatment, these differences disappear and changes in cast steel hardenability are caused by the same factors that have been confirmed in forged steels.

The main task of alloying elements during the heat treatment of castings is to increase the hardenability. The increase in the hardenability enables exchange of the cooling medium, *e.g.* water into oil, which reduces the risk of quenching cracks. Selection of the chemical composition of steel for thick-walled castings, in which the required strength at a given hardness is provided, is subject to certain additional effects.

Above all, martensite in carbon steel of the same hardness as alloy steel has a lower reduction of area and impact toughness. To achieve equivalent hardness while quenching and tempering alloy steels, a higher tempering temperature than for carbon steel is required. This reduces stress, but increases energy expenditure.

Manganese (at constant carbon content) is the cheapest alloying addition that increases the hardenability. In many cases, application of more expensive, but more effective elements, such as chromium or molybdenum, is justifiable. The introduction of nickel is the most expensive, but in the circumstances where high plasticity is required in thick-walled castings, its addition may not be replaced by any other element.

Selection of the chemical composition for thick-walled castings with increased hardenability and homogeneous structure should meet the following requirements in relation to individual elements:

– **carbon:** in structural castings, its concentration varies in the range of 0.25 to 0.45%. The upper concentration range is limited by the slower growth of YS (constant value of the ratio YS to UTS), by a tendency to quenching cracks and the low plasticity of the cast steel. The lower value of the carbon concentration provides good weldability and plasticity of cast steel. Keeping the carbon concentration below 0.25% reduces hot and quenching cracks, facilitates removal of risers and improves weldability. Therefore, most often castings of machine parts contain from 0.25 to 0.40% carbon. The increase in carbon content in this range leads to changes in the quantity of martensite after quenching.

Carbon has a double impact on the hardenability of alloy steel: it controls the final hardness and contributes to the increase in hardenability by its synergistic effect with other alloying elements in the alloy matrix. An example of an increase in alloy steel hardenability in the presence of various carbon contents is given in (Fig. **3**). In the presence of molybdenum, in particular chromium and molybdenum, the combined effect of C + Cr + Mo is much stronger than in the case of the introduction of only one component into steel. The most visible effect of carbon on steel hardenability results from measurements of the ideal diameter of quenching D_i. The largest impact is observed in Ni-Cr-Mo steels. After determining the proper carbon content, an optimum concentration of alloying elements should be selected. The primary objective of selection of the chemical composition is a homogenous structure throughout the casting wall cross section (most often tempered martensite), providing the best combination of alloy strength and plasticity.

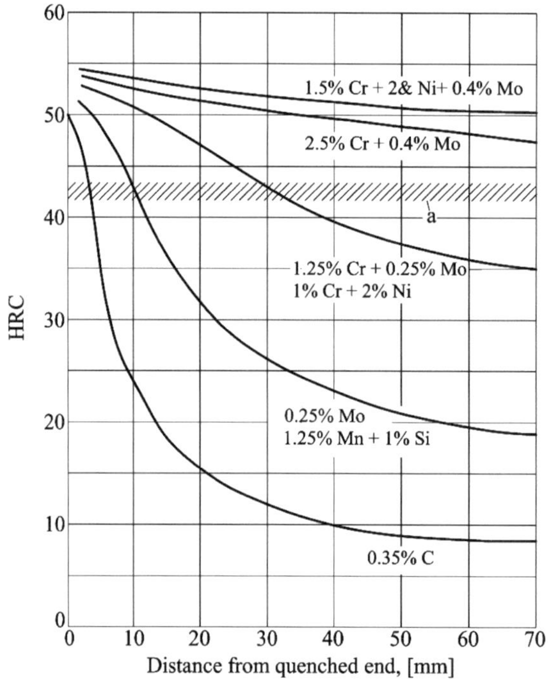

Fig. (3). Effect of chromium and carbon on reduction of the critical quenching rate [5]; a - min. required hardness (43HRC).

Manganese (T_m = 1517 K): as an alloying addition (over 0.9%), it is a cheap and effective means of increasing hardenability. Hence, there are manganese cast steels with manganese contents up to 2%. At manganese contents above 2%, segregation is increased in castings, and therefore, more complex cast steel grades, such as Mn-Si-Mo, or Cr-Mn-Mo, became more applicable;

- **silicon (T_m = 1683 K):** used as an alloying addition; if its concentration exceeds 0.5%, it increases the strength of steel (silicon cast steel typically has a strength of 1000-1200 MPa). It reduces quenching cracks, because silicon does not affect the temperature of formation of martensite and decreases phase volume changes during quenching. It increases hardenability slightly and reduces the quantity of retained austenite in Cr-Ni-Mo steels. Large concentrations of silicon (above 1-1.5%) significantly reduce weldability;
- **chromium (T_m = 2193 K):** used as an alloying addition with a content of more than 0.3%; increases hardenability, especially in combination with nickel and molybdenum. Due to the presence of hot and quenching cracks, its concentration in structural cast steel does not exceed 1.25% in practice. An increase in the chromium content of steel reduces the critical quenching rate (Fig. **4**). At a content of 3.5% Cr, a homogeneous structure and a strength of over 700 MPa are reached in castings, even those with a wall thickness up to 500 mm [5, 6]. The presence of molybdenum and nickel enhances the effect of chromium (reduces the critical quenching rate even more);
- **nickel (T_m = 1726 K):** used as an alloying addition over 0.3%; it has a moderate effect on hardenability (significant effect in the presence of chromium). It reduces the temperature of retained austenite formation, particularly in the areas with increased segregation (grain boundaries, interdendritic areas). It aligns mechanical properties in castings with wall thicknesses above 200 mm, increases the impact toughness (especially at temperatures below freezing). The basic element in heat-resistant cast steel;
- **molybdenum (T_m = 2883 K)**: used as an alloying addition above 0.1%; increases hardenability, strength (especially heat-resistance). It eliminates temper embrittlement, hence additions in the range of 0.2-0.4% are often used. The basic element in cast steel resistant to corrosion in a chloride ion environment (seawater, mine water) and creep resistant cast steel. It increases the yield strength in the presence of chromium and nickel;
- **vanadium (T_m = 2103 K):** used as an alloying addition at a content over 0.1%, it decreases the primary austenite grain (modifier). At a content of up to 0.1%, it increases hardenability, and above this concentration, it promotes secondary hardening of the alloy matrix during tempering. It forms long-lasting carbides, increases the UTS, YS and the heat resistance of steel, especially in combination with chromium after the steel's quenching and tempering. A preferred addition in tool steels;
- **boron (T_m = 2313 K):** used in cast steel in thousandths of a percent; one of the most effective elements enhancing hardenability (at concentrations up to

0.008%). In comparison to manganese, chromium or nickel, it prevents the formation of quenching cracks and residual austenite;

- **copper (T_m = 1357 K):** used as an alloying addition at over 0.3%; increases resistance to atmospheric corrosion, enhances austenite in the alloy matrix. At content levels above 1%, it enhances dispersion strengthening at temperatures of 523-623 K [6, 7]. It increases the UTS and YS of cast steel and the tendency of aluminium nitride to precipitate at the prior austenite grain boundaries;
- **titanium (T_m = 2085 K):** above 0.1%, it is used as an alloying addition, highly carbide-forming. It is used as a modifier in cast steel (similarly to vanadium, niobium and boron). It is a structural stabilizer in austenitic cast steel that is resistant to corrosion (% Ti = 5 x %C). It has a high affinity with oxygen, nitrogen and carbon;
- **tungsten (T_m = 3653 K):** used as an alloying addition above 0.1% (also similarly to Al, Co, V, Nb and Ti). It is carbide-forming and increases the hardness and strength of cast steel (1%W increases the UTS and YS in steel by 40 MPa). This is the basic component of heat-resistant steel due to the increase in high temperature resistance and improvement of temper embrittlement resistance (two times weaker than molybdenum).

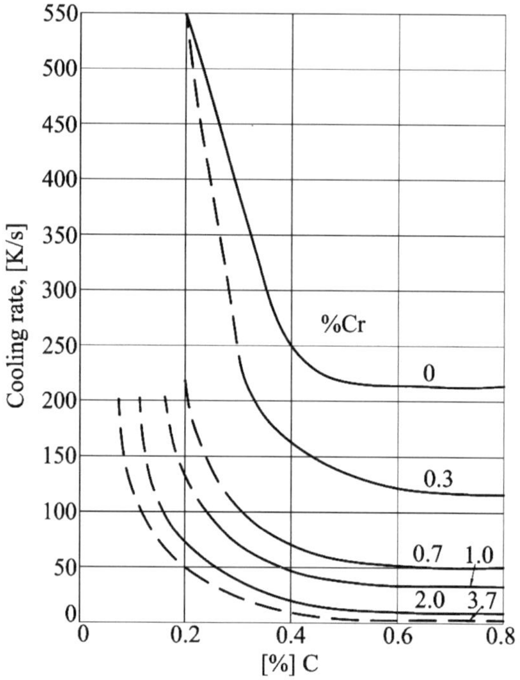

Fig. (4). The quenching depth of carbon and low-alloy cast steel containing 0.35% C [1].

Using several elements together is the most advantageous, and also the least expensive, way to increase the hardenability of steel. This is illustrated clearly by the research of K. Roesch [6], carried out for several cast steel grades. Compared to carbon cast steel, introduction of molybdenum, silicon and manganese, or chromium and molybdenum allows for an increase of the quenching depth from 5 to 60, and even up to 100 mm, (Fig. **3**). When introducing 2.5% Cr and 0.4% Mo into steel, an evenly hardened structure is obtained in castings with a wall thickness of more than 100 mm. These data are the basis for the optimum selection of steel grades for machine castings.

2. CASTINGS OF WALL THICKNESS UP TO 100 MM

The low hardenability of carbon cast steel (Fig. **3**) indicates the possibility of its application only to thin-walled castings, below 30 mm. As the castings used in machine and equipment construction are shaped castings for design reasons, they therefore have various wall thicknesses, typically from 20 to 80 mm. Such castings should have a relatively low quenching rate, because the wall connections of various thicknesses create heat distributions of even larger diameters, and in addition, they are subjected to quenching cracks or heterogeneous structures and properties. A low quenching rate also enables an exchange of the cooling medium, from water to oil. For these reasons, there are two groups of cast steel identified after quenching and tempering: those with a thickness of up to 100 mm and those above 100 mm. Their chemical compositions and equivalents in other countries are presented in Table **1**.

The first important difference between the cast steel designed for thin-walled and that for thick-walled castings is the much lower content of sulphur and phosphorus in the latter - in thick-walled castings each is below 0.025%. Therefore, segregation and the tendency to cracking are reduced. The second difference is the presence of nickel in castings with a wall thickness above 100 mm. Two fundamental characteristics of nickel, namely, its increasing hardenability and its ability to cause an equalisation of mechanical properties throughout the wall thickness of a casting, are applied here. In thick-walled castings, in particular in castings with a wall thickness above 150-200 mm, increased concentrations of molybdenum can be observed. The fact that the ratio of chromium to nickel most commonly assumed in practice (1: 1) is different from that in some known types of cast steel (1: 2), must also be taken into account [8].

The analysis of the data in Table **1** shows that after quenching and tempering in Cr-Mo cast steel, a UTS of over 950 MPa can be obtained in castings with a wall thickness of up to 100 mm. Suitable values for yield strength and elongation are 540 MPa and a minimum of 9%, respectively. The introduction of vanadium to this grade of cast steel further enhances YS to over 635 MPa and UTS to over 880-1030 MPa, at an elongation of at least 10%. These values are obtained not only by the cumulative action of Cr + Mo + V, but also as a result of the modification of the microstructure by vanadium.

Good mechanical properties of cast steel in thick-walled castings are therefore achieved by a much lower content of sulphur, phosphorus and non-metallic inclusions. To ensure a UTS above 950 MPa, a concentration of sulphur and phosphorus below 0.020% is required. These conditions are not too strict, because in almost every cast steel foundry that retains a good quality of product, concentrations of sulphur below 0.015%, and of phosphorus below 0.020%, are achieved. This is a result of the application of modern methods of deoxidation and desulphurisation of steel (blowing of powders with argon, double deoxidation, argon treatment in the ladle and other secondary metallurgy processes), which allow steels with even less than 0.010% of sulphur or phosphorus to be obtained. The introduction of these smelting methods for low-alloy steels, along with vacuum processes (LF, VOD, VAD and AOD), enables the gas content, *i.e.* oxygen, nitrogen and hydrogen, to be reduced, which is very important, as well as giving a reduction in non-metallic inclusions. These processes retain the properties given in Table **1**.

An important advantage of Cr-Mo cast steel is the control of its properties by the use of an appropriate tempering temperature. An example of such a technology is presented in Fig. (**5**). After normalising, quenching in oil and tempering in the range of 500-650°C, a tensile strength from 890 to 1100 MPa, and an elongation from 5 to 16%, respectively, were achieved. After tempering at 500°C, an impact toughness in the range of KCU = 59 - 63 J/cm^2 was obtained, and after tempering at 600°C, 100 to 103 J/cm^2. The hardness over the entire range of tests ranged within 270 - 325 HB. In addition to superior mechanical properties, Cr-Mo cast steel also has good abrasion resistance. Therefore, it is used for mining machinery and parts, *e.g.* on mine scrapers.

Low-alloy cast steel grades are introduced, not only because of the need to increase hardenability, but also to provide other operating parameters, such as: resistance to sudden load, high strength at elevated temperatures, high impact toughness at sub -zero temperatures and increased resistance to abrasion and erosion. Usually, there is no clear separation between the working conditions; the given grade of cast steel must combine several of these requirements. Modern industries want to use lighter machines (thinner walls of castings with higher strength) with an extended scope to their work, *i.e.* adaptation to higher loads and a wider range of temperatures. In practice, they usually meet some of these parameters, and such requirements determine the trends of development of structural cast steel.

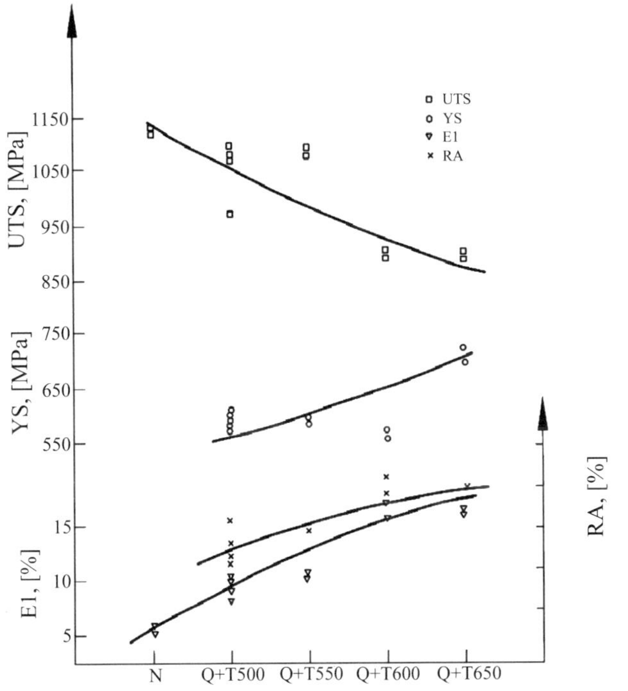

Fig. (5). Effect of heat treatment on mechanical properties of GS-42 CrMo 4 cast steel.

To meet these working conditions, many low-alloy cast steel grades have been developed and classified according to certain groups. Characteristics of each are presented below.

Low-Alloy Manganese Cast Steels

At a carbon content from 0.2 to 0.5% and manganese from 1 to 1.8%, these cast steel grades comprise the cheapest material with increased hardenability and simple heat treatment (normalising or normalising with tempering). These are castings with a pearlitic structure in which martensite is only obtained after a rapid cooling from the austenite temperature. Their advantage is low production cost and good mechanical properties after quenching and tempering, combined with good abrasion resistance [9]. However, their unfavourable parameters include a coarse-grained as-cast structure (Fig. **6**), a tendency to overheating and low hardenability.

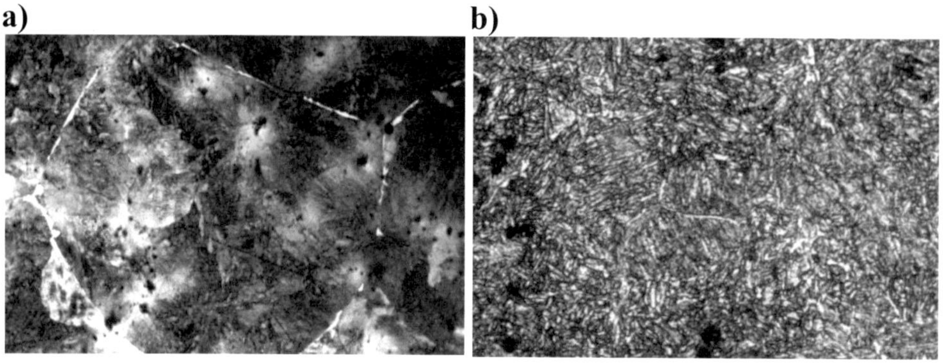

Fig. (6). Microstructure of cast steel (0.5%C-1,4%Mn) in as-cast condition: a) pearlite and ferrite at grain boundaries, above 100x, and after quenching and tempering: b) sorbit, above 500x [9].

In the as-cast condition, this structural cast steel has grain from 2 to -2 (according to the ASTM method, ferrite network, industrial heats), and requires modification of the structure while pouring the steel into the ladle. In order to eliminate coarse-grained structure, two methods are applied: addition of vanadium into the liquid steel or control of pouring conditions and cooling rate [9]. For castings with a thickness of 50-70 mm for the majority of walls, it is necessary to use a pouring temperature of 1560-1580°C and a pouring rate of about 40 kg/s. It is also important to minimise the time of pouring and the pouring distance.

The mechanical properties of this manganese cast steel range from: UTS = 950-1100 MPa, YS = 750-850 MPa, El = 8-14 and impact toughness KCU = 30-50 J/cm^2. These values are obtained [9] after quenching, tempering and the properly conducted refining of the steel. In the heat carried out in this manner, the following gas content was obtained: oxygen below 60 ppm, nitrogen below 70 ppm and hydrogen below 3 ppm.

The hardenability of pure manganese cast steel is low. However, it is higher than carbon cast steel, so giving a yield strength of over 700 MPa in castings with a wall thickness of 60-70 mm. Only after the introduction of other elements, such as silicon, chromium and molybdenum, the hardenability significantly increases. Since the addition of 2% Cr and 3.5% Ni contributes to a significant increase in hardenability (Fig. **3**), these grades are very often introduced instead of pure manganese steel.

To obtain specific mechanical properties, the optimum ratio of manganese to carbon should be kept. Requirements for good hardenability tend to maintain the upper limits of manganese content, while for conditions of good strength, the upper ranges of carbon are targeted. The higher the carbon concentration, the lower the content of manganese should be. For these reasons, in order to keep a good strength of castings, the following ranges of concentrations are applied [8]: I - 0.25-0.35% C and 1-1.4% Mn; II - 0.3-0.4% C and 1.4-1.6% Mn; III - 0.3-0.4% C and 1.6-1.9% Mn.

At low carbon concentration (up to 0.25%), with an increased manganese content to 1.4-1.6% and small additions of niobium, vanadium or titanium, it has been shown that the cast steel is a good material to use at low temperatures [10, 11]. Thus, according to [10], cast steel containing 0.17% C and 1.43% Mn reveals impact toughness from 35 to 58 J/cm^2 at a temperature of 40°C below zero. These values were obtained by double deoxidation of steel smelted in an induction furnace, modified with ferro–vanadium and normalised. The use of dual normalising, or quenching and tempering, improves the impact toughness at 60°C below zero to KCU = 65-80 J/cm^2.

Results of other studies [12] indicate that the combined introduction of vanadium (0.04-0.05%) and niobium (0.03-0.04%) gives an impact toughness at 60°C below zero from 48-56 J/cm^2 in manganese cast steel (0.18% C and 1.7 - 1.8% Mn). Similar data were obtained in reference [13]: after modifying the manganese cast steel with additions of vanadium and titanium (after homogenisation and normalising), at 60°C below zero, the impact toughness KCU = 46-50 J/cm^2.

Manganese-Nickel Cast Steel

Cast steels of this group are among the most common materials used in sub-zero temperatures (valves, heads and pump housings for gas and oil pipelines and the oil refinery industry). Their characteristic feature is their high impact load resistance and very good plasticity at temperatures less than 40°C below zero. Machine components made from these types of cast steel also have a high strength, along with good plasticity and abrasion resistance [4, 14]. This is confirmed by the results of industrial research presented in Figs.(**7** and **8**) [7, 15, 16]. The entire "family" of Ni-Mn cast steels ensures good plasticity and impact toughness at 40°C below zero (in cast steel grades with 0.25%C +Mn, Ni) or very high strength combined with good plasticity (in cast steel grades with 0.35%C+Mn, Ni).

The introduction of nickel into manganese cast steel eliminates its major disadvantages, *i.e.* its tendency to overheat during heat treatment and the irregularity of its properties with respect to the wall thickness of the casting. With a content of 1% Mn and 1% Ni, the mechanical properties in castings with a wall thickness of 200 mm are not significantly different from those in castings with a wall thickness of 40-60 mm [8]. The ratio of Ni: Mn, typically assumed to be equal to 1, is increased to 2 for thick-walled castings, with wall thicknesses of

300-400 mm. The main disadvantage of Mn-Ni cast steel is a tendency to temper embrittlement. However, after the introduction of small quantities of molybdenum and vanadium into Mn-Ni cast steel (Fig. **7**), the strength properties are clearly improved. For the grade 0.25%C + Mn, Ni, V, the minimum value of YS exceeds 650 MPa, with an elongation of 10% and impact toughness KCU = 400 kJ/m^2. A characteristic feature of this grade of cast steel is the ease of achieving the required strength parameters through small changes in the heat treatment. For example, application of quenching and tempering brings a significant improvement in the strength properties in relation to the condition after normalisation (Figs. **7** and **8**). At average tempering temperatures (450°C), a YS above 1000 MPa, and an elongation over 9%, were obtained.

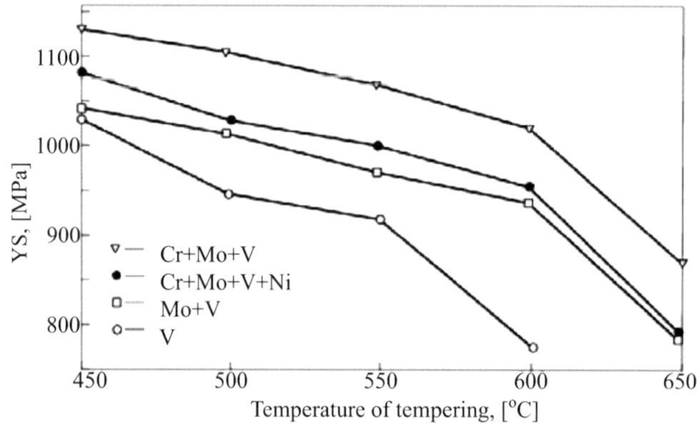

Fig. (7). Yield strength in 0.25%C +Ni,Mo,V cast steel [7, 15].

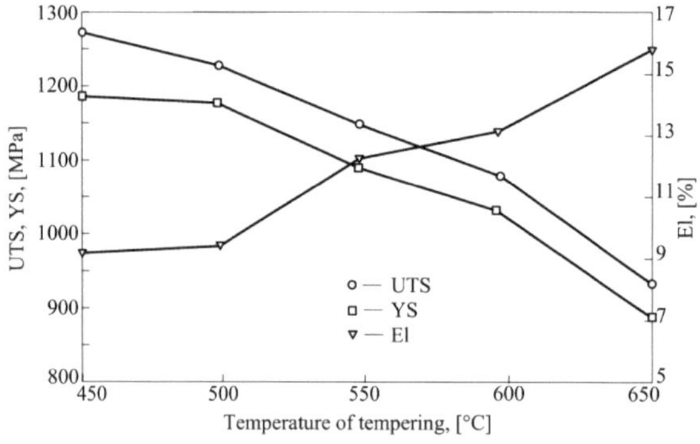

Fig. (8). The impact of tempering temperature on the properties of cast steel (0.35%C + Mn, Ni, Mo, V) [15, 16].

Mn-Ni cast steel with low carbon content has a very good weldability. All grades are cold-welded with EB 160 electrodes. These two features, good weldability and a favourable set of strength and plastic parameters, are extremely helpful for designers.

Mn-Ni cast steel with a higher carbon content (0.35% C) comprises another class of alloys. It is designed for very difficult working conditions, especially where high strength parameters, combined with good resistance to abrasion (mining, energy and machinery manufacturing industries), are required. Essential for this cast steel, tempered at 550 °C, are very good plastic properties, along with high strength. The ratio YS/UTS for this grade exceeds the value of 0.9 for all tempering temperatures after quenching.

A comparison of the abrasion of Mn-Ni grades, along with other grades of low-alloy cast steels, is presented in Fig. (**9**). Particularly favourable results (less weight loss) are obtained after a lengthy time of work on the cast steel. Under metal-non-metal abrasion conditions (aqueous suspension of SiC) after 12 hours of abrasion, Mn-Ni cast steel has smaller losses than Mn-Mo or Cr-Mo cast steel. It can be expected that cast steel with higher carbon content would bring even better results.

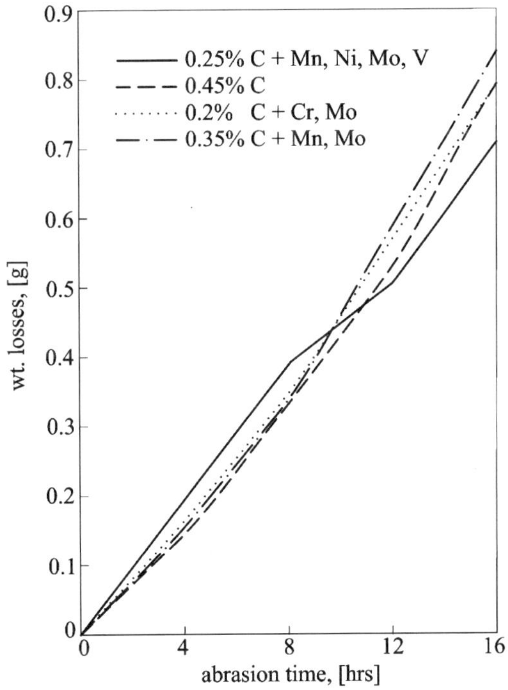

Fig. (9). Comparison of the wear rate by abrasion of various types of cast steels; research conditions: Miller apparatus, abrasive - SiC + water [15].

Manganese-Molybdenum Cast Steels

They are widely used, not only due to increased hardenability, but primarily for their good ratio of YS/UTS, increased strength at elevated temperatures and the reduced temper embrittlement found with the addition of molybdenum. At the usually applied carbon contents of 0.2-0.35%, two subgroups of cast steel can be identified, one with 1-1.35% Mn and 0.1-0.30% Mo and another with 1.4-1.80% Mn and 0.25-0.55% Mo.

Due to the good hardenability of this cast steel, very good mechanical properties can be obtained after applying normalisation combined with tempering. For example, after the double normalisation (930°C / over +820°C / over +565°C/ furnace) [8], a YS of over 820 MPa and an impact toughness of over 90 J/cm^2 are obtained.

Owing to their very good hardenability, Mn-Mo cast steels can be hardened in media with a cooling rate lower than water, *e.g.* air [8]. In combination with a second favourable factor, which is a low susceptibility to temper embrittlement (which allows for slow cooling after tempering from a temperature of 650°C), this cast steel is suitable for use in massive castings.

For these reasons, this cast steel is often used for castings for agricultural and construction machinery (tractors, excavators and dredgers), hammers and presses, and also gear wheels.

In medium-carbon cast steels containing 1.3-1.6% Mn and 1-1.4% Ni, additions of chromium (up to 0.8%) and molybdenum (up to 0.5%) are often found. Such a good hardenability is then achieved that castings with wall thicknesses of 50-70 mm can be hardened in oil or even in the air. After quenching and tempering from 500°C, a UTS$_m$ above 1100 MPa, a YS above 970 MPa, and an elongation of at least 12% are obtained [8]. To achieve such good properties in thick-walled castings with a complex shape, vanadium is also introduced for the fragmentation of the primary structure and a further increase in the hardenability.

Table 2. Characteristics of high strength cast steels.

Grade	Chemical composition, % weight									Mechanical properties				
	C	Mn	Si	P	S	Cr	Ni	Mo	Others	g, mm	UTS,, MPa	YS MPa	El, %	KCV J
GS-24Mn 6	0.24-0.25	1.50-1.80	≤ 0.60	≤ 0.02	≤ 0.015	≤ 0.30	–	–	–	50	700 - 800; 650 - 800; 600 - 750	550 500 400	12 15 18	40 50 60
GS-17CrMnMo 5 5	0.15-0.21	1.20-1.60	– < < –	– < < –	– < < –	1.20 - 1.50	–	0.45 - 0.55	–	50 80	820-970; 730-880	700 600	12 12	50 40

(Table 2) contd.....

Grade	Chemical composition, % weight									Mechanical properties				
	C	Mn	Si	P	S	Cr	Ni	Mo	Others	g, mm	UTS,, MPa	YS MPa	El, %	KCV J
GS-19CrMo4	0.15-0.22	0.60-1.00:	–<<–	≤ 0.015	–	2.00 - 2.50	–	0.70 - 1.10	–	150	700-850; 750-900	500 650	18 15	50 40
GS-42CrMo 4	0.38-0.45	0.60-1.00	0.30-0.50	<0.02	≤ 0.015	0.80 - 1.20	–	0.20 - 0.30	–	150 50	700-859; 900-1100	550 800	10 10	27 27
GS-30CrMoV 6 4	0.27-0.34	0.60-1.00	^0.60	–<< –	–<< –	1.30 - 1.70	–	0.30 - 0.50	V = 0.05- 0.15	100	850-1000; 900-1000	700 750	14 12	45 32
GS-35CrMoV104	0.32-0.39	0.60-1.00	–<<–	–<< –	–<< –	2.20 - 2.70	–	0.30 - 0.50	–<<–-	100	850-1000 1050-1250	700 850	15 10	45 27
GS-18NiMoCr36	0.17-0.22	0.8	0.60-1.20	–<< –	–<< –	0.40 - 0.90	0.60 - 1.00	0.04 - 0.70	–	80	850-980	700	12	–
GS-22NiMoCr 5 6	0.18-0.24	0.8	0.60-1.20	–<< –	< 0.005	0.50 - 1.00	0.80 - 1.30	0.50 - 0.70	–	50	980-1080 1050-1200	825 950	10 10	–
AISI/SAE 43 40	0.38-0.43	0.20-0.35	0.60-1.20	–	–	0.70 - 0.90	1.60 - 2.00	0.20 - 0.30	–	–	1980 1500	1860 1365	11 20	–
18G2ANb PN- 85/H-84018*	0.2	1.00-1.50	0.20-0.55	≤ 0.040	≤ 0.040	≤0.30	0.3	–	Cu = 0.30 Nb = 0.02- 0.05	–	540 680	400	19	–
18G2AV *	0.2	1.20-1.65	0.20-0.60	–<< –	–<< –	–<<–	–<< –		Cu= 0.30 V= 0.05- 0.15 N= 0.01-0.02	–	570 720	440	17	–
DIN 17102-83*	0.2	1.00-1.70	0.10-0.60	<0.030	<0.030	-<<-	1	-	Cu = 0.20; V = 0.20 Al. <. 0.02	–	560-730	440	17	–
Soldur 700"	0.04	1.8	0.25	–	–	–	–	0.4	Nb= 0.05; Ti= 0.03 B= 0.002	–	>760	>700	> 13	–
Chromansil	0.20-0.45	0.90-1.80	0.60-1.20	<0.030	<0.030	0.50-1.20			Mn: Cr: Si = 1.0:0.9:0.75	–				–
L35HGS	0.40	1.50	0.80	<0.04	<0.04	0.60-0.90	<0.30	–	–	–	600	350	14	KCU3-29 J
	0.29 0.28 0.27	1.22 1.18 1.26	0.38 0.81 1.04							–	510 560 660	280 320 380	30 30 22	– – –

g- wall thickness, * HSLA steels and cast steels

Chromium and Chromium-Molybdenum Cast Steels

Castings of pure chromium steels are used for machine parts with wall thicknesses up to 60 mm. Chromium reduces the critical quenching rate, but this effect is visible only when the carbon contents are above 0.3-0.4 [8]. Both for this reason, and due to chromium's strong tendency to form carbides in chrome steel castings, the total effect of carbon and chromium should be always considered. 0.7-0.9% Cr is usually used at contents of 0.3-0.45% C, obtaining a YS after quenching and tempering above 480 MPa, and a UTS above 650 MPa. The corresponding values of elongation and reduction of area are El = 16-18% and RA = 35-45% [8] (Table **2**).

Depending on the content of manganese, defects can occur in castings. When the content exceeds 0.6% Mn, cracks may appear during quenching.

An increase in the carbon content above 0.4% causes higher wear resistance in chromium cast steel, but rapidly lowers the plasticity. Although the strength increases (UTS above 1600 MPa), in castings with wall thicknesses over 60 mm that require a good plasticity, other alloying elements, such as molybdenum, vanadium and tungsten, should be introduced.

Cast steels with additions of Mo, V and W are mainly used in refineries, for gas extraction (casings and valves operating at high pressure), on turbines and as components of machines operated at temperatures up to 650°C. The concentration of chromium in these grades usually equals 0.8-1.1%, and that of molybdenum from 0.3 to 0.6%. These cast steels have a good set of strength properties, in contrast to their plastic parameters (Figs.**7** and **8**). Since these cast steel grades are used for operation at elevated temperatures, and additionally very often work under heavy stresses, their chemical composition must be closely matched to the operating conditions and the heat treatment applied. Under these conditions, the molybdenum content is increased to 0.5-0.7% and the chromium up to 1.5%, at carbon concentrations within the range of 0.3-0.4%. To improve the plasticity of these cast steel grades, vanadium (up to 0.2%) and tungsten (up to 1.2%) are introduced [8]. The introduction of molybdenum removes the tendency to undergo temper embrittlement, increases the creep strength and impact toughness, and also improves the machinability of castings. However, vanadium gives a finer grain during the primary crystallization, and thus raises the strength of the matrix. The effect of Cr + Mo + V is to give uniform mechanical properties in castings with a wall thickness up to 120 mm. With contents of 0.25% C, 1.5% Cr and 0.4% Mo, the following properties are obtained after normalisation at 900°C and tempering at 650°C: UTS above 700 MPa, YS above 550 MPa, El above 25% and RA above 40% [8].

After an increase in carbon content up to 0.7-0.8% (or the introduction of tungsten up to 1.2%), the abrasion resistance (at the expense of significant reduction in plasticity) is explicitly improved. Hence, cast steels containing 0.3-0.8% C, 1.8-2.1% Cr and 1-1.2% W are used in the production of pipes, rollers and the components of rolling mills [8].

3. CAST STEEL FOR THICK-WALLED CASTINGS

The grading of cast steels due to their intended use is extremely rare. Such a cast steel grade is practically established in Germany [1, 5]. In substantive and utility terms, it is fully justified, because making castings from one type of cast steel, but with various wall thicknesses (*e.g.* 30 or 350 mm), gives completely different technological problems. Most often, this is demonstrated when placing orders for large castings. The contractor then states, 'We do not have the conditions to make such a large casting,' which should be understood as, 'We do not have enough practice nor experienced staff to ensure the required quality of such a casting.'

An analysis of Fig. (**10**) shows that in castings with a wall thickness up to 100 mm, a tensile strength in the range of 850-950 MPa can be achieved by using Cr-Mo cast steel. At higher requirements relating to the strength and thickness of the walls, the cast steel grade should be changed by introducing additional nickel, or nickel and vanadium.

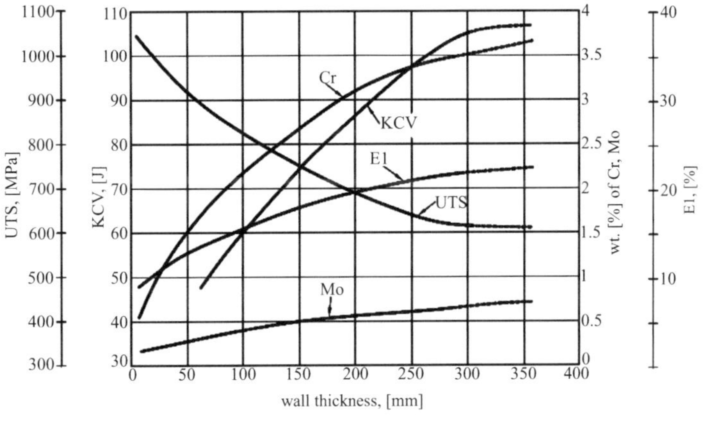

Fig. (10). Selection of Cr-Mo cast steels for thick-walled castings [5].

A similar procedure is recommended if castings have a wall thickness of more than 100-150 mm. High chromium (2.5-3.5%) and molybdenum (over 0.5%) contents then provide a good tensile strength (which can be more than 1000 MPa) and the appropriate yield strength (over 650 MPa) at an elongation of 6-10%. The more massive the casting, the higher the concentration of chromium and molybdenum required. However, with the increase in wall thickness, the strength properties (YS and UTS) decrease. Reducing the solidification rate, along with the increase in the wall thickness of the casting and an increase in the segregation of elements in massive parts, necessitates the introduction of modifiers (most often vanadium), performing double steel deoxidation or deep desulphurisation and dephosphorisation of steel (a sum of S and P less than 0.025% is often recommended).

Nonetheless, a heterogeneous structure, different in the outer layer of the casting and in its inside, is met in castings with a wall thickness above 150-200 mm. In the as-cast condition, there is typically a certain amount of ferrite inside the wall, and after quenching and tempering martensite (M) + bainite or bainite + pearlite.

The direct cause of this phenomenon is the uneven cooling rate within a single cast wall. This significantly affects the reduction of yield strength and impact toughness of the cast steel. This is also highly indicated by segregation processes of, *e.g.* oxygen and sulphur (in the case of type II sulphides), or carbon and chromium (during carbide precipitation at primary grain boundaries). Technological problems in the production of the thick-walled castings result from the difficulty in ensuring good material quality (low concentrations of oxygen, nitrogen, hydrogen and non-metallic inclusions), uniform solidification of the casting in the mould and the properly selected heat treatment. The last aspect is related to the information shown in the TTT chart, for a given cast steel grade. Examples of such charts are given in Fig. (**11**). Measurements of the cooling rates inside walls with a thickness of 150, 300 and 500 mm indicate that these rates pass through the formation of bainite (B), and under certain cooling conditions, ferrite (F) and pearlite (P) also occur.

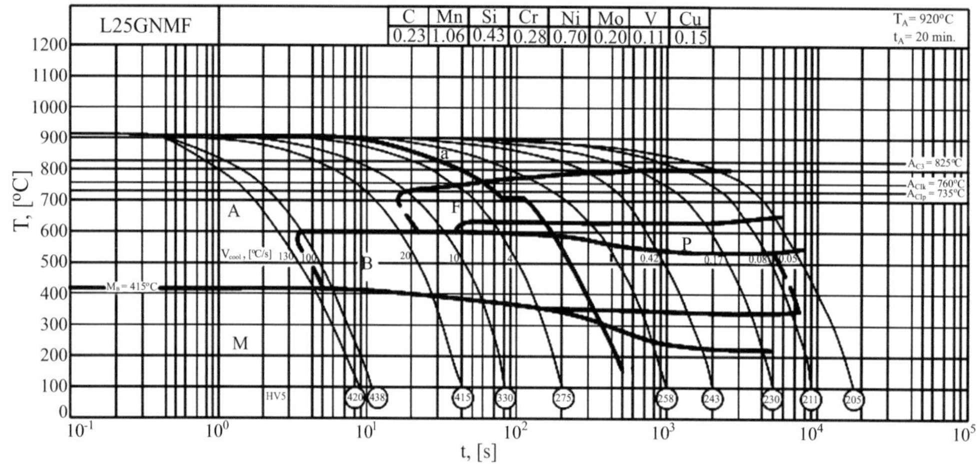

Fig. (11). TTT diagram for low alloy cast steel [15].

It can be assumed that the maximum permissible quantity of ferrite or pearlite, *e.g.* 10%, is a measure of the adjustment of the given cast steel grade intended for thick-walled castings, arranged according to the ease of the heat treatment, *i.e.* a certain kind of hardenability (Fig. **12**). These types, for which curves for the appropriate wall thickness are exceeded to the right, will contain a certain quantity of ferrite and pearlite in their structure.

The change of the cooling medium brings visible practical effects: the introduction of quenching in oil allows a homogeneous structure to be obtained in thick-walled castings. However, the characteristic feature of the production of castings with a weight of 30-40 tons is the necessity of quenching them in air, which results from limitations in the dimensions of quenching tanks, the possibilities of castings cracking during quenching and the difficulties in performing the quenching process.

If this is not possible, larger quantities of elements increasing hardenability, such as nickel, chromium and molybdenum, should be introduced. Fig. **(12)** gives information on making other types of cast steel, with higher concentrations of these elements.

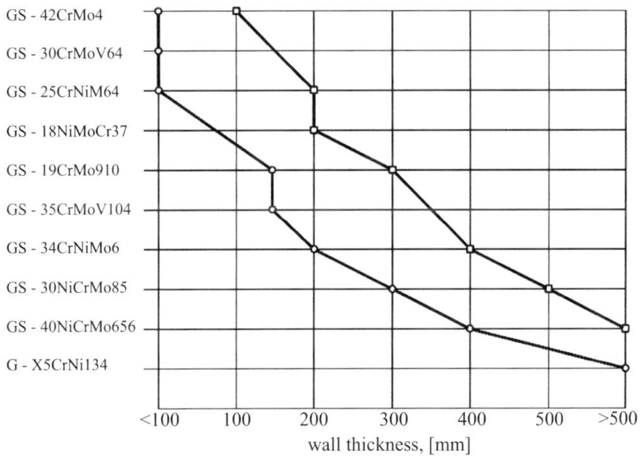

Fig. (12). Selection of low-alloy cast steel for thick-walled castings according to their hardenability [17].

The available data [1, 5] indicate that the following types of cast steel can be used for castings with a wall thickness of 500 mm:

– 0.30% C, 1.2% Cr, 2% Ni, 0.35% Mo with quenching in oil.

– 0.40% C, 1.4% Cr, 1.5% Ni, 0.6% Mo with quenching in the air.

– 0.05% C, 13% Cr, 4% Ni quenching also in the air.

The cast steel grades listed reach a high yield strength in the thickest parts of the casting as a result of limited quantities of ferrite and pearlite. Castings of steel containing 13% Cr-4% Ni, and below 0.05% C, tend to have lower concentrations of sulphur, phosphorus and gases (N, O and H). This is made possible by the use of advanced smelting technologies, *i.e.* AOD, VOD or VAD processes, which can give less than 3 ppm of hydrogen, less than 40 ppm of oxygen and less than 50 ppm of nitrogen. These processes also allow concentrations of sulphur or phosphorus below 0.020% to be obtained. Favourable parameters therefore lead to an improvement in the uniformity of the structure and to a reduction of hot cracking, which is important in the technology of making castings with walls of considerable thicknesses (over 150 mm).

Cast steel containing 13% Cr and 4% Ni was developed in the 50s in highly industrialized countries [2, 5]. It is applied in the energy industry: approximately 95% of Pelton and Francis wheels are made from this cast steel. Castings operating at elevated temperatures (for example gas turbine casings, pump components, *etc.*) are also an important area of application.

Another extremely important parameter of 13% Cr-4% Ni cast steel is good hardenability, allowing for the use of this composition at temperatures from 60 below zero to 120°C below zero. In these operating conditions, a very good impact toughness is achieved, and the threshold of brittleness of the cast steel is at 70°C below zero. The average value of impact toughness of this cast steel at a temperature of 70°C below zero equals over 55 J/cm^2. For this reason, this grade of cast steel is widely used for castings of cooling pumps for the primary circuit of nuclear reactors (instead of 18% Cr-9% Ni cast steel or 18% Cr-11% Ni cast steel). A good resistance to ageing during the long-term operation of the component at temperatures of 500 - 600°C, and good resistance to corrosion, are substantial advantages for the operation of this type of cast steel.

4. HIGH STRENGTH CAST STEEL

A lot of castings constituting elements of machine parts are faced with demands significantly exceeding the UTS values achieved for carbon cast steel (700 MPa) or Cr-Mo cast steel after quenching and tempering (1000 MPa). In such situations, high strength cast steel, for which the UTS exceeds 1200-1500 MPa, is applied. Obviously, these are castings operating at high stress, and they must simultaneously possess high yield strength and sufficient plasticity. Data presented in Tables **1** and **2** indicate that these conditions are met by several types of Cr-Ni-Mo cast steel. Their high ratio of YS/UTS= 0.8-0.9 and good plasticity (El above 10%) are worth emphasizing.

The following requirements must be met to ensure such parameters:

– keeping the proper content of Cr, Ni and Mo, the concentrations of sulphur and phosphorus should be reduced to 0.015% (for each),

– ensure low concentrations of oxygen (below 30 ppm), nitrogen and hydrogen (below 2 ppm), and non-metallic inclusions,

– eliminate any microshrinkage and microporosity in the alloy matrix,

– use more complex heat treatment (normalisation, quenching and tempering),

– provide the ability to weld castings and the possibility of easy machining.

Selection of alloying elements, in particular Cr, Ni, Mo, while also controlling the amounts of Al, N and oxygen, should provide a completely hardened structure for the wall thickness. Therefore, quenching and tempering are most commonly used, but due to the final properties, low and medium tempering are distinguished. Depending on which of these heat treatments are performed, castings of two ranges of strength are obtained [4, 5] (Fig. **13**):

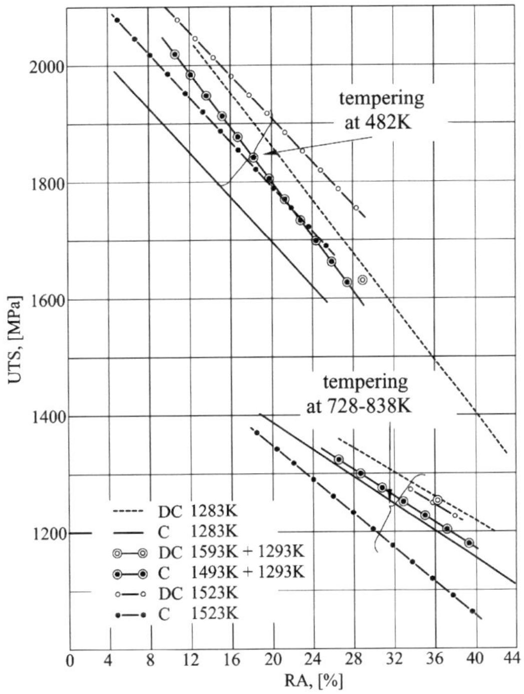

Fig. (13). Mechanical properties of high strength cast steel [1].

– UTS = 1600-2000 MPa, RA = 10-25% and El = 4-8%; after quenching and tempering at 473K,

– UTS = 1100-1350 MPa, RA = 28-45% and El = 10-12%; after quenching and tempering at 450-550K.

Under these conditions of heat treatment and a concentration of sulphur and phosphorus below 0.015%, a ratio of YS/UTS > 0.85 can be obtained. The use of tempering temperatures in the range of 350- 450°C causes an occurrence of lower plasticity range (Fig. **14**). To avoid this, the cooling rate must be increased while passing through this temperature range, or the addition of molybdenum must be increased.

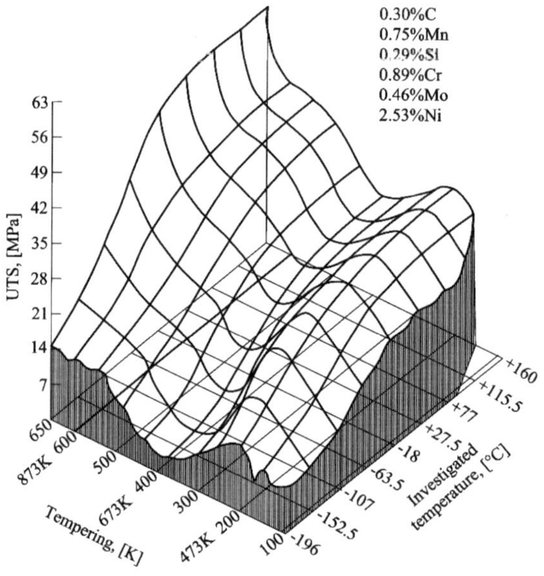

Fig. (14). Transition temperature of high strength cast steel, Cr-Ni-Mo [1].

In addition to the requirements for high-quality steel for castings from this cast steel type, it is also important to maintain an appropriate concentration of carbon. At low concentration of nickel (up to 0.6%), the strength parameters increase continuously with increasing carbon content. However, in cast steel containing approximately 1.8% Ni, the yield strength increases only up to 0.40% C, and then decreases. This is important information when selecting the chemical composition of cast steel as a construction material with a certain required ratio of YS/UTS.

4.1. High Strength Low-Alloy Cast Steels

This type of cast steel is based on previously developed HSLA steels (High Strength Low-Alloy Steels) [18, 19], meeting the conditions: UTS = 1200 - 2000MPa, YS = 950 – 1100 MPa and elongation above 8%. Keeping the ratio of YS/UTS above 0.8 is important for this group of (cast) steels. Such requirements must be met by elements of machines and equipment for the arms industry (tank hulls, muzzle brakes and suspension arms), aviation and astronautics (rocket sled parts, crank arms and support housings), but also in power generation and the petrochemical industry. Good wear resistance and toughness, combined with resistance to high stresses, allows for the use of this cast steel in earthmoving equipment (dipper teeth, mine scrapers, gears, axle housings and spindles).

The main types of steel that achieve the above-mentioned requirements are Cr-Mo cast steel and Cr-Ni-Mo cast steel, modified with vanadium, boron and titanium [20 - 23].

The microstructure of high strength cast steel is composed of ferrite and precipitations of carbide and carbo-nitrides. Depending on the applied heat treatment, the type of precipitate formed is subject to significant changes. In Fe-C-Mo, under equilibrium conditions, with an increase in the ratio of Mo/C the resulting carbide sequence is as follows: Fe_3C, $Mo_{23}C_6$ ($Fe_{21}Mo_2C$), Fe_2MoC, Mo_2C and Mo_6C [24 - 29].

CONFLICT OF INTEREST

The author (editor) declares no conflict of interest, financial or otherwise.

REFERENCES

[1] K. Roesch, and K. Zimmermann, *Stahlguss.* Verlag Stahleisen GmbH: Düsseldorf, 1966.

[2] C. McMahon Jr, "Intergranular fracture in steel", *Mater. Sci. Eng.,* vol. 25, pp. 233-239, 1976. [http://dx.doi.org/10.1016/0025-5416(76)90075-6]

[3] J. Svoboda, and A.S. Handbook, *Casting.* vol. 15. ASM Intern, 1988, pp. 722-735.

[4] M. Blair, *Steel Castings, Metals Handbook.* vol. 1. 10th ed. ASM Intern, 1990, p. 363.

[5] J.A. Niechendzi, *Stalnoje Litjo, Mietalurgizdat.* Moskwa, 1948.

[6] K. Roesch, H. Zeuner, and K. Zimmermann, *Stahlguss, Verlag Stahleisen.* GmbH: Duesseldorf, 1982.

[7] A. Maciejny, *Kruchość metali, Wyd.* Śląsk: Katowice, 1973. (in Polish)

[8] W. Gysel, and E. Gerber, "Kaltzaeher martensitischer Stahlguss", In: *Konstruieren und Giessen,* 1978, no. 2, pp. 17-25.

[9] J. Głownia, R. Bujas, and L. Młynarczyk, "Effect of heat treatment and deoxidation on the impact toughness of Mn-Ni cast steel", *Przegl. Odl,* vol. 1, pp. 1-6, 1995.

[10] E.R. Kuch, *Metals Handbook.* vol. 1. 10th ed. ASM Intern, 1990, p. 483.

[11] G. Kniaginin, "Staliwo,Metalurgia i Odlewnictwo, ed.Slask", *Katowice,* 1977.

[12] J. Głownia, B. Kalandyk, and A. Dzieja, "Technologia staliw o obniżonym progu kruchości, Project of Univ.of Sc.and Techn., no. 5.5.170.35", *Cracow,* 1998.

[13] M.M. Salama, *Metals Handbook.* vol. 1. 10th ed. ASM Intern, 1990, p. 662.

[14] G.F. Vander Voort, ibid, Metallography, Principles and Practise, ASM International, 1999, 689-736.

[15] T. Wachelko, M. Soiński, and A. Nowak, "Improvment of mechanical properties of low-alloyed cast steel through V and Nb additions", *Krzepnięcie metali i stopów, ed. PAN, Dep. Katowice,* vol. 3, p. 245, 1980.

[16] P. Leviček, and K. Stransky, "Oxidative remelts of Hadfield steel and its effect on the quality of castings", *63th World Foundry Congress, ref,* vol. 4, 1986.

[17] J. Głownia, B. Kalandyk, and L. Młynarczyk, "Constructional manganese-nickel cast steel of elevated yield point", *Przegl. Mech,* vol. 4, pp. 18-22, 1999.

[18] J.A. Rimebolt, and W.J. Harris, "Harris, Effect of Alloying Elements on Notch Toughness of Pearlitic Steels", *ASM,* vol. 43, pp. 1157-1214, 1951.

[19] L. Młynarczyk, J. Glownia, B. Kalandyk, E. Wychota, and A. Dzieja, Dzieja, Low-alloy constructional cast steel, Patent, Poland, no.PL 174868 B1, 1998.

[20] J. Głownia, Patent, Cast steel with microadditions, Patent Poland no.309839, 1995.

[21] F. Henke, "Low and high alloyed quenched and tempered wear-resisting steel castings", *Giesserei – Praxis,* vol. 23/24, pp. 377-407, 1975.

[22] R. Leveque, "Colloque international sur les alliages ferreux a hautes teneurs en chrome et en carbonde", *St.Etienne,* pp. 379-410, 1973.

[23] K. Achtelik, and J.M. Motz, "Vergütungsstahlguß für Gußstücke mit Wandstärken über 100mm", *Giesserei,* vol. 65, pp. 475-486, 1978.

[24] C. Richard, and G. Nectoux, Stainless Steel Castings, Editors Behal/Melilli, ASTM, STP 756, 201-221.

[25] T. Gladman, "Precipitation hardening in metals", *Mater. Sci. Technol.,* vol. 15, pp. 30-36, 1999. [http://dx.doi.org/10.1179/026708399773002782]

[26] Z-K. Liu, "Thermodynamic calculation of carbonitrides in microalloyed steel", *Scr. Mater.,* vol. 50, pp. 601-606, 2004. [http://dx.doi.org/10.1016/j.scriptamat.2003.11.032]

[27] T. Siwecki, "Physical metallurgy a.modelling of thermomechanically processed microalloyed steels", *Inż.Materialowa,* vol. 3, pp. 163-170, 1998.

[28] T. Wachelko, A. Nowak, and M. Soinski, "Einfluss des Mikrolegierens mit Vanadium und Niob auf das Gefüge von Stahlguss mit erhöhtem Mangangehalt", *Giesserei-Forschung,* vol. 38, no. 4, pp. 137-145, 1986.

[29] B.D. Jana, A.K. Chakrabarti, and K.K. Ray, "Study of cast microalloyed steels", *Mater. Sci. Technol.,* vol. 19, pp. 80-86, 2003. [http://dx.doi.org/10.1179/026708303225008644]

CHAPTER 8

Steel Castings for Low Temperature Application

Abstract: In engineering structure castings play important role, spatially in minus temperature. These conditions arise with additional stresses cracks and defects in steel castings. Since different properties of materials show different crack propagation, it is necessary to find the limiting parameters, which ensure the proper work of machines. For this reason, this chapter describes the fracture toughness, transition temperature, brittle and ductile fractures, chemical composition of cast steels and effect of heat treatment.

Keywords: Brittle fracture, Effect of alloying, Inclusions, Micro-porosity, Transition temperature.

1. INTRODUCTION

The vast majority of casting alloys and also other materials are characterized by a change in the plasticity at a certain range of temperatures. This means that the material loses its ductility and cracks in a brittle manner. The transition from ductile to brittle fracture occurs at a reduced temperature (Figs. **1** and **2**), especially in conditions of a multi-axis state of stress and dynamic loads. This state is well-illustrated by an attempt to test the impact toughness at temperatures of 223-323 K, which shows that more energy is needed to break the sample at higher temperatures than at sub-zero temperatures. At higher temperatures, the fracture of the sample has a plastic character (Fig. **1a, b**), with a clearly visible deformation on the shear lips of dimples.

Inside each dimple, there are visible inclusions: oval round particles are silicates and oxy-sulphides; square particles are vanadium nitrides and carbides. The occurring inclusions usually initiate the formation of defects and are the cause of further cracks.

Brittle fracture in the sample at low temperatures occurs on preferred crystallographic planes through the matrix and grain boundaries, (Fig. **2**). It can be noticed that the transition from the plastic to brittle fracture occurs for the same chemical composition of cast steel, at a certain temperature range, specified as the transition temperature (or brittleness threshold).

Jan Głownia

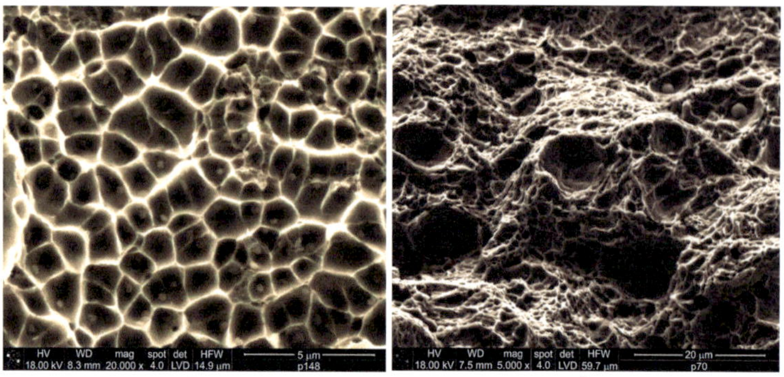

a b

Fig. (1). Ductile fractures in low alloy cast steel with vanadium additions; **a** – Magn. 20 000X; **b** – 5 000X.

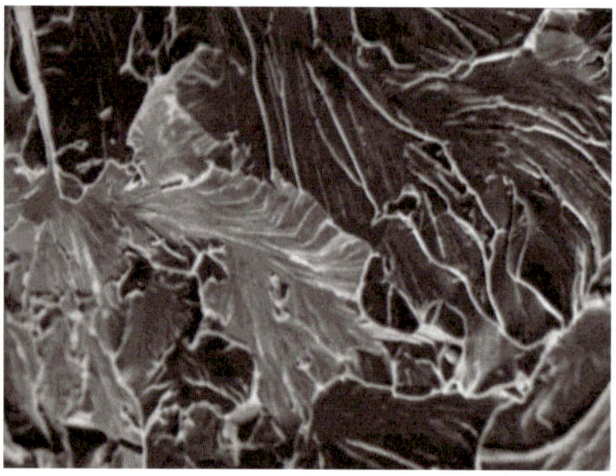

Fig. (2). Brittle fracture in the low alloyed cast steel (0.25%C – 1,4%Mn).Magn.600X.

Knowledge of the impact toughness at low temperatures is important in industries in which machines and devices work in the Arctic region (ships, tankers, offshore structures, (Fig. **3**), cryogenics (liquid nitrogen and oxygen vessels, pressure vessels, cooling systems), chemical and petrochemical industries, aviation and astronautics. Also, sudden weather changes in a temperate climate are the cause of destruction of a lot of machines, for example excavators, tractors, bulldozers and pumps. Ensuring safety (building constructions - structural frameworks, valve bodies, compressors, housings) and reducing failure of these machine components are extremely essential for the operation of machines in these conditions. Above all, this applies to sudden load changes, or the introduction of a tri-axial state of stress.

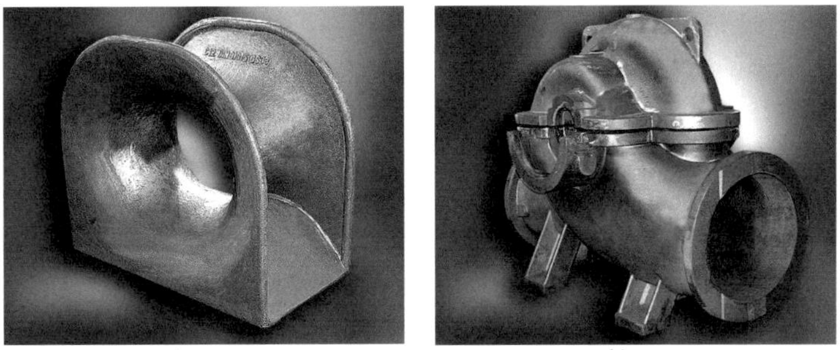

a) b)

Fig. (3). Examples of castings used in off-shore conditions: **a** - chain inlet for ships (weight 270 kg), **b** –casing pump (weight 2300kg); Courtesy Metalodlew S.A., Cracow, Poland.

During tensile testing, it can be seen (Fig. **4**) that along with decreasing temperature, the yield strength YS and the tensile strength UTS (also hardness, modulus of elasticity, and fatigue strength [1]) increase, while elongation, reduction of area and the impact toughness decrease. At a certain temperature, the yield strength YS reaches the value of ultimate strength UTS and elongation decreases to zero. The temperature at which both conditions are met is called the transition temperature T_k. As it is apparent from Fig. (**4**), such an ideal situation does not always happen and determining the temperature T_k, often specified as the brittleness threshold, presents some difficulties. Therefore, the definition of the temperature, at which the transition from the plastic state into the brittle state occurs, is based on the criteria: fracture energy (absorbed energy level), type of fracture (FATT - Fracture Appearance Transition Temperature), the drop-weight test (NDTT - Nil Ductility Transition Temperature), value of K_{Ic} and others (deformation behaviour of the impact specimen, Dynamic Tear Test) [1, 2].

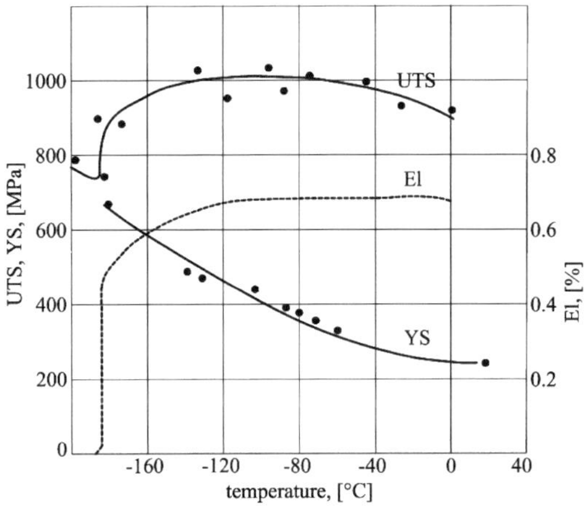

Fig. (4). Effect of temperature on the selected parameters of the mechanical properties of steel [3].

According to the first criterion, the embrittlement temperature is the temperature at which: impact energy is characteristic for a given temperature, *e.g.* of energy, corresponding to the value of 27 J (or conventional values of impact toughness equals 35 J/cm^2), obtained at the investigated temperature [4 - 7], or 20 J [4]. At this temperature, the ductile fracture is changed into brittle fracture, (Figs. **1**, **2**). According to the second criterion, transition temperature (FATT) is characterized by the conventional amount of brittle fracture. Usually, it is assumed that this is the temperature at which 50% of brittle fracture and 50% of ductile fracture occur [5, 7, 8]. This criterion is more difficult to perform, as at mixed fracture exist.

It is difficult to assess the exact value of the surface of the completely plastic or brittle fractures. According to the second criterion, the brittle temperature is the temperature at which the brittle fracture was observed for the first time, or it was stated that at this temperature, 50% of the surface is taken by the brittle fracture. Thc third criterion refers to the sample of drop-weight test, characterizing the temperature T_k with regard to service conditions [9]. In particular, it characterizes the temperature below which brittle cracking is easily initiated and propagated in the presence of small scratches or cracks.

There is a good correlation both for steel and cast steel among these criteria [10, 11]. Assessment of the brittleness of cast steel is performed in practice on the basis of the first criterion, examining the temperature T_k on "V"-shaped and sometimes "U"-shaped (for cast steel with high impact toughness values, *e.g.* high-manganese Hadfield cast steel) impact tests.

For alloys intended for operation at temperatures below zero, the transition temperature is the basic criterion for the assessment of these materials and their resistance to brittle fracture. The value T_k depends on a lot of factors, such as: state of stress, type of stress, applied stress rate, and mainly on the chemical composition of steel and cast steel, degree of deoxidation of steel and applied modification of the structure [5, 6, 12]. Hence, a certain temperature range in which there is the highest likelihood for brittleness occurrence is considered.

In industrial practice, the transition temperature is determined through impact toughness testing, selecting one of a number of U or V-shaped samples [1, 13]. Due to a large number of factors affecting this temperature (also - type of notch, grain size, type of heat treatment), a certain value of impact toughness is assumed, which characterizes the conventional transition temperature. This value is a comparison criterion for determining the resistance to cracking of various grades of steel and cast steel. In a lot of countries, it is assumed that the value of impact toughness equals 35 J/cm^2 (impact energy of 27 J DIN) is this criterion.

In cast materials, any kind of structural discontinuities, hard detectable by non-destructive methods, can affect the transition temperature. In the processed materials such defects include non-metallic inclusions and micro-cracks, macro- and micro-shrinkage (Fig. **5a**, **b**) non-metallic inclusions (Fig. **6**).

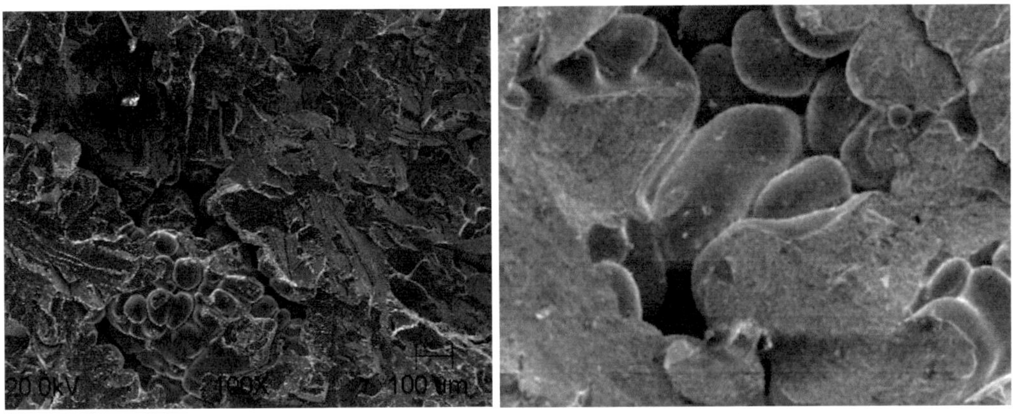

a b

Fig. (5). Microporosity in carbon cast steel; a- Magn.100X; b- Magn.1000X.

Testing the fracture surface gives relevant information about the reasons for the occurrence of cracks, indicating the impact that the presented defects had on the type of micro-cracks and the place of their creation. Based on fractography [14], the presented examples of fractures can be classified as intergranular (Figs. **1**, **5**) and transcrystalline cleavable fractures (Fig. **2**).

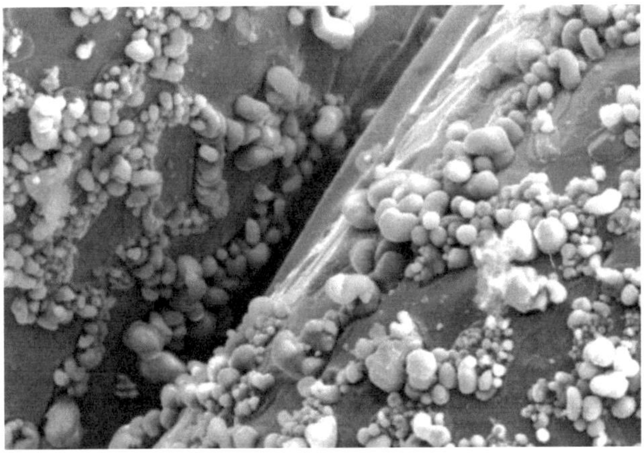

Fig. (6). Non-metallic inclusions in the steel castings Cr-Ni-Mo (on the surface of micro-shrinkage in 10-ton casting) Magn.1350x).

The existence of such discontinuities (defects) reduces the allowable load during long-term operation of the casting. Stresses are focused on these microdefects (equivalent to the width of micro-cracks 2c - in Orowan theory). For this reason, the concept of the stress intensity factor K was introduced, which means that the larger the crack (internal discontinuity, micro-shrinkage), the lower the stress required to cause cracks. In order to destroy the material with micro-shrinkage, extremely different stresses are necessary: reduction of the ductile cross-section by 20%, leads to brittle fracture creation already at about 65% stress, corresponding to a perfect defect-free material. For brittle alloy, fracture occurs even at stresses of 20%, in relation to defect-free materials.

In fracture mechanics, an important role is played by determination of the maximum stress, which the defected material of a specific size and geometry can withstand. The stress required for crack propagation in the presence of an existing defect is associated with the stress intensity factor K, which is expressed by the equation [11, 12]:

$$K = \sigma\sqrt{\alpha \cdot \pi \cdot c}\,,$$

where: σ – the existing tension, **(1)**
α - sample shape factor (1 - 4),
c - size of the defect.

If the value K is equal to or higher than the critical value K_c (G. Irvin's concept), the defect increases, and the material cracks.

The assessment of the safety level of the casting (the structure in which the casting works) is associated with two methods of treatment [15]: the adoption of the critical load criterion or of the critical crack size criterion. In the first case, it is necessary to determine the load necessary to cause a crack of a known size that does not cause damage. This means that the casting with an existing defect is working reliably. In the latter case, it is necessary to determine the size of the defect, which does not interfere with reliable operation of the casting under specific load conditions. This case refers to any technological defects, including micro-shrinkage that is so characteristic for castings (Fig. **5**).

Under conditions of occurrence of crevice critical length 2c and critical stress σ_{cr}, the value of K_c becomes critical K_{Ic} and is called the resistance to cracking. It depends on the load conditions, conventional yield strength, thickness of the component, geometry of crack and geometry of component. Under plane strain conditions, factor K_{Ic} does not depend on the geometric parameters, but only on the microstructure and it is therefore used by designers as a material constant.

In actual operating conditions of the casting, the material's ability to resist the cracks propagation depends mainly on:

- cleanliness of steel intended for castings, with a specified size and amount of non-metallic inclusions, admixture concentrations, gases, presence of carbides and brittle phases. Hence, the constant striving for the use of vacuum to melt steel (VOD, VAD processes), filters for gating systems, secondary metallurgy of steel (ladle furnace, steel argon treatment),
- the type of alloy (metals and alloys with a crystallographic network A1 have a brittleness threshold shifted to lower temperatures),
- wall thickness (thicker components have lower factor K),
- strength of the material (the increase in alloy strength leads to the reduction in impact toughness).

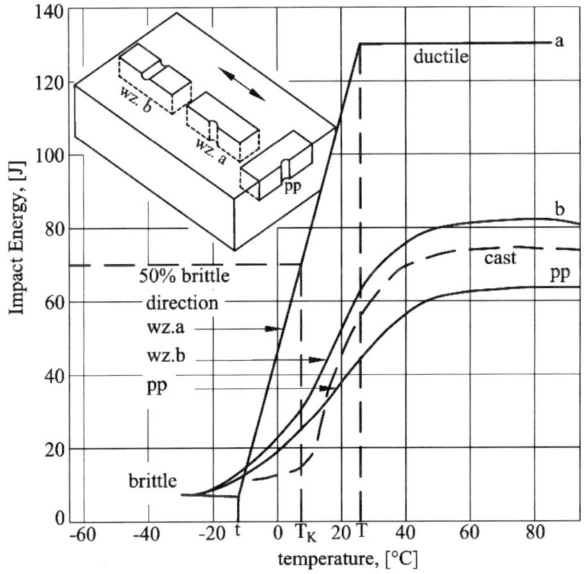

Fig. (7). The impact toughness value depending on the direction of plastic deformation [12].

Although steel castings are characterized with plasticity that is worse than steel, however, owing to their isotropic properties under certain conditions, they can be compared with one another. This applies the impact toughness of the cut-out material depending on the rolling direction (Fig. 7) and its comparison with the impact toughness of cast steel. Assuming that the sample cut out along the rolling direction is indicated by "wz.a" and "wz.b" and the sample transverse to the rolling direction "pp", it can be concluded that the impact toughness (of the sample) $K_a > K_b > K_{pp}$. However, impact toughness K_{cast} is even and higher than

K_{pp}, but lower than K_b (Fig. **7**).

Change of fracture energy thus shows that in a very narrow range of temperatures, a change in behaviour of the alloy may be caused, that is, the transition from the plastic to the brittle state. Casting crack and damage to the entire machine is the consequence of this. It means, therefore, that the casting is subjected to dynamic loads and must have a transition temperature lower than the temperature at which the machine works.

2. THE EFFECT OF THE CHEMICAL COMPOSITION ON THE IMPACT TOUGHNESS

Increasing the carbon concentration causes an increase in strength, but a reduction in plasticity. In carbon steels, an increase in carbon concentration also causes a shift of the brittleness threshold to higher temperatures, with simultaneous reduction of the impact toughness values.

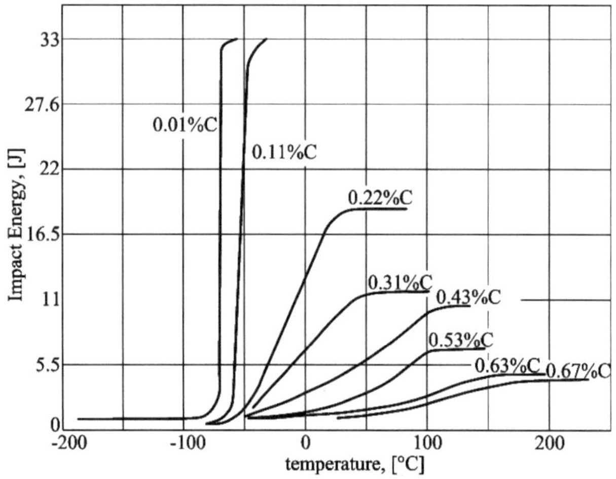

Fig. (8). Dependence of impact energy and transition temperature on the carbon content in steel [12].

The effect of carbon on the impact toughness is so significant (Fig. **8**) that in the steels intended for operation at low temperatures, its concentration does not exceed 0.20 - 28%. Therefore, the carbon content should be kept as low as possible under service conditions.

According to [12], the increase in the carbon content by 0.01% shifts the brittleness threshold to temperatures higher by 1 - 2°C. Therefore, in steels with good strength (0.3-0.4% C), other elements should be introduced to maintain good impact toughness (above 35-40 J/cm^2). The choice of the chemical composition is based on the studies [8], which said that in steel containing 0.30% C + 1.0% Mn and 0.3% Si, the transition temperature is increased by: copper (over 0.8%), chromium (over 0.4%), silicon (over 0.4%), phosphorus (in the whole concentration range) and vanadium (up to 0.15%). However, manganese, nickel and aluminium shift the brittleness threshold to lower temperatures (Fig. **9**).

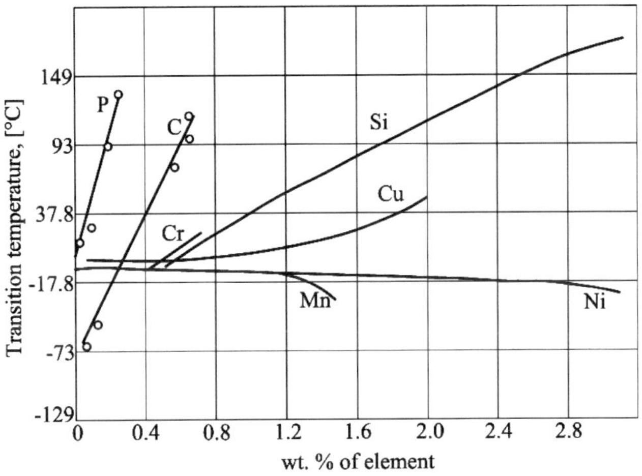

Fig. (9). The influence of alloying elements on the transition temperature [8].

The effect of **manganese** is best reflected in Fig. (**10**). The increase in manganese content by 0.1% decreases the transition temperature by 2-5 K [12]. The effect of manganese is reflected by fragmentation of carbides, reduction in the amount of cementite and increase in the solubility of carbon in ferrite. The beneficial effect of manganese is higher, the smaller the amount of nitrogen is in steel. The effect of the influence of manganese as a single element is often replaced with cumulative effect of carbon and manganese. In these cases, the ratio of Mn/C should be considered. In low carbon steels (up to 0.2% C), the ratio Mn/C is closely associated with a tendency of the steel to brittleness (Fig. **11**) [16]. Low values of Mn/C lead to an increased impact toughness and shifting the transition temperature to higher temperatures.

This effect is more visible for normalized rather than tempered steel. According to the data from [16], the growth in the ratio Mn/C also increases the certainty of obtaining the ductile fracture (characterized by 90% share of ductile fracture). Data mentioned here are valid only for steel with comparable strength, because at a high strength, the effect of Mn/C is variable.

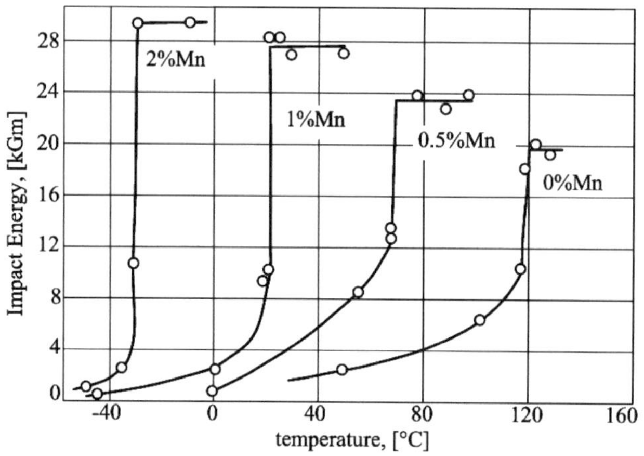

Fig. (10). Change the transition temperature of steel due to additions of manganese [8].

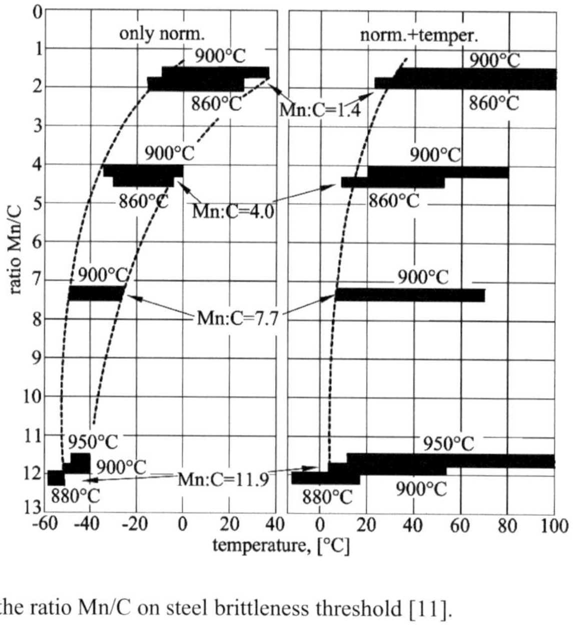

Fig. (11). The effect of the ratio Mn/C on steel brittleness threshold [11].

Nickel significantly reduces the tendency of steel to brittleness. The influence of nickel is particularly visible in its concentration range of from 2 to 5% (Fig. **12**). Higher concentrations of nickel do not bring such a positive effect.

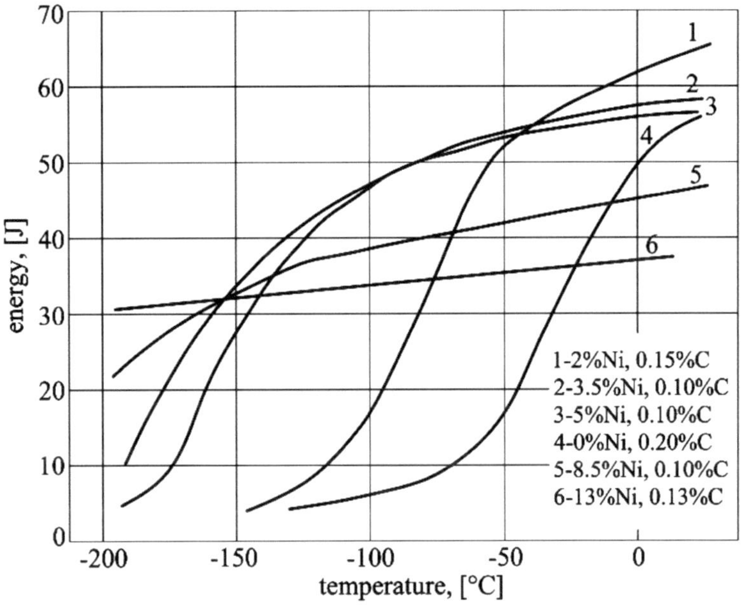

Fig. (12). The influence of nickel on the steel impact toughness at low temperatures [16, 17].

Aluminum reduces the brittleness of steel (with its content up to 0.04% Al). It is associated with fragmentation of grains and better degree of steel deoxidation. However, it is necessary to remember that a too high concentration of aluminium leads to formation of aluminium nitrides AlN, which cause brittle fracture (Fig. **13**). It has been frequently found in practice that at aluminium contents higher than 0.04÷0.045%, there is an increase in the transition temperature [5]. Aluminium is also an element modifying heavily the type of inclusions in cast steel, since it causes a change of the second type of inclusions into the third type. In this way, the cast steel plasticity is improved [18].

The importance of the impact of the final aluminium content is indicated in Fig. (**13**). Exceeding 0.05% of Al is not only associated with low impact toughness at room temperature, but also with low impact toughness at sub-zero temperatures. Obtaining stable values of aluminium in castings is associated with the proper control of oxygen and nitrogen contents in steel.

Only correct performance of deoxidation and refining process of steel during its smelting allows to obtain constant concentrations of aluminium, *e.g*. at 0.025 ÷ 0.035%. The use of oxygen sensors to control oxygen activity in liquid steel facilitates the task.

Chrome, dissolved in pure iron, reduces the transition temperature. But at its content of up to 5%, each percent of chromium increases T_k by 3 K for each percent.

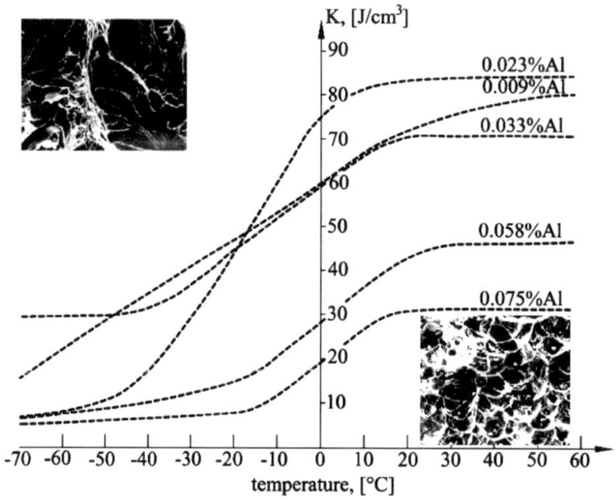

Fig. (13). Impact toughness of carbon cast steel (0.3%C) after deoxidation by Al [19].

In steels containing carbon, silicon, manganese and other elements, the effect of chromium is more complex. In Cr-Ni-Mo-V steel containing 1% Ni and 3% Cr the temperature T_k decreases with an increasing chromium concentration [6]. With the content of 1- 2% Ni, steel containing 2% Cr has the lowest transition temperature; at the nickel content up to 3.5%, the lowest T_k is obtained at the content of 1% Cr. Increase of the nickel content in steel containing 1.3% Cr leads to a continuous reduction of the transition temperature [6].

Addition of **titanium** in steel affects the impact toughness by interaction with nitrogen. By creating stable nitrides, titanium clearly raises the temperature T_k [6]. For example, in 0.4%C-Cr-Mn cast steel, the highest impact toughness with 0.07% Ti, and hot tearing resistance - with 0.15% Ti, were obtained [6, 13]. Due to high affinity with carbon and ability to create carbides, titanium causes a decrease in quantity of pearlite, which affects the brittle fracture [6, 5, 18].

Vanadium, similarly to titanium, is an element with a high affinity for oxygen, nitrogen and carbon. For this reason, it is introduced into steel after deep deoxidation and at the lowest possible nitrogen content. Vanadium, titanium, niobium, and also boron and aluminium are the basic micro-alloyed additions in low-alloy steels, which improve the yield stress (in steels and cast steels with YS above 800-850 MPa), weldability, impact toughness at low temperatures and fine ferrite grains. At carbon contents below 0.2% and about 1.5% of manganese, it is recommended that the sum of Ti + V + Nb ranges from 0.1-0.15%. However, contents higher than that results from the effect of strengthening the matrix, deteriorating the plasticity.

Similar results are achieved in cast steel. A skilful introduction of vanadium into 0.15%C-1.4%Mn cast steel allows to shift the transition temperature even up to minus 40°C (Fig. **14a**). At the optimum deoxidation of this cast steel with aluminium (Al final concentrations in the range of 0.02-0.04%), the lowest impact toughness at minus 40°C equals over 40 J/cm^2 (Fig. **14b**). The combined effects of vanadium and niobium are more efficient than the effect of one element. The data presented in Fig. (**14**) refers to wall thickness of 45 mm, but also in wall thickness of 120 mm, secure exceed impact toughness at temperatures of minus 40 and minus 60°C was obtained.

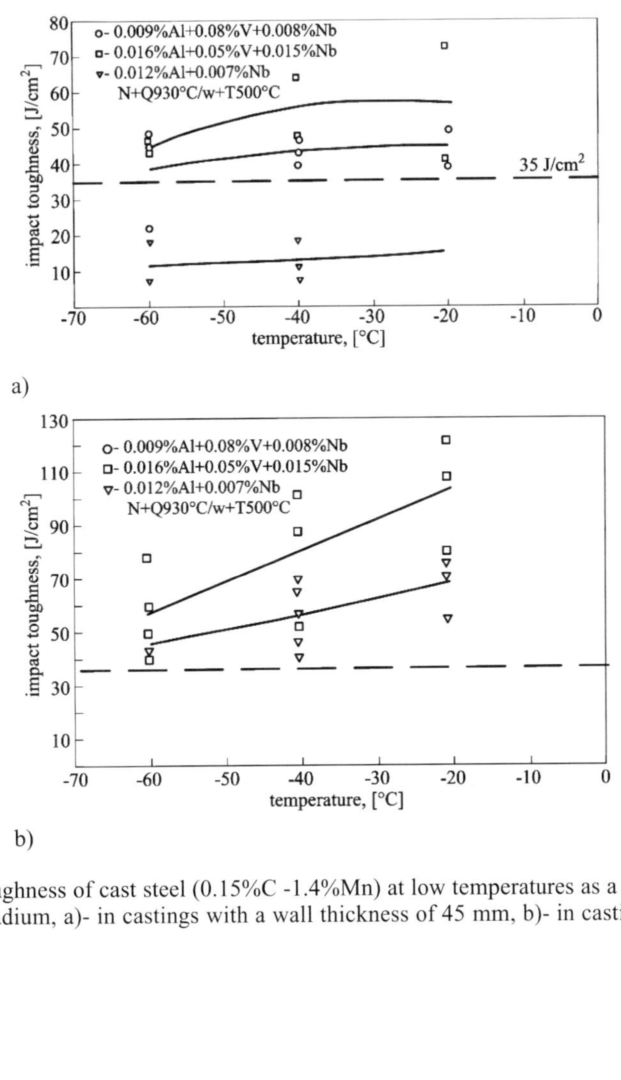

Fig. (14). Impact toughness of cast steel (0.15%C -1.4%Mn) at low temperatures as a result of modification of niobium and vanadium, a)- in castings with a wall thickness of 45 mm, b)- in castings with thickness of 120 mm [20].

The change in the share of ductile fracture surface in relation to the brittle fracture is the explanation of such significant differences in the impact toughness at temperatures ranging from 0° to minus 60°C. The share of brittle fracture with characteristic cleavage planes (Fig. **2**) increases with the transition temperature from 0°C to minus 60°C. Low impact toughness values under these conditions are caused not only by a small share of the ductile fracture, but also by the occurrence of micro-cracks, especially in the presence of non-metallic inclusions. Changing their form of occurrence (polygonal inclusions of the 3rd type), and leading to obtain a spherical form, significantly improves plasticity of cast steel at low temperatures.

Molybdenum has a beneficial effect on the impact toughness of cast steel at low temperatures. Its operation is second only to the effect of nickel, and is associated with the elimination of temper brittleness. However, due to the formation of carbides with increasing carbon content, the beneficial effect of molybdenum on the temperature T_k visibly decreases [6].

Boron is used in cast steel as a modifier in less than thousandths of a percent. Its presence in the matrix of low- and medium-carbon cast steel also increases the impact toughness. The maximum impact toughness is obtained with the addition of 0.001-0.005% of this element.

Another parameter significantly affecting the impact toughness, especially at low temperatures, is **cleanliness** of steel. This term is meant not only by the amount of non-metallic inclusions, but also by the presence of elements such as phosphorus, sulphur, tin, zinc, antimony, arsenic, bismuth and lead. The effect of phosphorus is associated with hardening of ferrite and its tendency to segregate at the grain boundaries. The effect of phosphorus at the temperature T_k is stronger when the steel contains more carbon and alloying elements. Steels containing 0.02% P may have a mixed fracture (ductile-brittle) at room temperature. Reduction of phosphorus content to 0.01% balances the effect of 1% nickel [6].

Other low-melting elements (Sn, Zn, Sb, As, Bi and Pb) are introduced into steel and cast steel with commercial scrap. Pressed scrap of galvanized sheet metal, not sorted machine scrap containing certain quantities of brass or bronze, causes rapid reduction in the plasticity of cast steel. The detrimental effect of these elements is due to their segregation at the grain boundaries, particularly in the presence of other elements, such as phosphorus. It is characteristic that already in the presence of 0.002% of each of these elements, there is a significant reduction in the impact toughness at low temperatures [5, 13].

Sulphur reduces the plasticity of cast steel particularly strongly. This is visible when its content exceeds 0.02%. In connection with the existing effect of oxygen on the final concentration of sulphur, the process of deoxidation of steel definitely affects the quality of the casting. At high contents of sulphur and oxygen, formation of the 2nd or the 2nd and 3rd type of oxy-sulphides can occur, which lead to reduction of El, RA and K_{Ic}. Thus, steel is not only where impact toughness depends on inclusion morphology dependent on the plastic work direction [5], but also in cast steel, there are such compound deoxidisers used, which lead to the formation of spherical inclusions (Fe-Ca-Si, mischmetal, Al-Mn-Ca, rare earths alloys, wired powders of Fe-Ca-Si, secondary metallurgy processes). A necessity to introduce a two-step deoxidation is visible in cast steel containing less than 0.15% C. In these cast steel grades, deoxidation with only aluminium is insufficient and must be supplemented with Fe-Ca-Si or alloy with the cerium [21, 22].

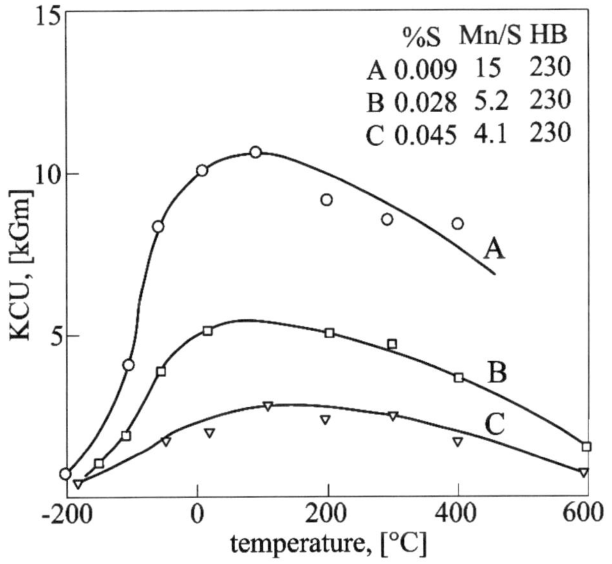

Fig. (15). The effect of sulphur on the impact toughness of cast steel containing 0.15% C – 5% Ni – 1% Cr [22].

There are a lot of examples indicating the adverse effect of sulphur. For example, G. Goutenoire [22] showed that in quenched and tempered cast steel, impact toughness at low temperatures is strongly dependent on the concentration of sulphur (Fig. **15**). When reducing the concentration of sulphur to about 0.010%, the transition temperature has been moved to minus 100°C. Similar examples are provided by D. Wittekopf [21]: an increase in impact energy by 20-60 J for any reduction in sulphur by 0.010% can be obtained in the carbon and low-alloy cast steel Cr-Mo. In this paper, it is stated that in carbon cast steel containing 0.20% C, impact energy can be given by the following formula:

$$KU3\ [J] = 157.2 - 280.72 \times \%C - 16.80 \times \%Si + 25.11 \times \%\ Mn + \\ - 732.9 \times \%P - 1488 \times \%S + 11.06 \times \%Cr - 9.97 \times \%Mo - 9.30 \times \\ \times \%Ni + 7.31 \times \%V + 22.42 \times \%Cu + 218.6 \times \%Al$$

These types of effects are associated not only with a reduction in the sulphur content, but also a change in the morphology of non-metallic inclusions.

3. THE EFFECT OF DEOXIDATION OF STEEL

Oxygen and nitrogen increase the tendency of steel to brittleness. When oxygen is below 0.003% in steel, there are no changes in the shifting of the transition temperature observed. If the oxygen content in liquid steel is lower than 0.003%, the impact toughness can even reach a value of 130 J/cm^2. However, the increase in the oxygen concentration to a value of 0.057% shifts the T_k from 258 to 618 K [10]. The oxygen concentrations in steel higher than 0.01% refer to rimmed or semi-killed steels. In steels intended for castings, the acceptable oxygen concentration should not exceed 0.007%.

The influence of oxygen on the impact toughness of steel is generally known [5, 23]. Non-oxidized steels practically reveal the transition temperature already at 20°C. Deoxidation with silicon shifts the T_k by 20 K and double deoxidation with silicon and aluminium, by further 20-30 K.

The effect of aluminium is directly related to the activity of oxygen in the metal bath before deoxidation. Additions of aluminium as a deoxidizer (in an amount of 1 kg/t) result in its diverse final content. A high concentration of aluminium is obtained after the introduction of a usual amount of deoxidizer 1kg/t, but at low oxygen activities in liquid steel. Such conditions exist at the initial deoxidation of steel Fe-Mn or Fe-Si. Therefore, lack of usual steel smelting technology may lead to failure to keep the required impact toughness values. Only using oxygen sensors during refining period, and varying amounts of deoxidizer, dependent on the oxygen activity in the bath, give beneficial impact toughness values, and also bring savings in the quantities of consumed deoxidizer.

This effect of reduction in the plasticity due to unstable aluminium content after deoxidation of steel is presented in Fig. (**13**). At ambient temperatures, the impact toughness of carbon cast steel containing about 0.06% of Al does not reach the required value of 35J/cm^2. Smelting should therefore be performed in a manner so that the final analysis revealed the optimum aluminium content, ranging from 0.015-0.040%. At these concentrations of aluminium, the transition temperature was shifted from the temperature of 273 K to 253 K [23]. To ensure such optimum concentrations of aluminium, oxygen sensors should be used to check the oxygen activity in the liquid steel, and not only the five basic elements (C, Mn, Si, P, and S). The proper heat control should contain also the contents of aluminium, gases (O_2, N_2, H) and some alloying elements: Cr, Ni, Ti, Mo. This is due to the fact that aluminium, titanium and vanadium are not only good deoxidizers, but also are easy to combine with nitrogen, creating brittle fracture owing to the formation of nitrides (Fig. **14**).

The importance of deoxidation of steel for castings is presented in Fig. (**16**). Depending on the deoxidizer applied, in low carbon Mn-Ni cast steel, the impact toughness can be obtained above 70 J/cm^2 at double deoxidation or only 40 J/cm^2 - at deoxidation only with aluminium. It is also important that the transition temperature is shifted from the temperature of +10°C to minus 40°C (Fig. **16a**). For economic reasons, it is better to conduct double deoxidation and normalization rather than deoxidation with aluminium and quenching and tempering.

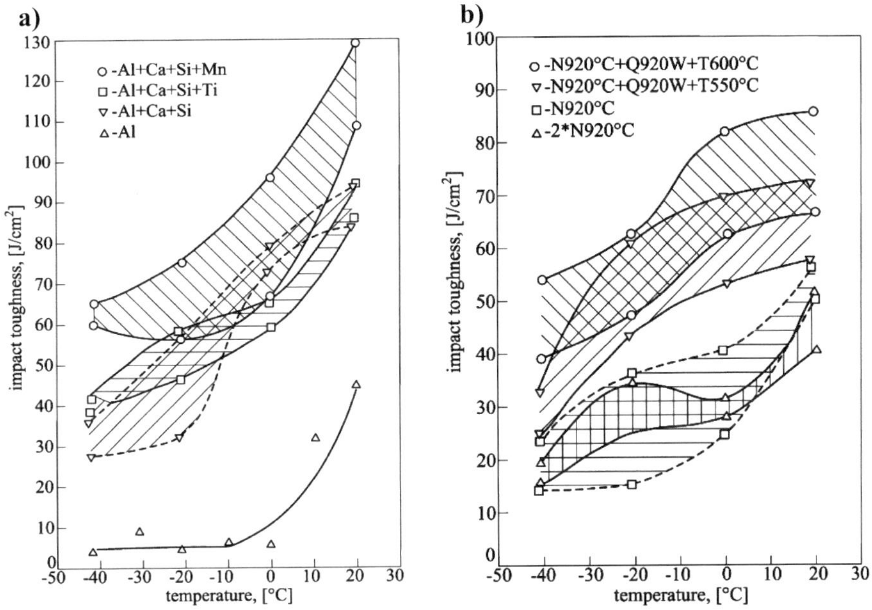

Fig. (16). Change of the impact toughness of Mn-Ni cast steel as a result of:
a) various method of deoxidation of 0.25%C-1.4%Mn-0.7%Ni and
b) heat treatment of 0.35%C-1.4%Mn-).7%Ni cast steel, [24].

The changes in the impact toughness of Mn-Ni cast steel with higher carbon content are presented in Fig. (**16b**). After deoxidation with aluminium (1 kg/t) and normalization, the impact toughness at ambient temperatures exceeds a value of 40 J/cm^2. An introduction of double normalization does not bring significant benefits. Only an application of quenching and tempering increases the impact toughness of cast steel twice. These data also indicate that an improvement in impact toughness can be obtained by changing normalization into quenching and tempering, or in addition - by raising the tempering temperature. Thus, after tempering at 550 and 600°C, over the whole temperature range of testing (up to minus 40°C), the impact toughness exceeds 40 J/cm^2.

4. CHEMICAL COMPOSITION OF CAST STEELS FOR OPERATION AT LOW TEMPERATURES

There are three groups of cast steel used in these conditions [4]:

– carbon cast steel (below 0.25% C) with ferritic-pearlitic structure and strength UTS = 380-600 MPa. Depending on the heat treatment, castings made of this cast steel can be used for operation at temperatures of minus 40°C (after their normalization) or up to minus 50°C (after their quenching and tempering).

– quenched and tempered cast steel with the strength of 650-750 MPa. It is used for operation at temperatures below minus 120°C.

– austenitic cast steel. After solution annealing, it is applied on castings operating at temperatures up to minus 196°C.

Carbon cast steel has impact toughness higher than 35 J/cm^2 both after quenching and tempering, and after normalization. Shifting the T_k to minus 50°C after conducting heat treatment (Fig. **16**) is caused by fragmentation and homogenisation of the structure. However, this effect is only visible in thin-walled castings because thick-walled castings have a heterogeneous structure (except for tempered martensite, where bainite may occur partially). Hence, low-alloy cast steels, especially the chromium-molybdenum or chromium-nickel-molybdenum grades, are more frequently used. In addition to increasing hardenability by the combined effect of Cr + Mo, molybdenum removes the effect of temper brittleness. Therefore, Cr-Mo cast steel is a grade competitive to Cr + Ni + Mo cast steel, especially after taking into account the following: material costs, tendency of nickel steel to increase hydrogen content and more complex heat treatment.

There is also an important condition limiting the application scope of Cr-Mo cast steel. It is the lowest operational temperature, within the temperature between minus 80° to minus 100°C. Castings operating at lower temperatures require the addition of nickel (3 to 5% Ni, Fig. **17**). With additions 2-2.5% Ni, the transition temperature is at a temperature of about minus 70°C, and with additions 3-3.5% Ni - at minus 100°C. The growth in the nickel content admittedly increases the amount of the retained austenite, but the introduction of quenching and tempering with hardening in water allows to obtain a homogeneous structure in castings with wall thickness of up to 80 mm.

Under conditions, which require good impact toughness at temperatures below minus 120°C), cast steel 13% Cr + 4% Ni or austenitic cast steel 18% Cr + 9% Ni is applied. The impact toughness of the former depends on the content of nickel, Fig. (**17**). Due to high chromium content and low carbon concentration, it is advisable to use higher nickel content (over 3 - 4%). Impact toughness higher than 35 J/cm^2 can be obtained in thick-walled castings made of this cast steel (above 150 mm), even after hardening in air. The other, not a less important advantage of cast steel, is its good weldability. Therefore, it is also used for casings and bodies of turbo-compressors operating at temperatures below 100°C.

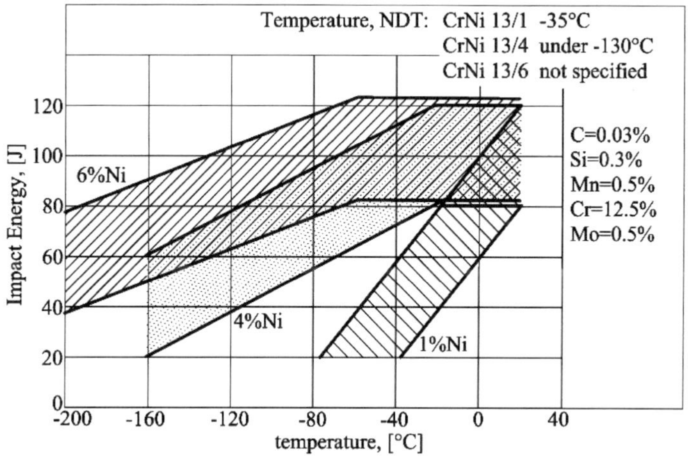

Fig. (17). The influence of nickel on the impact toughness of cast steel containing 13% chromium [12].

At temperatures lower than minus 120°C, only austenitic cast steel (cryogenics, chemical industry, astronautics, *etc.*) can be used. This is mainly cast steel grade containing 18% Cr and 9 or 11% Ni. At similar concentrations of chromium and nickel, the reduction of the carbon concentration is the factor deciding about an improvement in the impact toughness at low temperatures, (Fig. **18**). Reduction of the carbon content to 0.06% ensures obtaining impact toughness at minus 200°C above 90 J/cm^2.

Depending on the carbon content and wall thickness in the casting, δ-ferrite and a certain amount of carbides can occur in the structure. In order to ensure good impact toughness, both these factors should be eliminated. Delta ferrite is removed by increasing the concentration of nickel and decreasing the concentration of carbon (advisability to introduce AOD, VOD, VAD processes or processes related to smelting of the steel for castings). Due to the diverse wall thickness in castings, austenite stabilization is performed by introducing elements with higher affinity to carbon than chromium. These include additives of niobium (in the amount % Nb = 8 × %C) or titanium (in the amount %Ti = 5× %C). Introduction of niobium or titanium into steel eliminates the formation of chromium carbides at grain boundaries.

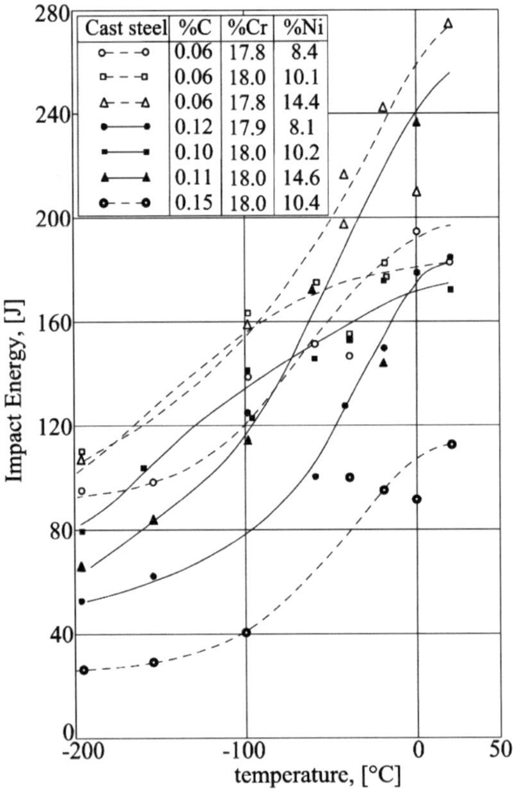

Fig. (18). The impact toughness of cast steel 18% Cr-9% Ni after solution annealing [25].

To eliminate delta ferrite and increase in impact toughness, a reduction of carbon content is more effective than increase in the concentration of nickel. As an example, here are studies of G. Mayer and K. Balayiva [25]. At the temperature of minus 100°C and comparable nickel content (10.1-10.4%), a reduction of carbon concentration from 0.15 to 0.06% increases the impact toughness from 40 to 120 J. Such an effect is a significant cause of rapid development of the converter processes of smelting steel for high alloyed castings.

5. THE INFLUENCE OF STRUCTURE AND HEAT TREATMENT

Resistance of steel with ferritic - pearlitic structure to brittle fracture is dependent on the grain size of ferrite, ferrite distribution in the matrix, the quantity of pearlite, and carbon concentration in the matrix [6] (Fig. **19**). Fragmentation of grains lowers the transition brittleness temperature, which is very significant in these types of steels.

With an increase in the content of pearlite, elongation and impact toughness decreases and T_k temperature increases. At high contents of pearlite, its morphology has a decisive influence on the impact toughness and temperature T_k. Reducing the distance between the plates of cementite lowers impact toughness of steel, while reducing their thickness increases it significantly. This is due to the fact that generation of cracks in pearlite is the effect of simultaneous dislocation and formation of micro-cracks at the "ferrite - cementite" boundary. Increasing the volume fraction of pearlite, assuming a constant carbon content, promotes raising the transition temperature. The highest brittle temperature is in steel, in which cementite is in the form of thick plates, and the lowest transition temperature occurs when carbides are spheroidal and are evenly distributed in the matrix. The fine pearlite is dependent on the applied overheating temperature above Ac_1 and the cooling rate of the element from these temperatures.

Normalizing is the most common heat treatment of low carbon steels, intended for operation at low temperatures. It causes a significant reduction in the transition temperature through the fragmentation of the structure. Resistance to brittle cracking of bainite structures also depends on the concentration of carbon (Fig. **19**). Additional parameters affecting the impact toughness are also here: the amount and type of carbides and the size of original austenite grains [6]. There is data [5, 6] indicating that the structures with lower bainite lead to the reduction of the transition temperature as compared to the structures with the upper bainite. This is due to a more favourable distribution of carbides, their smaller size and thus smaller tendency to crack formation.

Resistance to cracking of tempered martensite depends largely on the austenite grain size. Transition temperature of structures comprising of fine-needle shaped martensite is approximately 40°C lower than that of coarse-needle shaped martensite.

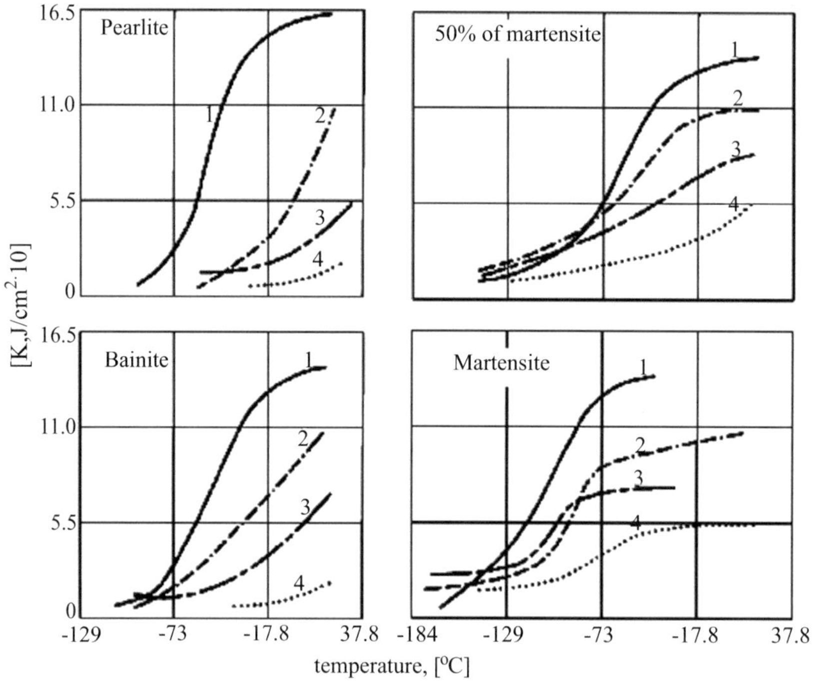

Fig. (19). The effect of steel microstructure on its impact toughness [3]; 1 – 0.17% C; 2 – 0.278% C; 3 – 0.40% C; 4 – 0.54% C.

One method of reducing the brittle fracture of low and medium-carbon steels is the introduction of several types of heat treatment. For example, the introduction of quenching is related to the dissolution of harmful admixtures in austenite, where their solubility is higher (this concerns, among others, P, Si, Al). However, a large role is played by the tempering temperature: in low-alloy steels containing 0.3÷0.4% C, increased plasticity is achieved only when the temperature Ac_1 exceeds 30÷50°C [1, 2, 6]. Heat treatment also brings a significant improvement in the impact toughness in cast steel. Change of normalizing into quenching and tempering of carbon cast steel shifts the T_k temperature to lower values and increases the impact toughness in the whole temperature range. This is a result of homogenization and fragmentation of the structure. This is clearly illustrated by the studies of H. Zeuner and K.-E. Höhner (Fig. **20**), which show that at sub-zero temperatures, the impact toughness increases several times at quenching and tempering. An additional element here is the possibility of improving the impact toughness by using higher tempering temperatures.

In some kinds of cast steel, an increase in brittleness during tempering can be observed. This effect is visible in low-alloy cast steel Cr-Ni and other grades of construction cast steel. This is about the reversible temper brittleness of steel and cast steel hardened and slow cooling, in the temperature range of 350÷550°C. The characteristic features of reversible temper embrittlement are: rapid decrease in the impact toughness and formation of a brittle intergranular fracture [6]. The reason for this phenomenon is the segregation of the surface active elements of steel, such as Sn, Sb, Pb, As, P and Bi at boundaries of primary austenite grains. It was found that, among elements dissolved in solid solution during austenitization, chromium, nickel, manganese and silicon increase the tendency of steel and cast steel to the reversible temper embrittlement. However, reducing the tendency is possible after the introduction of additions of molybdenum (0.2÷0.4%), tungsten (up to 1.5%) into cast steel or the use of rapid cooling of castings during their cooling in the temperature range of about 450°C. Under these conditions, it is recommended to cool the castings in the air, or even introduce cooling with a water mist. Cooling rate control brings significant shifting of the transition temperature to lower temperatures (Fig. **21**). The use of cooling with air instead of cooling castings along with the furnace shifts the T_k temperature up even to 80°C (Fig. **21**).

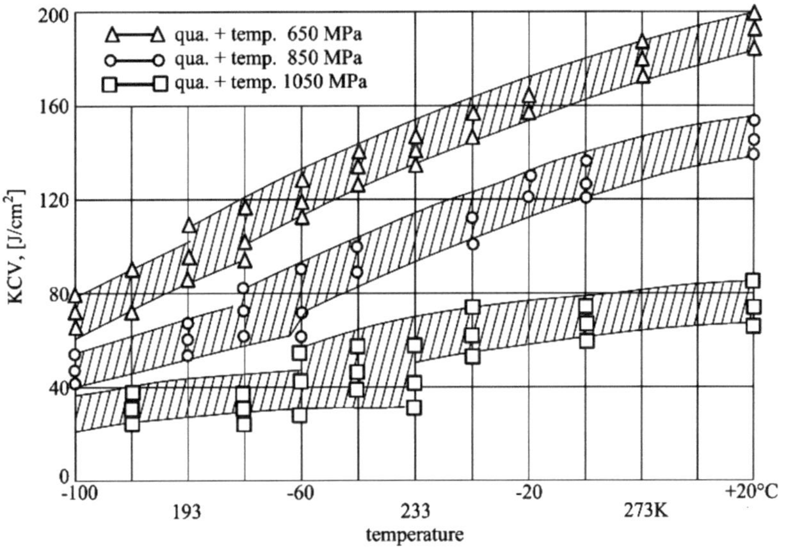

Fig. (20). Shifting the transition temperature of carbon cast steels after quenching and tempering [4].

6. THE INFLUENCE OF WALL THICKNESS OF THE CASTING

Changes in the chemical composition of the alloy, its microstructure and smelting process, have the same effect on the impact toughness of steel and steel castings. In relation to the anisotropic properties of steel, castings are characterized by even properties of the casting wall thickness, but they may have other disadvantages.

Due to the presence of micro-shrinkage and micro-porosity, cast steel has a lower impact toughness than steel. Most often, a wider range of the impact toughness at ambient temperature, and a rapid reduction in impact toughness at sub-zero temperatures are demonstrated. These effects are due to a greater number of inclusions of a larger size, higher concentration of gases and segregation of elements during solidification. Each of these parameters reduces the plasticity of the material and their combined activity is even more dangerous. This is visible and known in steels and particularly in cast steels. In low manganese cast steel (0.2%C-1.4%Mn), the impact toughness in the middle of the wall with a thickness of 75 mm is 30% lower than in the middle of the wall with a thickness of 45 mm, (Fig. **22**). In castings with a wall thickness of 120 mm, these trends are even stronger. The importance of the impact of the final aluminium content is indicated in Fig. (**13**). Exceeding 0.05% of Al is not only associated with low impact toughness at room temperature, but also with low impact toughness at sub-zero temperatures. Obtaining stable values of aluminium in castings is associated with the proper control of oxygen and nitrogen contents in steel.

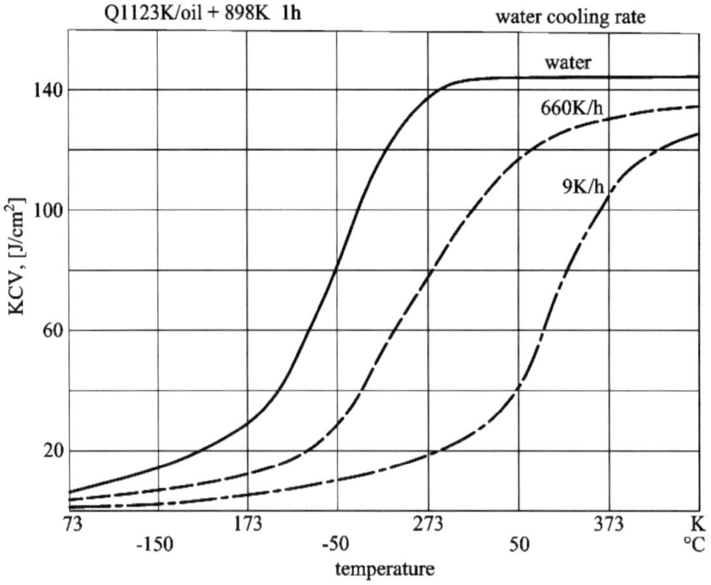

Fig. (21). The effect of increasing the cooling rate of castings after annealing on the impact toughness of cast steel [4].

Already at 20°C, the impact toughness value is determined by the manner of cutting-out the samples. At the edge of the wall, the impact toughness is higher than the value required (35 J/cm^2), and in the middle of all the walls (45-120 mm), it is below 20 J/cm^2. The negative effect of wall thickness can be reduced by modification of steel in the ladle. This has been shown in Fig. (**14**), indicating the beneficial effect of vanadium and titanium on the impact toughness of low alloyed cast steel at temperatures up to minus 60°C. In all wall thicknesses (from 25 to 100 mm), not only the effect of vanadium and titanium is visible, but also the final content of aluminium after deoxidation. At over 0.04% Al already, at minus 20°C, the reduction of the impact toughness is very significant. It should be emphasized that at concentrations of sulphur and phosphorus below 0.01% each, even aluminium content exceeding 0.04% is less harmful than at the optimum aluminium content (0.036%), and higher contents of sulphur and phosphorus (over 0.015%).

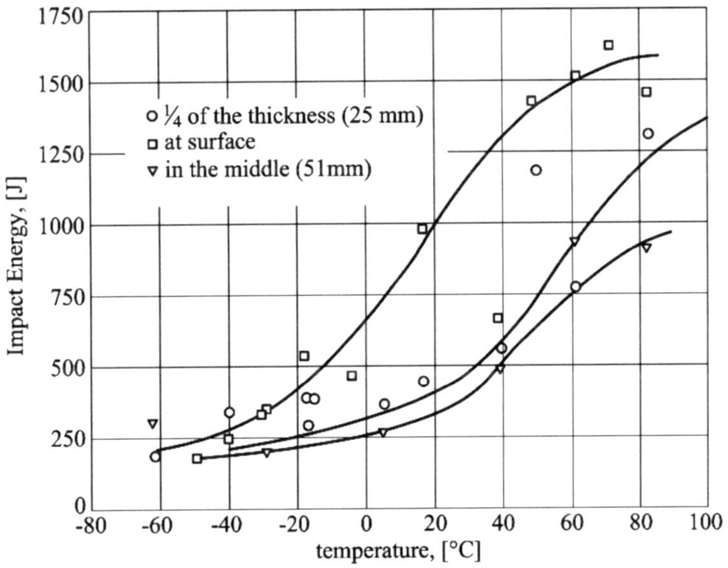

Fig. (22). Differences in impact toughness in the steel plates containing 2.25% Cr – 1% Mo [5].

Only the combined effect of niobium and vanadium provides the required value of impact toughness. A higher concentration of vanadium (above 0.05%) is more effective because it does not lead to a reduction in the impact toughness with an increase in the wall thickness of the casting (Fig. **14**). This effect is connected with easier nucleation of carbides and carbo-nitrides of vanadium and niobium in areas with a lower rate of coagulation.

The total effect of all the factors on the changes of the impact toughness in the temperature range is presented in Fig. (**23**). Its knowledge and compliance during smelting of the steel grade, making castings of it and subsequent heat treatment, provides the best basis to achieve good technical and economic results.

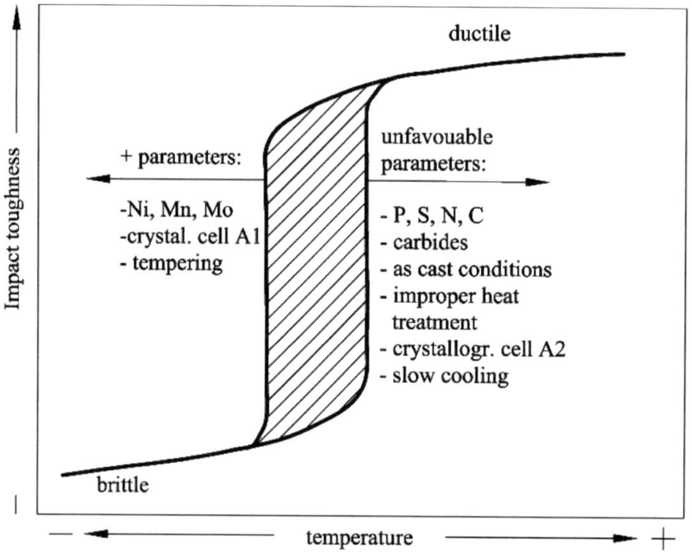

Fig. (23). Summary statement of all parameters affecting the impact toughness of cast steel [4].

CONFLICT OF INTEREST

The author (editor) declares no conflict of interest, financial or otherwise.

REFERENCES

[1] M.M. Salama, "Ambient and Low-Temperature Properties of Structural Steels", *Steel Casting Handbook,* chapter 20. 5th edSteel Founders Society of America, pp. 662-669.

[2] R.E. Smallman, *Physical Metallurgy and Advanced Materials Engineering.* 7th ed. Butterworth-Heinemann, 2007.

[3] *Fracture: An Advanced Treatise, Technology and Engineering.* vol. 5. Academic Press: N.Y., 1968.

[4] K. Roesch, H. Zeuner, and K. Zimmermann, *Stahlguss.* Verlag Stahleisen: Düsseldorf, 1982.

[5] G. Roe, and B.L. Bramfitt, "Notch Toughness of Steel, Metals Handbook", *ASM Intern.,* vol. 1, p. 737, 1990.

[6] A. Maciejny, *Kruchość stali,* Katowice Śląsk, Ed., , 1977. (in Polish).

[7] Standard Specification for Steel Castings, Ferritic and Martensitic for Pressure-Containing Parts, Suitable for Low-Temperature Service. Standard of ASM A352 M-06

[8] J.A. Rinebolt, and W.J. Harris Jr, "Effect of alloying elements on notch toughness of pearlitic steels", *Trans. ASM,* vol. 43, p. 1175, 1951.

[9] Odlewy staliwne do pracy pod ciśnieniem, EN 102 13:2007/AC,2008.

[10] W.P. Rees, and B.E. Hopkins, "Intergranular Brittleness in Iron-Oxygen Alloys", *JISI,* vol. 172, p. 403, 1953.

[11] J.H. Gross, "Comparison of Charpy V-Notch and Drop-Weight Tests for structural Steels", *Welding Research Journal,* 1956.

[12] W. Szczepanski, *The Brittlness of Steel.* J.Wiley and Sons: N.Y., 1963.

[13] N.S. Stoloff, Effects of Alloying on Fracture Characteristic. *Fracture of Metals.* vol. 6. Academic Press: London, 1969. [http://dx.doi.org/10.1016/B978-0-12-449706-1.50006-8]

[14] Fractography and Atlas of Fractographs, *Metals Handbook.* vol. 9. ASM Metals Park: Ohio, 1974.

[15] In [3] vol. 6

[16] W. Barr, and A. Honeyman, "Effect of the Manganese/Carbon Ratio on the Brittle Fracture of Mild Steel", *JISI,* vol. 157, pp. 239-242, 1947.

[17] H. Zeuner, and K.E. Höhner, "Die Kaltzähigkeit von niedriglegierten Cr-Mo-Ni und Cr-Ni-Mo Stahlguss nach Unterschiedlicher Wärmebehandlung", *Giesserei,* vol. 47, p. 623, 1960.

[18] C. Sims, and C. Briggs, "The Effect of Various Deoxidizers on Cast Steel", *25th Inter. Foundry Congress,* 1958, Brussels.

[19] M. Kowalówka, *"Effect of deoxydation of carbon cast steels by Al-addition on their mechanical properties"*, PhD.Thesis, Univ.of Sc.a. Technology, Cracow, Poland, 1984. (in Polish)

[20] J. Glownia, "Technology of cast steels for minus temperatures", *Project of Univ.of Sc.a.Technology,* 2000. no. 5.5.170.115

[21] D. Wittekopf, "Der Einfluss des Schwefelgehalts auf die Zähigkeitseingeschaften von unlegiertem und niedriglegiertem Stahlguss", In: *Giessereitechnik,* 1988, no. 9, pp. 267-270.

[22] G. Goutenoire, "Facteurs d'influence metallurgique sur les resultants de l'essai de resilience des aciers moules an carbone on faiblement allies", *Fonderie,* vol. 29, no. 336, pp. 285-297, 1974.

[23] K. Mamro, *Deoxidation of Steel,* Katowice Śląsk, Ed., Poland, 1976. (in Polish)

[24] J. Głownia, R. Bujas, and L. Młynarczyk, "Effect of heat treatment and deoxidation on the impact toughness of Mn-Ni cast steel", In: *Przegl. Odl.* vol. 1. , 1995, pp. 1-6. (in Polish)

[25] G. Mayer, and K. Balajiva, "The Low Temperature Mechanical Properties of some Cast Stainless Steels", *Metallurgia,* vol. 59, p. 221, 1959.

CHAPTER 9

Corrosion – Resistant Cast Steels

Abstract: Cast steels are chemically unstable in air-saturated water and other solutions at ambient temperatures. Their commercial applications therefore depend on protective mechanisms: passivation and formation of thin oxide films, prevention of passivity breakdown, cathodic protection, coatings, control of chemical composition and microstructure, along with their service conditions (stress, fatigue, voids and erosion-corrosion). The corrosion of steel castings is usually considered as a problem involving chemical (carbon content, alloying), microstructure (austenite, ferrite-austenite) and quality (non-metallic inclusions, gases, segregation, grain size) parameters. In this chapter, some topics concerning the manufacturing of corrosion-resistant steel castings are considered.

Keywords: Composition, Duplex structure, Passivation, Microstructure, Types of corrosion.

1. THERMODYNAMICS OF AQUEOUS CORROSION

In most working conditions of castings, the metallic state of their surface is not durable in the environment in which they are found. This is accompanied by the conversion of atoms from the metallic state into the ionic state, that is to say, corrosion.

At metal - water solution boundaries, there are two reactions: chemicals from the aquatic environment remove electrons from the metal, while on the other hand, the atoms on the surface of the metal are involved in the supplementation of lost electrons. On the surface of the metal, these two reactions form a current (cell), the potential value of which balances the exchange of electrons of both reactions.

At the appropriate potential value, the metal is dissolved and metal atoms diffuse into the solution (*i.e.* the metal corrodes). If dissolved oxygen is removed from the solution, the potential value is decreased. If the potential increases (by introduction of additives into the solution), solid reaction products form on the metal's surface, which comprise a protective coating.

Jan Głownia

The corrosion rate can be controlled either by controlling oxygen solubility in the solution, or by the introduction of elements that tend to inhibit the corrosion process into the iron. The corrosion process is characterised by a change in the free energy and is described mostly by electrochemical terms, which include oxidation and reduction reactions on the corroding surface.

The thermodynamic characteristics of chemical reactions are illustrated by Pourbaix diagrams (potential - pH diagrams), which give phase stability as a function of electrode potential E, and the pH of the solution, (Fig. **1**).

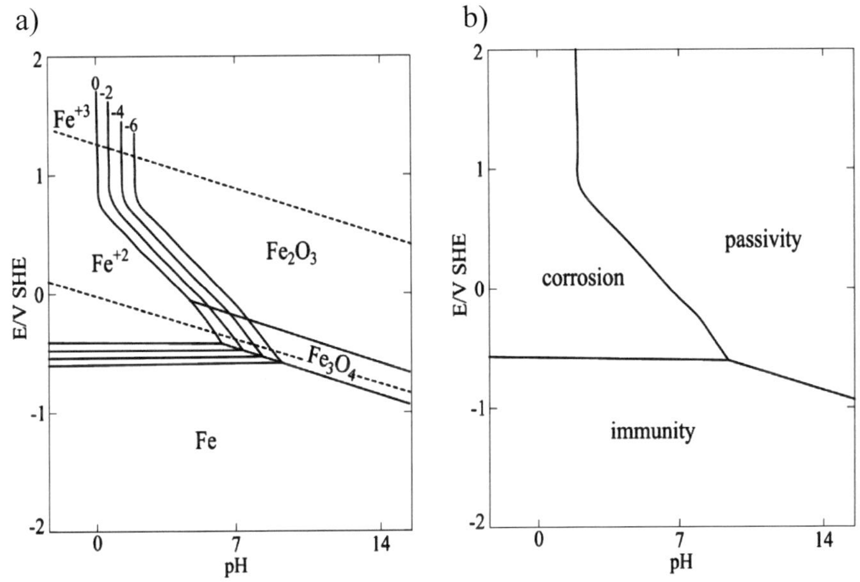

Fig. (1). Pourbaix diagram for the iron-water system at 25°C, a- nominal zones; labels: 0-$a_{Fe2+} = a_{Fe3+} = 1$; 2-$a_{Fe2+} = a_{Fe3+} = 10^{-2}$; 4- $a_{Fe2+} = a_{Fe3+} = 10^{-4}$; 6 – $a_{Fe2+} = a_{Fe3+} = 10^{-6}$ [1].

These three domains characterise metal behaviour under the conditions of:

- immunity - the metal is constant within the values given in the diagram;
- corrosion - metal corrodes in the form of dissolved ions;
- passivity - on the surface, there is an indissoluble, protective layer (adherent layer).

Designation of domains (conditions) of corrosion in Fig. (**1b**) is possible assuming [2, 3] that at a metal solubility lower than 10^{-6} mol/ dm^3, corrosion does not occur. There are then (as shown in Fig. **1**) corrosion resistant, subject to corrosion and passive domains. The scope of potentials E_{corr}-E_{pp} (see Fig. **2**) corresponds to the active dissolution of metal, and E_{pp}-E_p corresponds to the domain in which the protective passive film is formed; above $E_{p,}$ there is a breakdown of the passive film (pitting), and a continuous (catastrophic) corrosion process starts. The value of the potential E_{corr} means that, at this potential, the rate of all anodic reactions equals the rate of all reduction reactions (Fig. **2**).

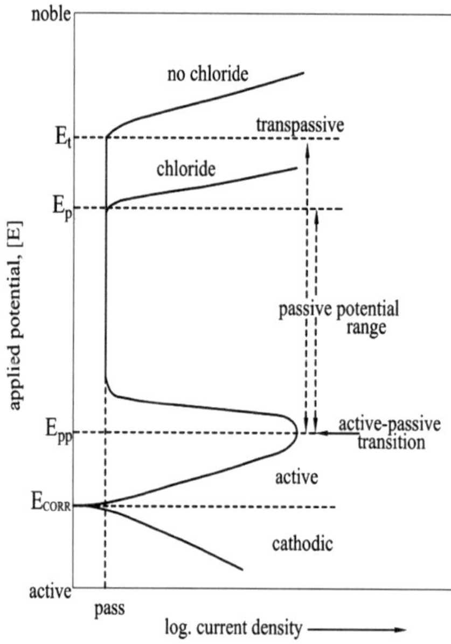

Fig. (2). Schematic polarisation curve for stainless steel in aqueous medium [1 - 3].

The selection of steel appropriate to the conditions of the chosen environment (including castings) consists of the largest possible reduction in the corrosion rate; that is, assuring a passive domain. The characteristics of the passive domain are visualised by making potential - current density diagrams (*i.e.* polarisation curves) for the given chemical composition of steel and the environment in which it is situated, (Fig. **2**).

In the active domain, metal atoms go into solution in the form of hydrated ions (Fe^{2+}, Cr^{3+}), in accordance with the reaction:

$$Fe = Fe^{2+} + 2e^- \quad \textbf{(1)}$$

At the critical potential value E_{pp} (which corresponds to the critical current value i_{crit}), there are thermodynamic conditions for oxide formation, and inhibition of the anodic process of iron dissolution begins. This is the beginning of the creation of a passive film. To create it, not only the correct parameters (pH and potential value) with regard to the Pourbaix diagram are essential, but also the coherence and adhesion of the film (chloride, sulphate or sulphide inclusions may affect it), passing from the alloy matrix to the solution.

Above the potential E_{pp}, the current drops rapidly to the value i_{pass} and remains constant over a wide range of potential value. The stability of the passive range is the basis for the selection of a steel appropriate to the working conditions of individual structural elements. At a potential value E_p, the passive range ends, and a sharp increase in corrosion (transpassive range) starts. The value of the transpassive potential E_t is dependent on many factors. These include:

- type of solution (aqueous, sulphuric acid, brine, *etc.*).

- chemical composition of the forged or cast steel.

- presence of defects in materials (matrix discontinuities, pores, micro-shrinkage, cracks), impurities (carbides, sulphides, non-metallic inclusions at the grain boundaries) and inhibitors (very low concentrations of solutes protect steel surfaces from corrosion).

The presence of chlorides in aqueous solutions (brine) or acids is reported as one of the most important factors, (Figs. **2**, **3**). They cause pitting due to the interruption of the passive film, after exceeding the pitting potential, E_p.

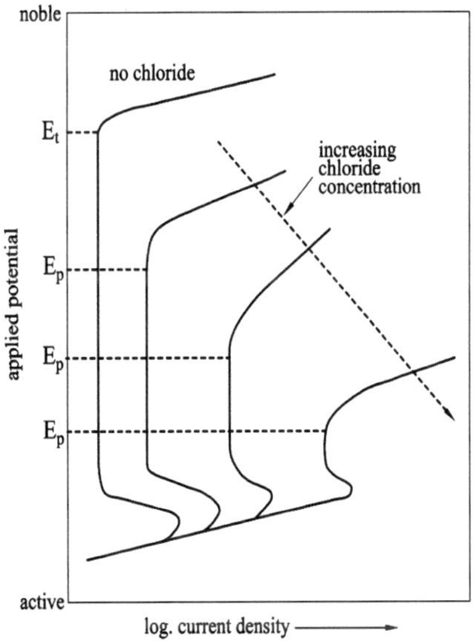

Fig. (3). Anodic polarisation curve of steel at corrosion and passivation conditions [4].

The mechanism of breakdown of durable passive films may be presented as shown in Fig. (**4**). It is based on the impact of chloride ions [4]. The model assumes that the passive film is composed of hydrated oxide, wherein metal ions created under anodic dissolution form ions MOH^+. They react with the surrounding water, releasing hydrogen ions (a). The presence of chloride ions in bridge sites H_2O-M-OH_2 (b) is then seen. The introduction of chloride ions at these bridge sites forms soluble complexes that are removed from the passive film (b`). This is one of the known mechanisms describing the beginning of pitting corrosion.

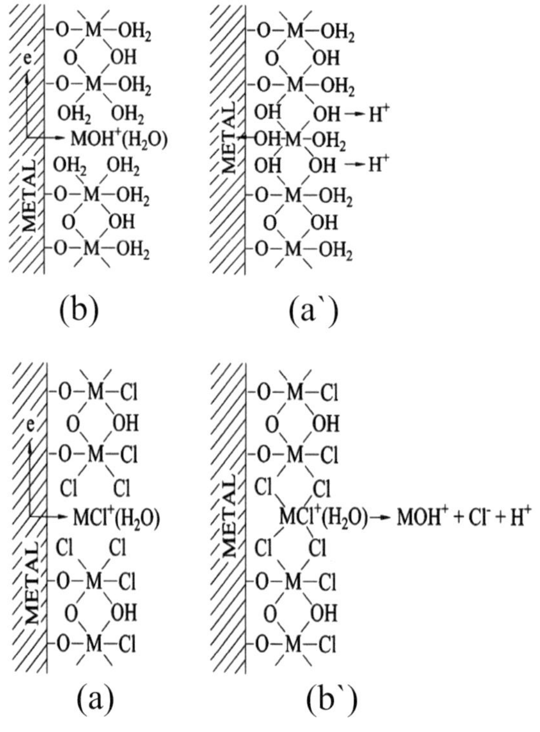

Fig. (4). Mechanism of breakdown of passive film [5].

The chemical composition of the selected steel (including cast steels) is another very important factor. Chromium is the most effective element in corrosion-resistant steels (Fig. **5**). This alloying metal decreases the current density i_{pass} and extends the passive range (E_p - E_{pp}). The effect of alloying elements in steel on the parameters of the passivation curve can be characterised as follows:

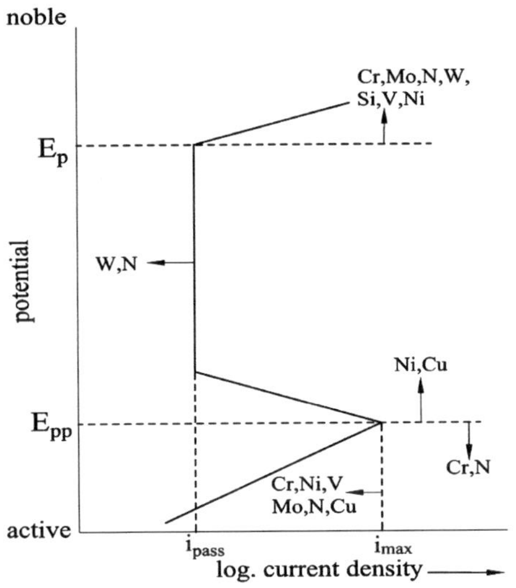

Fig. (5). Schematic summary of the effect of alloying elements on the anodic polarisation curve [3, 6].

Effect of chromium (Fig. 6): among the three elements Fe, Ni and Cr, chromium improves passivation conditions the most:

- reduces current density of passivation.
- increases the transpassive potential.
- reduces the critical anode current and passivation current density

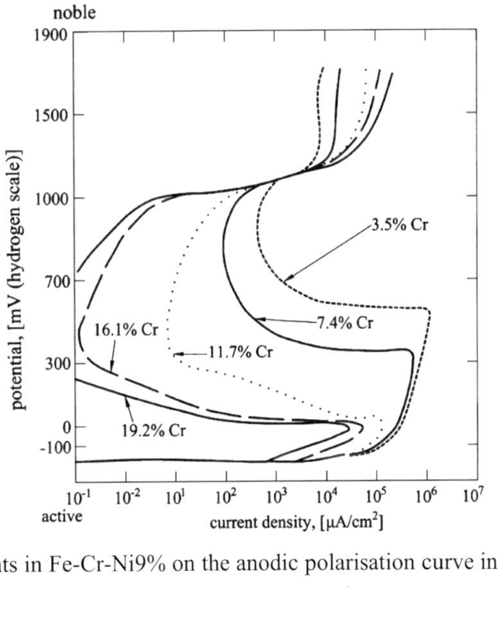

Fig. (6). Effect of Cr contents in Fe-Cr-Ni9% on the anodic polarisation curve in 2N H2SO4 solution at 90°C [7].

The positive impact of chromium is associated with the formation of amorphous hydroxide connections enriched with chromium, giving passive films with high corrosion resistance [2].

Effect of molybdenum: molybdenum and nitrogen act similarly to chromium. So in corrosion-resistant steels of ferritic and austenitic structure, molybdenum improves resistance to pitting corrosion, (Fig. **7** [2]), especially in the presence of chromium. The presence of Mo_4^{2-} ions in the passive film causes an increase in its stability [3]. According to J.R. Ambrose [8], these ions form bonds with oxides of chromium and iron, and according to N. Sato [9] they hinder the migration of ions and remove adsorbed chloride ions. The beneficial effect of the introduction of molybdenum into steel is limited by chi (χ) phases, which decrease the yield strength parameters (El, KCV) and reduce pitting resistance.

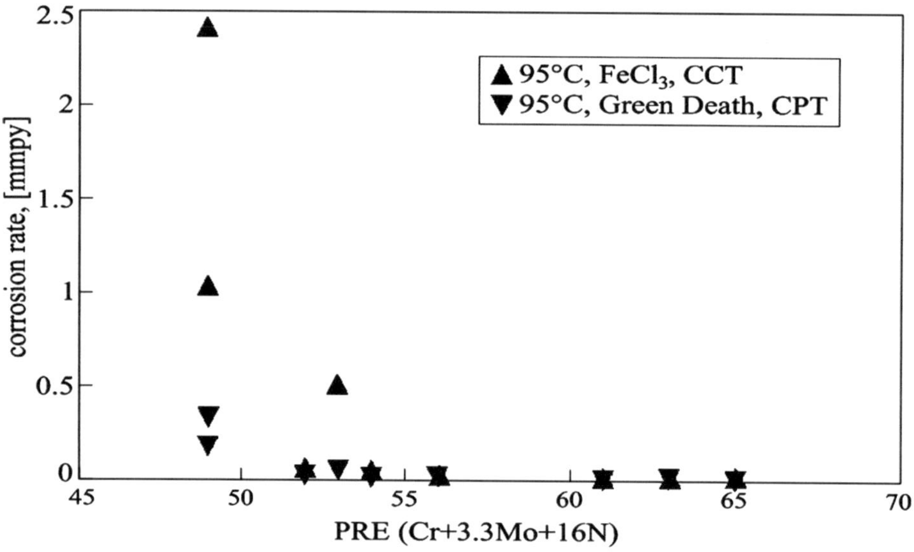

Fig. (7). Effect of Cr contents in Fe-Cr-Ni9% on the anodic polarisation curve in 2N H2SO4 solution at 90 C , [10 - 14].

effect of nitrogen: nitrogen dissolved in the steel matrix (including that of cast steel) increases yield strength [10], stabilises the austenite and increases resistance to pitting corrosion, especially in the presence of Cr and Mo. This effect is determined by the so-called 'pitting resistance equivalence', PRE, defined as:

$$PRE = \%Cr + 3.3(\%Mo) + 16(\%N) \quad [10, 13] \qquad \textbf{(2)}$$

If PRE equals 35 - 45, steel is characterised by good resistance in seawater; if it equals > 50, the steel is resistant to corrosion in aggressive chloride solutions [10].

The impact of PRE on the steel corrosion rate containing < 0.01%C, 24 to 28%Cr, 22% Ni, 4 to 8% Mo and 0.7 to 0.95% N is presented in Fig. (**7**).

To ensure low corrosion rate of the above-mentioned steel grades in samples of $FeCl_3$ and Green Death, the PRE value should be above 55.

The addition of nitrogen improves the corrosion resistance, particularly in the presence of molybdenum. The mechanism explaining the beneficial effect of nitrogen in steels is based on the enrichment of the matrix-passive film by chromium, the formation of a nitride film at the metal-film boundary and the formation of a ferrous molybdate film outside the passive film [14]. Thus, for example, in 0.03%C - 22%Cr - 5.5%Ni - 1.8%Mo - 1.4%Cu cast steel [15], higher contents of nitrogen result in changes to the pitting potential. Increasing the nitrogen content from 0.14% to 0,30% diminishes the susceptibility of cast duplex steel to pitting attack. Simultaneously with the increase in the nitrogen concentration up to 0.30%, the beginning of pitting corrosion occurs at higher temperatures, (Figs. **8** and **9**).

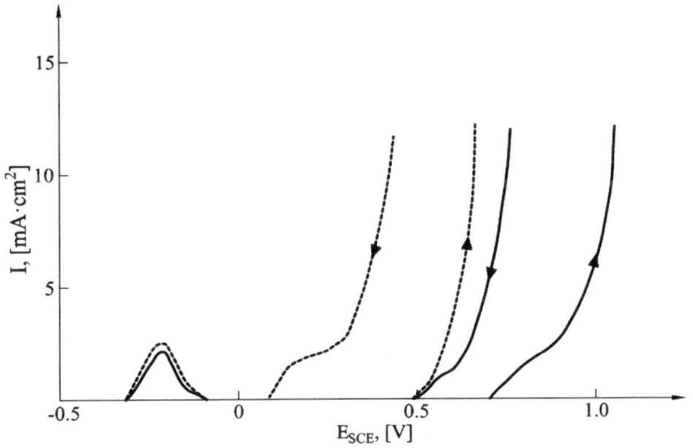

Fig. (8). The effect of nitrogen content on anodic polarisation in 1M H_2SO_4 – 1M NaCl at 50°C dotted line - 0.14%N; full line – 0.30%N, Scan rate 300mV/min [15].

In corrosion resistant cast steel, the addition of nitrogen changes the morphology of delta ferrite from the acicular to the nodular form, (Fig. **10**). This beneficial effect of increased nitrogen content in the cast steel is connected with the danger of the formation of gas porosity in castings. Although increased additions of manganese [10], chromium and molybdenum increase the solubility of nitrogen, its segregation in thick-walled castings can lead to the formation of gas bubbles.

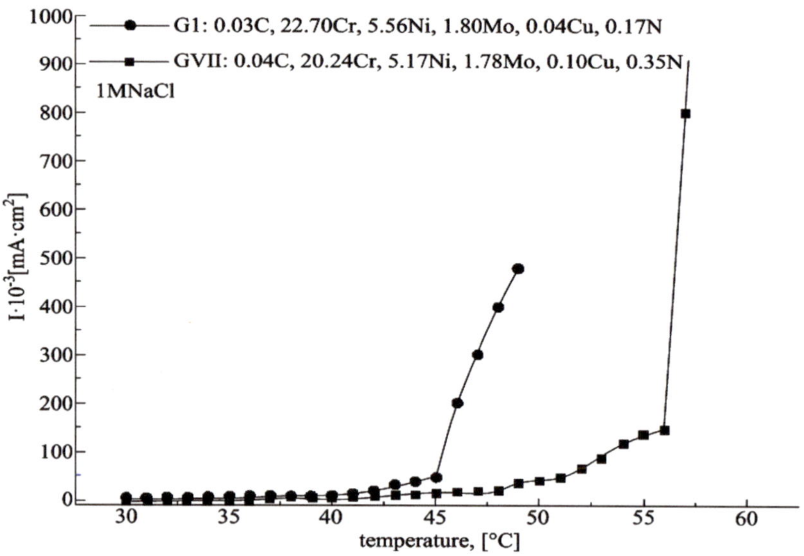

Fig. (9). The effect of nitrogen on critical pitting temperature CPT in 1M NaCl at 0.3V [15].

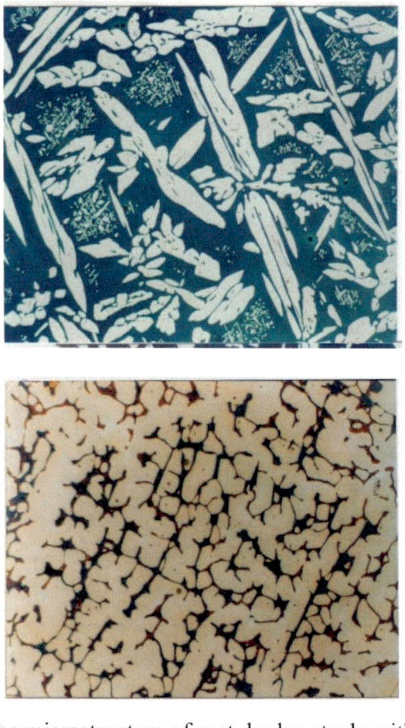

0,03%C-22.78%Cr-5.56%Ni-1.8%Mo- **0.17%**N.
50% δFe CPT 45°C

0.04%C-20.24%Cr-5.17%Ni-1.78%Mo-**0.36%N**
22.5% δFe CPT 54°C

Fig. (10). The microstructure of cast duplex steels with different nitrogen contents [16].

effect of carbon: It is particularly unfavourable, due to the formation of carbides of chromium, titanium and niobium. In aqueous solutions of chlorides, chromium depletion in the matrix near carbides destroys the passive film and leads to intergranular corrosion. The effect of carbon is particularly evident in its total action when added along with chromium.

The effect of other elements on the corrosion resistance of austenitic steel can be found in references [2, 17, 18], and for cast steel in references [19, 20].

In most cases, it is important to obtain passive conditions, that is, a stable passive film, for the selection of forged steels and cast steels to work in corrosive conditions in aqueous environments (diluted acids, presence of chloride ions). Passivation abilities determine: the passivation potential E_{pp}, the density of the passivation critical current i_{pas} and the range between the transpassive potential and the passivation potential.

The end of the metal dissolution reaction (point E_{pp} – Fig. **2**) means transition to the so-called 'active–passive' domain, at the end of which both the potential and the current (to the value of so-called passive current density i_{pas}) are stabilised. Under these conditions (E_{pp}, i_{pas}), there are thermodynamic conditions for the reaction of oxide formation:

$$nM + mH_2O \rightarrow M_nO_m + 2mH^+ + 2m\cdot e \tag{3}$$

However, under certain environmental conditions both the potential and the current can rise again, mainly owing to, for example, the presence of oxygen in the solution or the formation of pits in the protective layer. Above the potential E_p the formation of such pits occurs, and this is the potential limit value that should not be exceeded.

2. TYPES OF CORROSION

Protection of a given item or casting against corrosion depends, among other factors, on the stability of the passive film. In other words, knowledge of the properties of the film allows for the protection of the metal against various environmental conditions and operating parameters. Typically, the passive film is subjected to the following types of corrosion:

general (uniform) (Figs. **11** and **12**): in these conditions anodic and cathodic areas are evenly located over the entire surface, and the whole element is a compact corrosive cell consisting of local anodes and cathodes. The corrosion rate is usually measured in mm/year [mm/y], mils per year [mpy] or [$g/m^2\cdot h$] (grams per square meter per hour). There are also other units known: inches per year [ipy] and milligrams per square decimetre per day [m/dd] [2, page 367]. The alloy is resistant if it corrodes at a rate of less than 0.1 $g/m^2{\cdot}h$ (approx. 0.1 mm/year) [21]. In practice, there are alloys used whose corrosion rate reaches a maximum of $1g/m^2{\cdot}h$. The general corrosion rates can be obtained under the conditions characterised by the curves:

E_1 - i_1; $i_1 - i_2 - i_3$ and $E_4 - i_4$, Fig. (**11**). In the first case, these are the conditions of corrosion in dilute sulphuric acid, while the other represents the rates under the conditions of corrosion in the transpassive range. A rapid increase in corrosion in an active domain (i_1 - i_2 - i_3) is characteristic of the presence of the oxidants Fe^{+3} and Cu^{+2}. Atmospheric corrosion is a typical example of uniform corrosion.

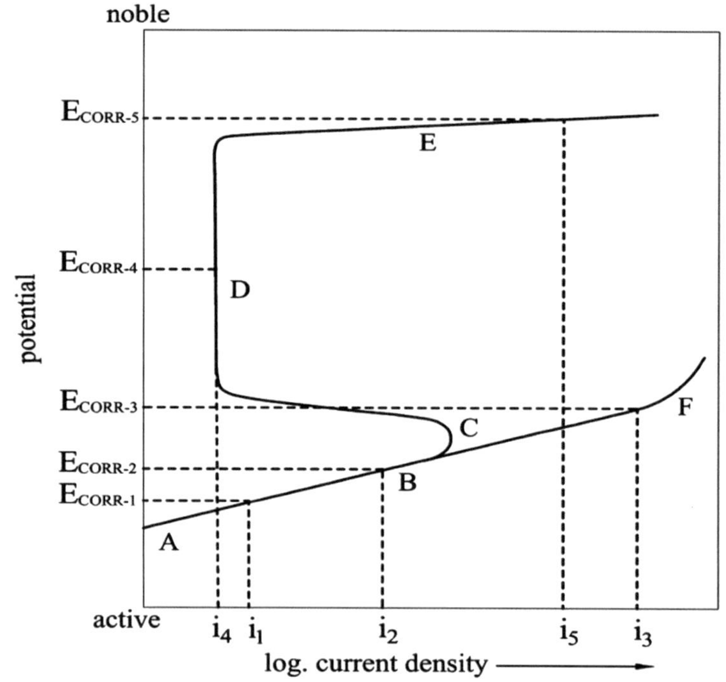

Fig. (11). Polarisation curves of general corrosion [22].

Pitting (Fig. **12b**), this type of corrosion depends on the stability of the passive film in a given environment. However, the presence of chloride, bromide or iodide ions in the solution results in local damage to the protective film and rapid oxidation of the metal. In such cases, thin-walled castings or elements can be damaged over the entire wall thickness. Therefore, it is more dangerous than uniform corrosion.

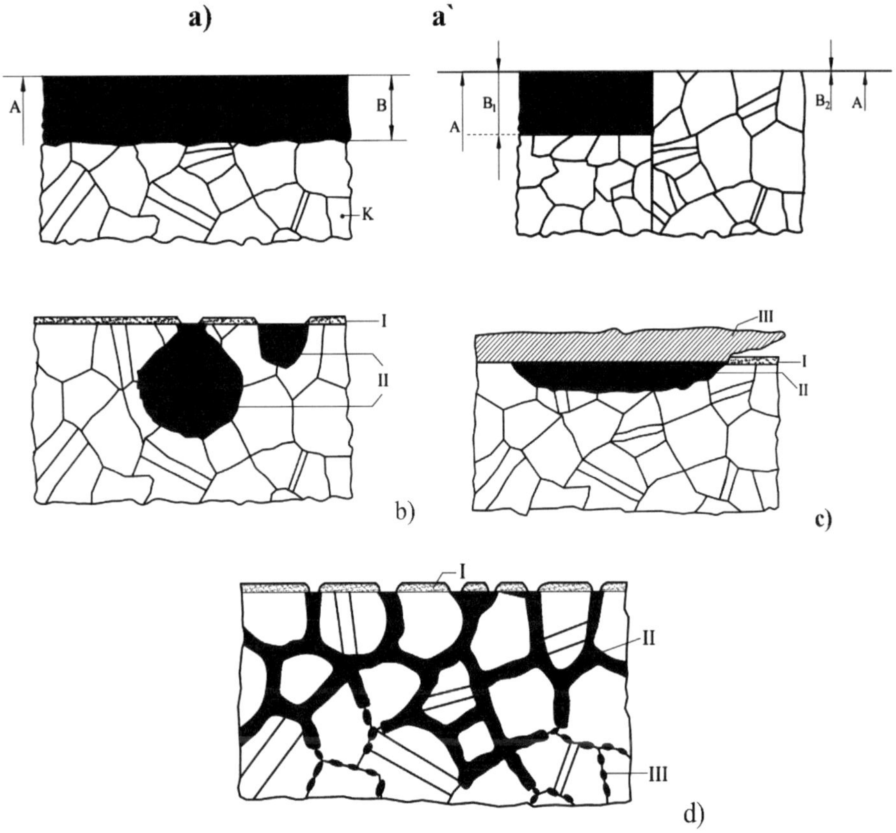

Fig. (12). General characteristics of corrosion types: a and a` - general corrosion; b- pitting; c-crevice corrosion; d- intergranular corrosion [22].

Pitting corrosion progresses faster in more concentrated acids, in solutions with more chloride ions and in environments more exposed to oxidation. Uneven surface features, such as cracks, scratches and sharp edges after machining, also accelerate pitting attack. Introduction of high-chromium (or chromium–nickel) cast steel and the addition of molybdenum are the main factors favourably reducing the corrosion rate. Conditions of pitting corrosion are observed mostly on the anodic polarisation curve (Fig. **2**) by:

– reduction of the potential value below the transpassive potential, E_p.

Introduction of elements shifting E_p to higher values (Cr, Ni, Mo and N) into the matrix may not ensure the required corrosion resistance. This applies to conditions in which the passive film thickness and its chemical composition change over time [2, 23]. However, introduction of molybdenum into Fe-18%Cr steel shifts the pitting potentials to higher values, (Table **1**).

Table 1. Pitting potentials (volts *vs.* S.C.E.) for Cr18-Mo steels in 1M NaCl at 25°C [24].

Chemical composition	260mV/h	Scratch technique
Fe – Cr18%	0.090	0.120
Fe – Cr18% - Mo1%	0.202	0.189
Fe – Cr18% - Mo2%	0.297	0.284
Fe – Cr18% - Mo3.5%	0.423	0.413
Fe – Cr18% - Mo5%	0.637	0.535

– reduction of re-passivation potential [25, 26]. Application of a 'protection potential', E_x, by reversing the direction of potential value beyond E_p, (Fig. **13**).

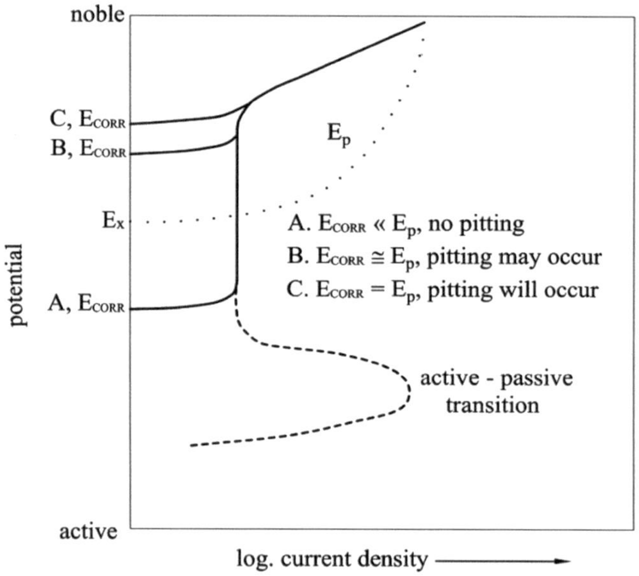

Fig. (13). Schematic polarisation curve with the idea of a 'protection potential' [2].

Pitting in castings from corrosion-resistant steel is especially dangerous. The first pits appear at the interdendritic boundaries, in regions adjacent to intermetallic compounds, non-metallic inclusions (Fig. **14b**) and carbides. Any defects that occur during solidification, such as shrinkage micro-porosity, or the segregation of species such as sulphur, phosphorus or carbon to the original interdendritic boundaries, can also be sites of pitting initiation. Examples of their occurrence are presented in Fig. (**14a** and **b**).

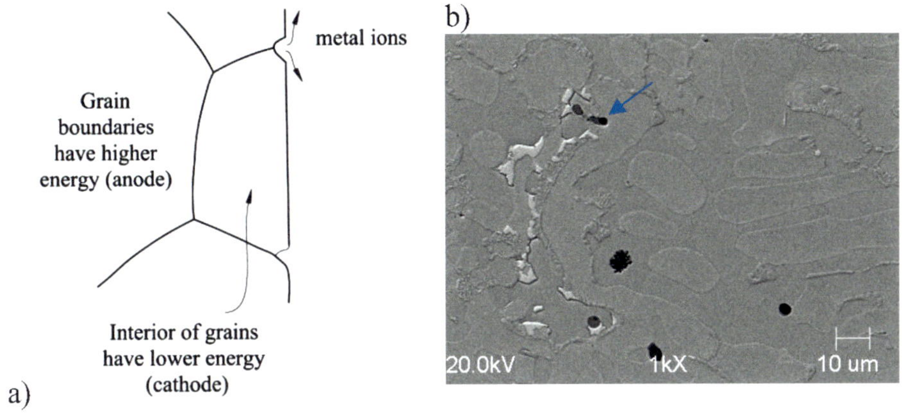

Fig. (14). First steps of pitting at grain boundaries (**a**) [2], and some examples in F –A cast steels between the matrix and oxy-sulphides (**b** - see arrow), Magn. 1000X

The mechanism of pitting corrosion involves two stages of this process: an incubation period and a period of pitting development. The start of the process takes place mostly in those alloy structures in which there are carbides, areas with a certain degree of segregation, dislocations or interdendritic boundaries. The presence of non-metallic inclusions (sulphides, oxy-sulphides or oxides) is also a possible reason for pitting formation, due to their dissolution in the corrosive environment and the creation of common hollows (pits).

It has been stated [8], that the pit bottom constitutes an anode, wherein dissolution of iron takes place according to the reaction:

$$Fe = Fe^{2+} + 2e \quad \textbf{(4)}$$

and the crater edge is a cathode where the oxygen reduction occurs:

$$O_2 + 2H_2O + 4e = 4OH^- \quad \textbf{(5)}$$

As a result of the course of these reactions, the concentration of chloride and iron ions increases in the pit bottom, and hydroxide ions raise the pH value locally at the crater edge.

– **Crevice corrosion** (Fig. **12c**) occurs in such places on the corrosive alloy surface to which access of oxidant (oxygen) is difficult (there is no metal passivation in these places). The lack of a protective film offers the possibility of a chemical reaction consisting of the continuous dissolution of metal in the crevice (anodic reactions $M = M^{n+} + ne$) and the disappearance of the cathodic reaction ($O_2 + 2H_2O + 4e = 4OH^-$, oxygen consumption). Iron ions form salts that penetrate into the interior of the crevice, where they are subject to hydrolysis, increasing the hydrogen ion concentration (even up to seven times). The area surrounding the crevice is covered with corrosion products (hydroxide ions).

– **Intergranular corrosion** (Fig. **12d**), is the result of changes in the chemical composition (mainly chromium) at the interdendritic boundaries or between carbide precipitates and the cast steel matrix, Fig. (**15**). The reduction in chromium concentration below the critical value is primarily due to too high a carbon content.

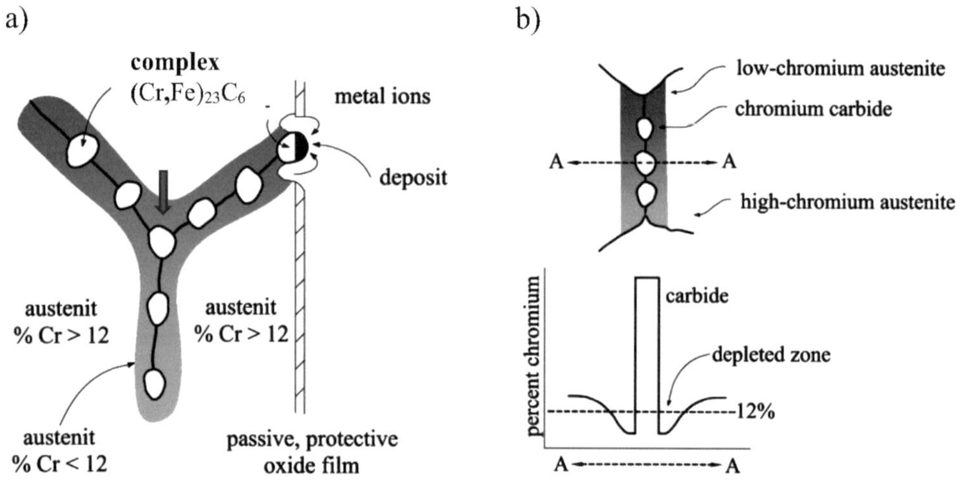

Fig. (15). Precipitation of chromium carbides at the grains of austenite (a) and chromium content at the carbide-matrix region (b) [2].

In the corrosion resistant steels, it is easier to fight intergranular corrosion, as processes of vacuum steel melting are commonly used, in which a carbon content below 0.02% (and even below 0.01%) is obtained.

In cast steel foundries, these processes are used less frequently and obtaining such carbon content is a rare case. Therefore, in castings with a wall thickness of 40 - 70 mm, at a low cooling rate and relatively high carbon concentrations (*e.g.* 0.03 to 0.04%), precipitation of carbides $(Fe, Cr)_{23}C_6$ can be observed at the primary austenite boundaries or at austenite - ferrite interphase boundaries, (Fig. **16**). The problem of changes in chemical composition during steel solidification is particularly important in castings. The effects of the segregation of carbon, oxygen, sulphur and chromium lead to the nucleation of precipitates, mainly chromium carbides $(Fe, Cr)_{23}C_6$, sigma σ and chi χ, along with other phases.

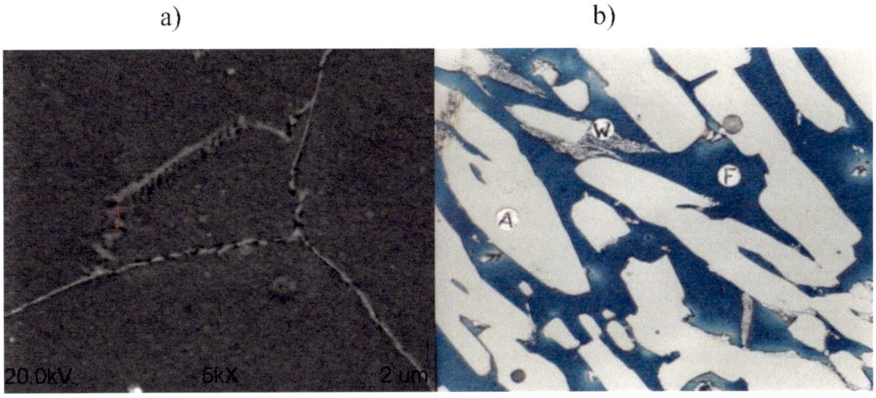

Fig. (16). Carbide precipitation in ferritic cast steel (0.21%C-14.7%Cr) - a, Magn.800X; and in cast duplex steel (0.04%C-22.6%Cr-6.1% Ni-1.4% Mo) – b (A – austenite, W - carbides, F – ferrite), Magn.400X

Heat treatment (solution annealing) applied in further stages of the process allows for the dissolution of these precipitates into the cast steel matrix, but during heating the castings to the austenite temperature, stresses are formed, which may cause cracks. For this reason, striving to achieve the lowest carbon concentration is economically justified.

In addition to chromium carbides, at the time of solidification other compounds are also released in the corrosion-resistant cast steel. Depending on the share of alloying elements and their concentration, they can be of a very complex type, (Fig. **17**).

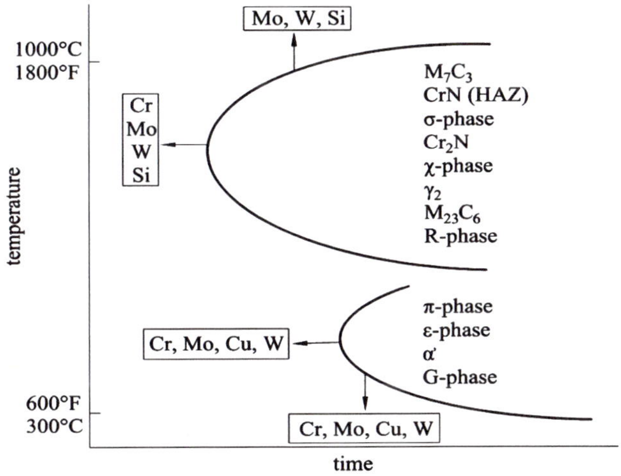

Fig. (17). The precipitation conditions of harmful phases in stainless steels [2].

The conditions for precipitate formation are presented in the form of a TTP (time-temperature-precipitation) diagram for the appropriate grade of forged steel or cast steel, Fig. (**18**). Chromium carbides are already released after several minutes, while other intermetallic compounds appear after several hours. These effects of changes in microstructure are found in castings kept at elevated temperatures while being worked upon over long periods. This also applies during the service conditions of castings (if working at, *e.g.* 300 – 400°C, with wear), during the welding of castings (Heat Affecting Zones -HAZ) and on places in castings which are subject to friction (*e.g.* when machining at incorrectly selected cutting conditions).

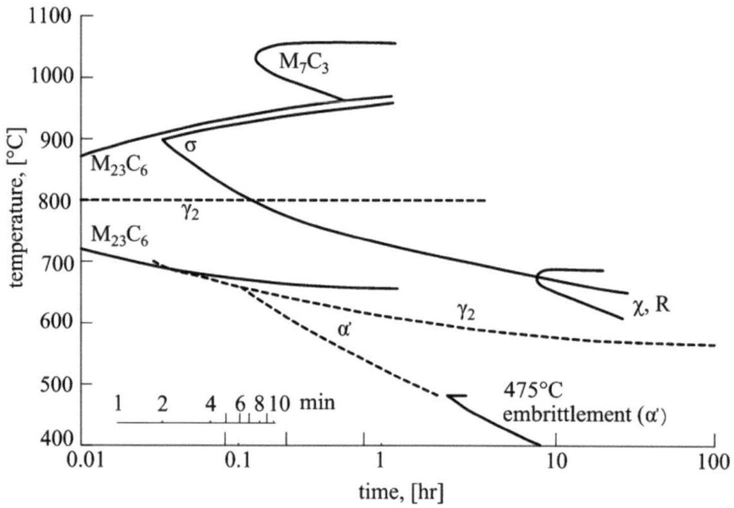

Fig. (18). Time-temperature – precipitation for grade 316 (according AISI) [27].

The second condition occurs more often, as chromium carbides are released at grain boundaries due to Cr–Ni cast steel staying at a temperature of 550-850°C over a long period of time. It is followed by the diffusion of chromium and carbon from the matrix, adjacent to the carbide, into carbide precipitates, and chromium depletion in the matrix. Since the rate of carbon diffusion is greater than the diffusion of chromium, carbides of $Cr_{23}C_6$ are first formed at the grain boundaries, and a continuous film of carbides is formed over a longer period of time. Inside the carbides, the concentration of chromium can equal more than 65%, and the depleted zone contains less than 8% chromium (anodic domains). These domains are not resistant to the corrosion of the surrounding environment (Figs. **15**, **19**). In corrosion-resistant castings, such conditions occur during long-term operation at elevated temperatures and during the repair of defects by welding. The increased carbon content in castings (above 0.03-0.04%), in combination with these castings' working conditions, is the most common cause of pitting initiation in areas of carbide precipitation.

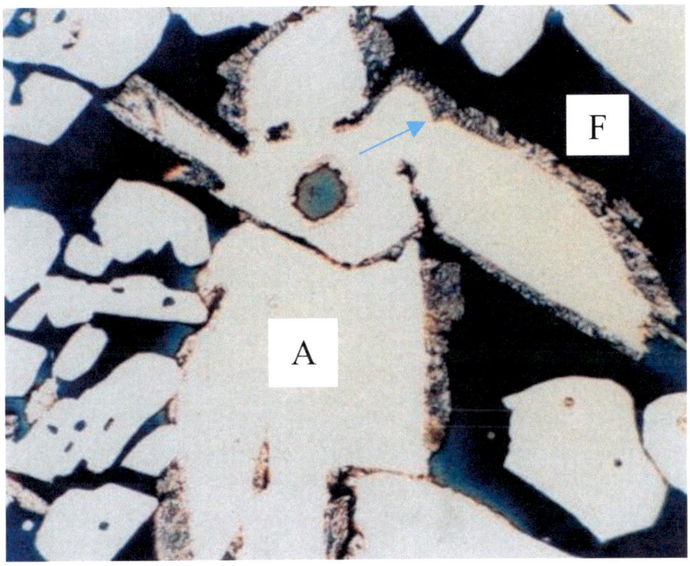

Fig. (19). Examples of $Cr_{23}C_6$ carbides (arrow) in F-A cast steel: 0.05%C-20.6%Cr-6.84%Ni- 4.6%Mo Magn.630X, Beraha II etching.

In cast duplex steel (ferrite-austenite), zones depleted in chromium near carbide, $Mn_{23}C_6$, are slightly different to those in austenitic cast steel. Because the rate of chromium diffusion in ferrite is 100 times faster than in austenite, carbide grows faster in the ferrite than in the austenite (see Fig. **19**). Hence, the zone depleted of chromium in ferrite is wider and deeper than in the austenite. Corrosion occurs faster at the F-A interphase boundary.

The slow cooling of castings at temperatures below 850°C, along with the necessity of repairing castings (welding and casting work at temperatures of 600-800°C) is the main cause of intergranular corrosion. Its prevention consists of:

1. solution treatment in which the carbides dissolve in the alloy matrix,
2. additions of niobium or titanium (stabilisers) that have a greater affinity for carbon than chromium; they are introduced into steel and by forming stable carbides (TiC, NbC) protect the matrix against depletion of chromium. Typically, the amount of stabiliser introduced depends on the final carbon concentration and equals:

$$\% Nb = \% C \times 5$$
$$\% Ti = \% C \times 8 \qquad \textbf{(6)}$$

3. production of steel with very low carbon concentration (nowadays, even below 0.01% can be obtained). In steels intended for corrosion-resistant castings, this requires the necessity of the introduction of AOD, VOD or VAD processes in converters of small capacity (up to 15 tonnes).

Other types of corrosion, such as galvanic, fatigue, stress corrosion cracking, hydrogen or selective corrosion of castings are found less frequently. Their characteristics can be found in references [2, 3, 20, 21, 27].

3. THE EFFECT OF ALLOYING ELEMENTS

The corrosion resistance of steel is associated with elements with a high capacity for passivation. Most often, these are metals with high oxygen affinity (low enthalpy of oxide formation) and the capacity to form stable passive films. In chromium-–nickel steels, these are primarily chromium, silicon and aluminium. Pure chromium shows a broad passive domain, and in Fe–Cr–Ni alloys, it is the most favourable element for steel passivation. The effect of chromium on atmospheric corrosion is presented in Figs. (**6**, **15** and **20**). It indicates that at a concentration of about 12% chromium, the Fe–Cr alloy is resistant to atmospheric corrosion. These include stainless steels, applied in corrosive conditions in water and vapour environments below a temperature of 315°C. Due to their good resistance to stress corrosion cracking, they are used for the castings of valve bodies and trim in the chemical industry and food processing. CC-50 cast steel (according ACI designation, with 18-21%Cr, max.0.05% C) is used for castings operating in a mine water environment, the pulp industry and the automotive industry (catalyst pipes and casings). Nickel (up to 2%) and nitrogen (up to 0.15%) are added to enhance the yield strength parameters in the ferritic cast steel.

To ensure corrosion resistance of Fe-Cr alloy (Figs. **15**, **20**), the chromium concentration should equal 12%. However, steels contain a certain amount of carbon, which causes formation of carbides $(Cr, Fe)_{23}C_6$. Since 1%C binds 14.5%Cr [21], ferritic steels with:

$$1\%C \text{ should contain } 12 + 14.5\%Cr = 26.5\%Cr$$
$$0.1\%C \ 12 + 1.45\%Cr \sim 13.5\%Cr \qquad \textbf{(7)}$$

So the clear effect of chromium is the basis for the division of corrosion-resistant cast steels into three groups:

- chromium cast steels (from approximately 12% Cr and 1% Ni),
- chromium-nickel cast steels (typical 18% Cr – 9% Ni; 18% Cr – 11% Ni – % Mo),
- nickel –chromium cast steels (*e.g.* 32% Ni – 20% Cr; 25% Ni – 20% Cr).

The beneficial effect of chromium occurs also in Cr18-Ni9% steels, Fig. (**6**). An increase in the chromium content from 3.5 to 19% decreases the current density in the passive range and extends the area of the passive range potentials [7].

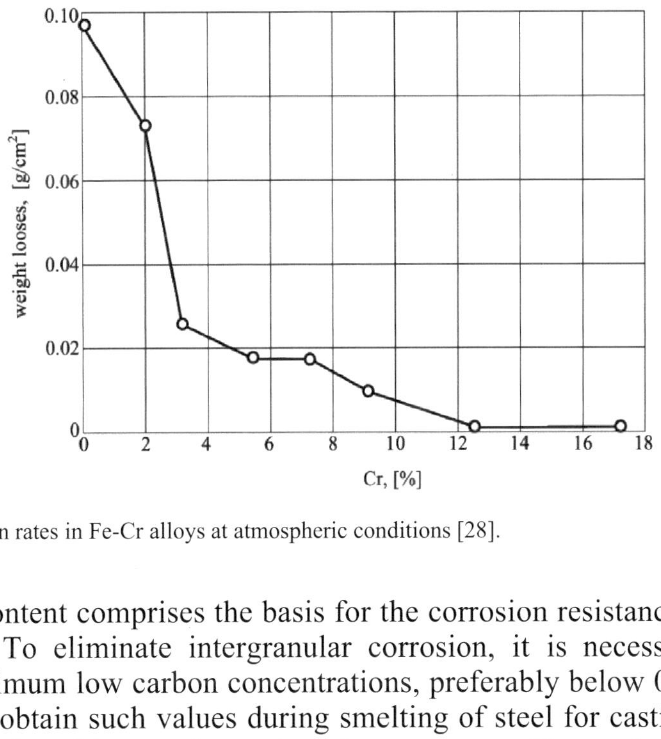

Fig. (20). Corrosion rates in Fe-Cr alloys at atmospheric conditions [28].

The carbon content comprises the basis for the corrosion resistance of these cast steel grades. To eliminate intergranular corrosion, it is necessary to aim at reaching maximum low carbon concentrations, preferably below 0.03%. Since it is difficult to obtain such values during smelting of steel for castings in electric arc furnaces, foundries produce steel for castings with much higher carbon concentrations. However, it is then difficult to obtain significant corrosion resistance, because the amount of carbides at grain boundaries changes unfavourably, and therefore the conditions of heat treatment and resistance to intergranular corrosion are worsened. The introduction of new techniques of smelting steel in AOD or VAD converters (including ladle furnaces) is advisable here, because they allow to a carbon content of even less than 0.01% to be obtained.

The effect of other elements on the corrosion resistance is well illustrated by Fig. (**5**), indicating the possibility of extending the passive domain. Nickel, the second basic element of corrosion-resistant cast steel, only slightly increases the stability of cast steel in the active domain and slightly shifts the transpassive potential in a positive direction. In cast steels with addition of nickel, only a slight improvement in the corrosion resistance was observed in low oxidizing centres and neutral solutions of chloride ions, but there was a clear increase in yield strength and especially in the large wall thickness of castings. The beneficial effect of nickel on the technological properties of cast steel, in particular in shaped castings (equalisation of the properties with respect to different wall thicknesses, improving castability) is also very important.

The addition of molybdenum significantly improves resistance to pitting corrosion. This is a result of modifications to the structure of the passive film, through enrichment in molybdenum of the matrix film directly adjacent to the oxide layer and the presence of MoO^{2-}_4 ions in the passive film [2, 13]. There are also suggestions that its anions form a barrier that reduces ion conductivity in the passive film, and therefore blocks the adsorption of chloride ions.

The effect of silicon is complex: at a content below 0.1% and above 2-3%, there is no intergranular corrosion observed (it improves the passivation properties of steel in highly oxidizing centres). However, the highest attack of pitting corrosion occurs at a content of 0.7% [2]. On this basis, forged steel and cast steel containing 4% Si have been developed (*e.g.* Uranus S [10]), which were made to work in concentrated solutions of nitric acid (in the presence of $Cr_2O_7^{2-}$, MnO_4^-, Fe^{3+}, V^{5+}, Ce^{4+} [29] and at elevated temperatures).

Nitrogen has a beneficial effect on the increase in corrosion resistance of the cast steel. As a component stabilizing austenite, it has a particularly beneficial effect in the presence of molybdenum. The direct impact of nitrogen is demonstrated by its shifting the transpassive potential to a higher value of E_p.

4. MICROSTRUCTURE OF CAST STEEL

An important division of corrosion resistant cast steel is based on the structure obtained in the finished casting. All standards (ASTM A 890M – 99, EN 10 088, EN 10 213 and EN 10 283) also take into account this factor, indicating that, depending on the chemical composition, steel grades with a purely ferritic (chromium ferrite), martensitic, austenite–ferritic or purely austenitic structure can be obtained.

The concentration of ferrite-promoting (Cr, Mo, Si and Nb) and austenite-promoting components (C, Ni, Mn, N and Cu) determines the final structure of the cast steel grades. This effect is usually presented in the Schaeffler diagram (Fig. **21**) [19, 20, 30], in which the participation of delta ferrite depends on chromium and nickel equivalents (Cr_{eq}, Ni_{eq}), included in the equations:

$$Cr_{eq} = \%Cr + 1{,}5\times(\%Si) + 1.4\times(\%Mo) + \%Nb - 4.99,$$
$$Ni_{eq} = \%Ni + 30\times(\%C) + 0.5\times(\%Mn) + 26\times(\%N - 0.02) + 2.77 \quad \mathbf{(8)}$$

Unlike forged steels, which are usually austenitic, cast steels of identical chemical composition contain a certain amount of delta ferrite. This is due to the slow cooling of castings and the segregation processes during their solidification. 5-25% of delta ferrite, arranged evenly in the alloy matrix, can be found in castings made from austenitic cast steel grades. The presence of delta ferrite is beneficial, because it improves weldability, resistance to stress corrosion cracking (SCC) and intergranular corrosion [31]. Islands of delta ferrite impede crack spreading and are domains favouring the formation of carbides in their interior, and not at the ferrite–austenite boundary. This effect, however, depends largely on the environmental conditions, stress-strain conditions, heat treatment and chemical composition.

There are also conditions in which presence of delta ferrite is detrimental. This relates to the casting operation at elevated temperatures (above 315°C). For these reasons, it is usually necessary to control the amount of delta ferrite in given operating conditions. For the castings of Cr–Ni steels (containing: 16-26% Cr, 6-14% Ni, up to 4% Mo, up to 2% Mn, up to 1% Nb, up to 0.2% C and up to 0.19% N) the amount of δ Fe is specified by the Schoefer diagram (Fig. **22**), determined from the previously calculated equivalents of Cr_{eq} and Ni_{eq}.

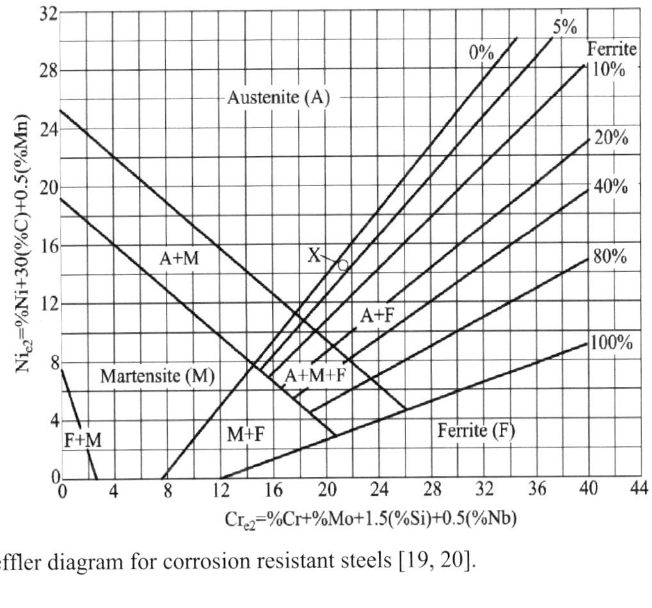

Fig. (21). The Schaeffler diagram for corrosion resistant steels [19, 20].

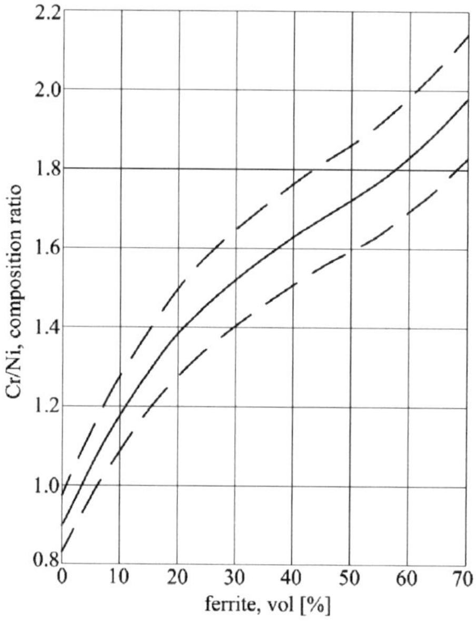

Fig. (22). Schoefer diagram for stainless cast steels [31, 32].

This dependence is the basis for prediction of the amount of δ Fe in a cast steel of known chemical composition.

Failure to comply with the required chemical composition, temperature changes and segregation in different walls of the casting are causes of certain differences in the calculated and measured amounts of delta ferrite. In the past 20 years, the beneficial effect of δ Fe has led to the use of a new austenitic – ferritic cast steels group, *i.e.* duplex steels (and cast duplex steels) in industrial environments.

They are characterised by a larger amount of ferrite (from 40-65%) in the matrix (Fig. **23**) and numerous additional advantages such as [29]:

– better resistance to intergranular corrosion and stress corrosion cracking (SCC) in many centres [33],
– better resistance (UTS, YS) than in grades with purely austenitic structure (YS even twice as high [27]),
– better weldability,
– less tendency to hot cracking [22],
– they are magnetic,
– no worse resistance to fatigue corrosion than austenitic steels.

Fig. (23). The microstructure of F-A cast steel in the as-cast conditions (a-Magn.800X – white areas are carbides) and after solution annealing (b- Magn.400X – yellow areas – austenite, dark - ferrite).

The Schoefer diagram is used to determine the amount of ferrite at a known chemical composition of the steel for castings, or to calculate the chemical composition of the heat for smelting, when the required amount of delta ferrite in the castings is assumed.

5. FERRITIC – AUSTENITIC, F-A (DUPLEX) CAST STEELS

One of the important reasons for the increasingly widespread application of F-A cast steel is the increase in the strength parameters. This applies to both UTS and YS over the entire range of participation of delta ferrite in the structure of cast steels. In relation to the cast steel grades with austenitic structure, in which up to 20% of delta ferrite is achieved, austenitic-ferrite cast steel grades have more than twice an increase in strength. This is very important for thin-walled castings, such as pump rotors and vanes, working in the mining (saline mine water) or chemical industries (acid solutions), (Fig. **24**).

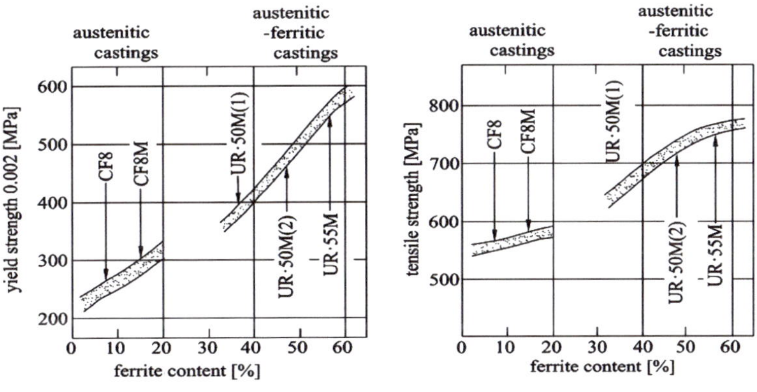

Fig. (24). The effect of ferrite contents on the tensile and yield strengths [6].

Due to segregation processes and differentiated cooling rates in particular walls of the casting, there can also be carbides in this group of cast steels in the as-cast condition, (Figs. **19** and **23a**). Their amount at grain boundaries depends on the carbon content. Hence, the smelting process and the necessity to use vacuum furnaces is so important. Only the technology of vacuum melting and casting (VOD, VAD, LD, LF) guarantees obtaining carbon below 0.01 - 0.02%.

6. HEAT TREATMENT

Although the overall heat treatment course of stainless steel castings is very similar to the heat treatment of forged steel, due to segregation and the dendritic microstructure, homogenization is used (*e.g.* 1100-1150°C) to obtain a uniform microstructure and uniform distribution of elements in the matrix.

Other treatments are associated with the presence of carbides $Cr_{23}C_6$ along the grain boundaries (or ferrite–austenite boundaries in cast duplex steels). They may be formed in the as-cast condition, in the case of high carbon content (above 0.06% [11, 34],), slow cooling of castings in the range of 425 to 900°C or castings staying over a long period of time at these temperatures.

To eliminate these conditions, castings are solution treated at temperatures of 1040-1200 °C and cooled in water. An example of eliminating carbides by applying such heat treatment is shown in Fig. (**23b**). Cast steel grades with larger chromium contents require a higher temperature for this heat treatment. The presence of brittle phases (σ, χ and α') creates similar requirements. In both cases, rapid cooling is essential (after solution treatment) in the temperature range of 500-900°C, so as to avoid brittle phases, (Figs. **17** and **18)**.

In some cases (thin-walled castings with low carbon content), keeping castings at 870-980°C for 24 to 48 hours and cooling them in air can be used instead of solution treatment. The resistance to intergranular corrosion obtained is, however, worse.

CONFLICT OF INTEREST

The author (editor) declares no conflict of interest, financial or otherwise.

REFERENCES

[1] D. Talbot, and J. Talbot, *Corrosion Science and Technology*. CRC Press: Boca Raton, 1998.

[2] A.J. Sendriks, *Corrosion of Stainless Steels*. John Wiley: N.Y., 1979.

[3] J. Horvath, and H.H. Uhlig, "Critical potentials for pitting corrosion in Ni, Cr-Ni, Cr-Fe, and related stainless steels", *J. Electrochem. Soc.*, vol. 115, no. 8, pp. 791-795, 1968. [http://dx.doi.org/10.1149/1.2411433]

[4] J. Banaś, *Corrosion resistant alloys fundamental aspect of material selection, Conf.of Polish Academy of Science* AKAPIT: Cracow, 2002, pp. 89-112.

[5] G. Okamoto, and T. Shibata, *Passivity of Metals, Corrosion Monograph Series,* R.P. Frankenthal, J. Kruger, Eds., The electrochemical Soc.: Pennington, N.Y., 1978.

[6] M.O. Speidel, *Handbook on Stress Corrosion Cracking and Corrosion Fatigue.* Published by Advanced Research Projects Agency: Washington, D.C., 1984.

[7] K. Osozawa, and H.J. Engell, "The anodic polarization curves of iron-nickel-chromium alloys", *Corros. Sci.,* vol. 6, no. 8, pp. 389-393, 1966. [http://dx.doi.org/10.1016/S0010-938X(66)80022-7]

[8] J.R. Ambrose, *Passivity of Metals* Electrochemical Society: N. Jersey, 1978.

[9] N. Sato, "Toward a more fundamental understanding of corrosion processes", *Corrosion,* vol. 45, pp. 354-368, 1989. [http://dx.doi.org/10.5006/1.3582030]

[10] G.O. Rhodes, and J.J. Convay, "High-nitrogen austenitic stainless steels with high strength and corrosion resistance", *JOM,* vol. 48, no. 4, pp. 28-31, 1996. [http://dx.doi.org/10.1007/BF03222915]

[11] *Annual Book of ASTM Standards* Easton, MD, ASTM, 1995, pp. 174-179.

[12] R. Kirchheimer, M. Köhler, and U. Hübner, *Conf. NACE International Corrosion/90* paper no. 90. Houston, TX, 1990.

[13] J.E. Truman, *Effects of composition on the resistance to pitting corrosion of stainless steels, U.K. Corrosion 87* Brighton, England, 1987, pp. 26-28.

[14] G.P. Halada, C.R. Claton, D. Kim, and J.R. Kearns, *Conf. NACE Corrosion 95, Preprint no. 531,* 1995.

[15] J. Banaś, J. Głownia, B. Kalandyk, and B. Stypuła, "Corrosion and passivity behavior of duplex stainless steels with high nitrogen content", *Proc.3rd Kurt Schwabe Corrosion Symp.,* 2000pp. 235-239 Zakopane, Poland

[16] Grant of the Polish State Committee for Scientific Research (KBN), *Contract no.7 TO8 C 021 14,* Cracow, Poland, . (in Polish)

[17] Z.Szklarska- Śmiałowska, "M.Janik-Czachor, Passivity corrosion of 13Cr-Fe alloy in Na2SO4 solutions containing chloride ions", *Corros. Sci.,* vol. 7, no. 2, pp. 65-70, 1967. [http://dx.doi.org/10.1016/S0010-938X(67)80103-3]

[18] Y.M. Kolotyrkin, "Electrochemical behaviour of metals during anodic and chemical passivation in electrolytic solutions", *Proc.1st Symp.Corrosion,* 1962.

[19] C. Houska, *Castings-Stainless Steels and Nickel-base* Nickel Development Institute Reference Book Series 11022, 2012.

[20] Steel Casting Handbook, *Steel Founders' Soc. of America.* 6th ed. ASM International, 1995.

[21] K. Roesch, and K. Zimmermann, *Stahlguss.* Verlag Stahleisen: Düsseldorf, 1982.

[22] J-H. Wang, C.C. Su, and Z. Szklarska-Śmiałowska, "Effects of Cl- concentration and temperature on pitting of AISI 304 stainless steel", *Corrosion,* vol. 44, no. 10, pp. 732-737, 1988. [http://dx.doi.org/10.5006/1.3584938]

[23] J-H. Wang, C.C. Su, and Z. Szklarska-Śmiałowska, "Effects of Cl- concentration and temperature on pitting of AISI 304 stainless steel", *Corrosion,* vol. 44, no. 10, pp. 732-737, 1988. [http://dx.doi.org/10.5006/1.3584938]

[24] E.A. Lizlovs, and A.P. Bond, "An evaluation of some electrochemical techniques for the determination of pitting potentials of stainless steels", *Corrosion,* vol. 31, no. 6, pp. 219-221, 1975.

[http://dx.doi.org/10.5006/0010-9312-31.6.219]

[25] L.L. Shreir, *Corrosion: Metal/Environment Reactions* vol. 1. Newnes-Butterworths: Boston, 1976.

[26] T. Suzuki, and Y. Kitamura, "Critical potential for growth of localized corrosion of stainless steel in chloride media", *Corrosion,* vol. 28, no. 1, pp. 1-6, 1972. [http://dx.doi.org/10.5006/0010-9312-28.1.1]

[27] R.N. Gunn, *Duplex Stainless Steels.* Abington Publ: Cambridge, England, 1997. [http://dx.doi.org/10.1533/9781845698775]

[28] P.Gümpel und 6 Mitautoren, Rostfreie Stähle, Expert Verlag, Renningen-Malmsheim, 2000.

[29] M.A. Streicher, "General and intergranular corrosion of austenitic stainless steels in acids. Effect of cations in the acids and the influence of heat treatment and grain size of the steel", *J. Electrochem. Soc.,* vol. 106, no. 3, pp. 161-180, 1959. [http://dx.doi.org/10.1149/1.2427304]

[30] *Metals Progress Data Book.* vol. 112. ASM, 1977.

[31] M. Blair, and A.S. Metals Handbook, *Metals Handbook* vol. 1. 10th ed.. ASM International, 1990.

[32] E. Schoefer, "Metals Progress Databook, vol.112, p.51, 1977", *Welding Research,* vol. 39, p. 10, 1974.

[33] V. Neubert, DSS castings-metallurgy, manufacturing, mechanical properties, corrosion resistance and welding of duplex stainless steels, 1994.

[34] J.J. Eckenrod, and C.W. Kovach, "Effect of nitrogen on the sensitization, corrosion and mechanical properties of 18Cr-8Ni stainless steels", In: *Conference: Properties of Austenitic Stainless Steels and Their Weld Metals* ASTM STP 679: Philadelphia, 1979.

CHAPTER 10

Heat – Resistant Cast Steels

Abstract: The high temperature oxidation of cast alloys represents an important problem in several industries: petrochemicals, chemical processing, the arms industry, gas turbines, aircraft engines and cement mill equipment. Many alloyed cast steels are used in these areas. The present chapter describes the alloys' resistance to corrosion at elevated temperatures, the stability of their microstructures and three group of cast steels: Fe-Cr; Fe-Cr-Ni and Fe-Ni-Cr.

Keywords: Cast grades, Elevated temperatures, Mechanism of oxidation, Microstructures, Oxidation, Oxide scale.

1. INTRODUCTION

The term 'heat resistance' describes alloys with a special resistance to the oxidising effect of gases at temperatures higher than 873K - 923K [1 - 3]. Weight loss from the matrix is the usual measure of resistance. Cast steel that has a loss of less than 1 g/m^2 h at a temperature *x*, and at a temperature (*x* + 50K) less than 2 g/m^2 h, is regarded as heat-resistant when these temperature changes occur within 120 hours [4].

Heat-resistant cast steel typically requires further features: primarily, it must have a resistance to reactions with chemical species within the environment in which it operates. Therefore, there is heat resistance in oxidising conditions (in the presence of SO_2, air), reducing conditions (in the presence of H_2S, sulphur in gaseous form), alkaline media (ash, clinker) or in flue gases, *etc.*

Castings operating at temperatures above 873K generally fulfil one of the two basic conditions:

- they are resistant to air, gas, salt, slag or metal, *etc.*, (heat-resistant cast steels),
- they are strong, *i.e.* have resistance to long-term mechanical stress (strength- and creep-resistant cast steels).

These two conditions also include other requirements: resistance to temperature changes, adhesion of scale, weldability, resistance to abrasion and mechanical influence of the environment. These three parameters (chemical composition,

Jan Głownia

temperature and the type of ambient environment atmosphere) determine the scaling rate. The chemical composition of heat-resistant cast steel is based on known heat-resistant steel grades. Compared to ordinary steel, heat-resistant cast steels exhibit some advantageous features:

1. they are cheaper (they eliminate forging, rolling, cutting and welding),
2. because they have a slightly higher carbon content than ordinary steels, they have higher strength parameters (creep strength at elevated temperature, tensile strength),
3. they allow for accurate mapping of the required element
4. they allow for the use of grades with high chromium contents, since these can only be cast, rather than forged.

Heat-resistant castings also have some unfavourable features: larger weight, tendency to brittleness, longer delivery time and a possibility of the occurrence of defects (micro-shrinkage, segregation of Cr, C, N and O).

The properties of the alloy in its operating conditions are dependent on the mechanism of casting surface wear. In one environment, there is a formation of a protective film (scale), and in others, a formation of volatile reaction products that subsequently evaporate from the surface (weight loss occurs under these conditions). It is therefore not possible to describe gas attack at a high temperature with a single mechanism. When selecting heat-resistant alloys for a particular technological process, the following operating conditions may be present: oxidation, sulphidation, carburising, nitriding or ash attack (low-melting eutectics of vanadium and sodium). Appropriate characteristics of alloys have been made for these separate areas of application.

2. OXIDATION - SELECTION OF ALLOYS AND MICROSTRUCTURE

This is described by the metal - gas reaction in which an oxide layer is formed on the surface. This reaction occurs at high temperatures in an atmosphere of oxygen, air, carbon dioxide, sulphur or sulphur oxide water vapour, or industrial gases containing an excess of one of these components.

Chrome forms a tight and thick layer of Cr_2O_3 on the casting surface, protecting the alloy matrix from further oxidation. It is the basic element providing heat resistance to iron alloys. The presence of silicon and aluminium enhances the effect of chromium. A beneficial effect under conditions of cyclic temperature changes is also caused by calcium, beryllium and lanthanum. Hence, the scale is built not of pure Cr_2O_3, but a mixture of oxides of these elements.

The positive effect of the oxide layer occurs when its thermal expansion coefficient does not differ from the thermal expansion coefficient of the matrix material. Another important factor is to avoid phase transitions in the alloy, since these can cause changes in the volume of the alloy and the chipping off of its protective film.

The operating conditions of the casting should ensure the stability of scale, so that the formed oxide layer is not damaged during operation. Therefore, in practice, the aim is not to exceed, either locally or throughout the casting, the highest recommended temperature for the casting, which is usually given in a standard or a catalogue for the steel grade. Local heating of the alloy by a flame is detrimental in this case. Contact of the casting (or part of it) with an element of a different chemical composition is also disadvantageous, because a chemical reaction may occur between the oxide layer and the material. Low-melting slags, low-melting metals, their salts and oxides have a similar effect. Gases, dust, vanadium oxides, salts and alkaline compounds are especially detrimental.

The tendency to oxidation is characterised by the free energy change ΔG^o (standard free energy of the reaction), accompanying the formation of oxide on the alloy surface. Changes in ΔG^o for individual elements, with respect to increasing temperature, are presented in chapter 2 (Fig. **2**). The more negative the value of ΔG^o, the higher the affinity of a given element for oxygen, and the easier the formation of oxide on the surface. All the elements whose ΔG^o is lower than the ΔG^o of iron (ΔG^o_{Fe}) will oxidise earlier than iron, and those whose ΔG^o is higher than ΔG^o_{Fe} will oxidise after iron is oxidised (Co, Ni, Cu are noble elements here - see red line in Fig. (**2**) in chapter 2). It should be emphasised that the chemical affinity of elements for oxygen decreases with increasing temperature. Only carbon has a different tendency, and at high temperatures it has a higher affinity for oxygen (it has a lower value of ΔG^o than the ΔG^o of silicon, manganese, vanadium, niobium and chromium). Under these conditions, it is a reducer of oxides of these elements.

2.1. Kinetics of Oxidation

Changes in the free energy indicating a possibility of the formation of an oxide film on a surface do not ensure that the oxidation reaction proceeds continuously to give a permanent layer. First, oxygen molecules are dissociated on the metal surface and are chemically combined with atoms of elements present on the alloy surface. After creating several monoatomic layers, nucleation of the oxide takes place in privileged places (dislocations, admixture atoms and point defects), until the formation of a uniform and stable oxide layer occurs. However, due to the occurrence of discontinuities on the metal surface (non-metallic inclusions, carbides and grain boundaries) or defects (*e.g.* casting defects, such as micro-shrinkage and micro-cracks), the increase in the oxide film may be heterogeneous. Oxygen molecules can diffuse through these defects to the metal - oxide surface. The oxide film is then porous. However, the oxidation kinetics of the alloy are based on a continuous oxidation of the metal due to diffusion of oxygen through the oxide film.

The most common methods of measuring the oxidation rate are based on changes in sample weight per unit area in a known time period. An example of such tests is presented in Fig. (**1**), indicating that the oxidation process is usually associated with weight gain. Depending on the chemical composition or oxidation conditions, weight gain is slow (OAB segment in Fig. (**1**) or rapid (OF segment). In the first case, these are slow heating conditions at low temperatures (100-200°C), and in the other there is rapid heating in a strongly oxidising atmosphere (rapid heating to the austenite temperature, welding). The OACD line is characterised by oxidation conditions in which the scale is unstable and there are spatters from the alloy surface, (on the oxidation curve, there are visible rapid changes in weight gain). An observation of the casting surface and control of the scale construction in a scanning microscope reveal such areas (Fig. **1b**). Of course, such scales do not constitute proper protection against oxidation, and the alloy cannot be regarded as heat-resistant in industrial conditions.

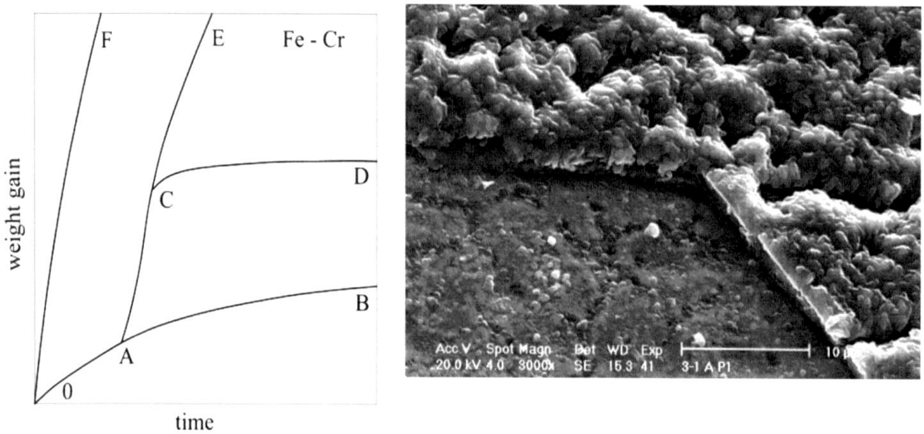

a) b)

Fig. (1). Diagram presenting examples of curves for the kinetics of Fe-Cr alloys [3] (**a**); and an example of the scale spatter on the surface of a single crystal of Fe-Cr17-Al5 (**b**); Magn.3000X [5].

Heat resistance is usually provided by even oxidation of the alloy surface and a stable binding of the oxide film with the matrix. To fulfil the latter condition, it is necessary that the linear expansion coefficients of the alloy matrix and the oxide film are similar. An important parameter of heat resistance is tight adhesion of the oxide film to the matrix and its low penetration by oxygen diffusion.

Appropriate vapour pressures are the condition for the formation of the oxide film. If the vapour pressure of the oxides formed is greater than the oxygen partial pressure in the ambient atmosphere, oxidation does not occur. If the vapour pressure of the oxide distribution is lower than the oxygen partial pressure, the metal is covered with an oxide film; this applies to such elements as Fe, Cr, Si, Al and Ni. Despite the low dissociation pressure, molybdenum is an exception, since it does not form oxides, because it evaporates at high temperatures.

The physical properties of the oxide film are essential for its durability. There are certain requirements that must be fulfilled so that the scale is a protective film:

- it should be compact, without discontinuities, allowing for easy diffusion of oxygen,
- it should adhere to the metal, so that it does not come off (chip off) the alloy surface,
- it should not form low-melting compounds or eutectics,
- it should not facilitate the diffusion of oxygen through the scale and the elements from the matrix to the scale.

The ratio of the molar volume of products formed to the volume of the reacted alloy is the criterion of compactness of scale, and it was named the Pilling-Bedworth ratio (P - B). For compacted scales, the P - B ratio is always higher than one. For example, the P - B factor equals: Cr (3.92); W (3.30); Mn (2.07); Fe (2.06); Co (2.10); Si (2.04); Al (1.28); Ni (1.68) and Cu (1.70) [3 - 5]. Similar data are shown also in Table **1**.

Table 1. The values of P – B ratios for some metals [6].

Metal – oxide	Density of oxide, [Mg/m^{-3}]	Pilling–Bedworth ratio
Fe – Fe_2O_3	5.3	2.1
Ni – NiO	6.9	1.6
Al – Al_2O_3	4.0	1.3
Ti – TiO_2	5.1	1.5
Cr – Cr_2O_3	5.1	2.1
Si – SiO_2	2.7	1.9

The P - B ratio can be only applied to conditions in which the diffusion of the oxidant occurs inwardly (through the reaction product layer), *i.e.* the process of scale accumulation is performed at the internal interphase boundary (important for the oxidation of Ti, W, Zr and Nb). If the process takes place outwardly, *i.e.* scales are formed on the external surface (such as for most metals and alloys), the criterion of P - B is not applicable.

Formation of a protective film does not form a complete protection against further oxidation. Its structure and chemical (and phase) composition depend on the oxidation conditions (time, chemical composition, the surrounding environment and alloy). Therefore, there are several types of scales, according to their compactness. They include:

- porous scales (Fig. **2a**): under these conditions, the chemical composition of the reaction product does not affect the oxidation process, since the flow of oxidant to the metal-scale reaction surface takes place at high speed and through discontinuities in the scale. An alloy forming this type of scale is not a heat-resistant material.
- single or multi-layered scales (Fig. **2b**): several layers in the scale can occur due to differences in the morphological structure (porous, compact) or the phase structure (metal oxides of different valence). For example, iron forms a triple film consisting of the following oxides: FeO - adjacent to the pure metal, Fe_3O_4 - central and Fe_2O_3 - external [7 - 9].
- homo- or heterogeneous scales: in two- and multi-component alloys, several components (metals) may form compounds with an oxidant. If two compounds dissolve mutually without limitation, then the single-phase scale remains. If there is no mutual solubility, the less noble metal becomes oxidised. In this case, we are talking about selective oxidation, in which a scale consists of:
    - metal oxides B (BX - less noble) and metallic precipitates of more noble metal A.
    - heterophase mixture of oxides AX + BX, if both metals are oxidised (Fig. **2c**). If the more noble metal passes into scale in smaller quantities than the other metal, *e.g.* B, it means that component B is subject to preferential oxidation.
    - scales with internal oxidation zones.

If oxygen passes through the scale and dissolves in the alloy matrix, it may react with an element with a high affinity for oxygen, forming areas with different chemical composition, *e.g.* oxides in the matrix. This process is called internal oxidation.

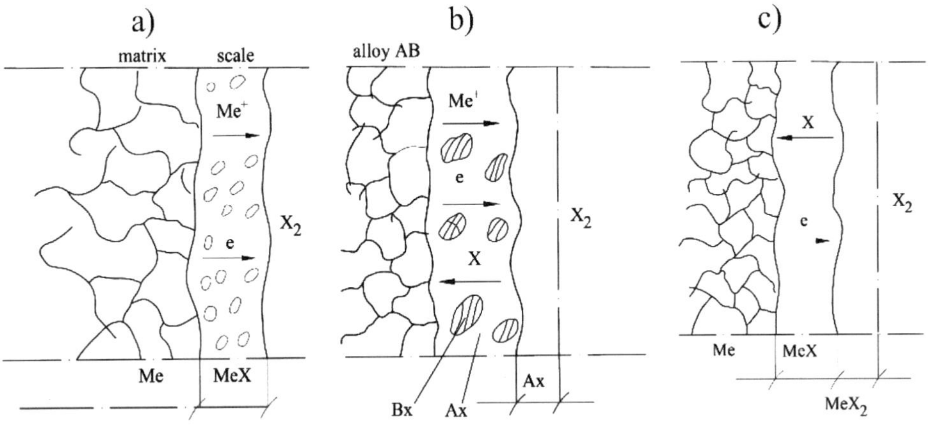

Fig. (2). General division of scales and the mechanism of their growth [5, 7]

2.2. The Mechanism of Scale Formation

Despite the existence of a stable and compact oxide film (scale) on an alloy surface, the corrosion process still occurs as a result of diffusion of the attacking gas through the scale. Since, at high temperatures, the diffusion rate of elements in crystalline solids is many times slower than the rate of chemical reactions forming the oxides, the rate of the whole corrosion process is determined by the transport of matter in the scale. The higher the rate of oxidant diffusion, the faster the oxidation of the alloy is. The whole process of formation of oxide film consists of several stages:

1. delivering the oxidant to the scale – atmosphere interphase (through mass transport),
2. incorporation of oxygen into the scale,
3. transport of metal from the alloy matrix onto the metal - scale surface,
4. incorporation of metal into the oxide scale,
5. transport of metal and oxygen through the scale.

The course of these partial processes is controlled by redox reactions:

$$Fe = Fe^{2+} + ne^{-} \quad \textbf{(1)}$$

$$2e^{-} + \frac{1}{2}O_2 = O^{2-} \quad \textbf{(2)}$$

These reactions occur rapidly and are difficult to control. Diffusion of oxygen or another element through the scale is assisted by defects in the crystal lattice of the oxide layer, which can be vacancies, micro-cracks or ions in the interstitial spaces (Fig. **2** and **3**).

According to C. Wagner's theory [9], diffusion of elements through the scale occurs at high temperatures as a result of ions passing, along with electrons, through ion and electron lattice defects. In the case when scale is formed by compounds of a defective cationic lattice (FeO, Fe_3O_4, NiO or Cu_2O), the oxidation process occurs due to an outward diffusion of ions and electrons from the metal. In the case of a defective anionic lattice, the oxidation is performed by inward diffusion of ions and oxidant electrons. There is also a possibility of scale formation as a result of the total diffusion of both reagents in both the opposite directions (*e.g.* TiO) [10].

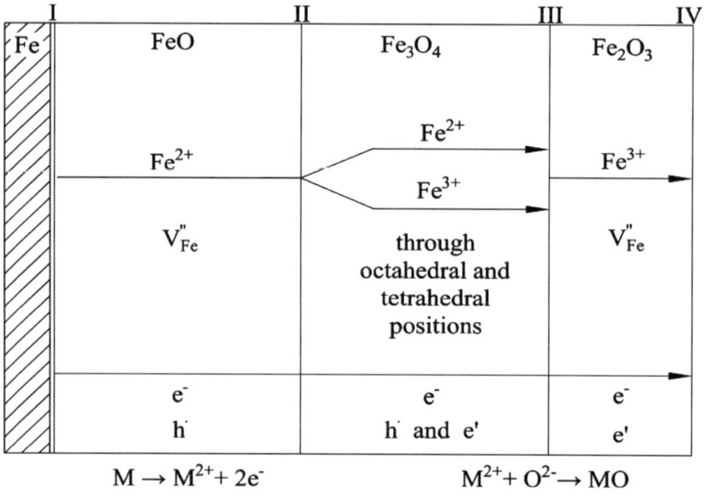

Fig. (3). Formation of oxide film on iron surface [10].

The analysis of the experimental results indicates that the higher the concentration of lattice defects in the scale, the faster the alloy is oxidised. This is also confirmed by the fact that the easier the formation of scale, the higher the deviation of the compound from the stoichiometric composition.

The above-presented mechanism of scale formation is more complex in practice [7], since many metals form several oxides (*e.g.* Fe - FeO, Fe_3O_4, Fe_2O_3, Cu – Cu_2O, CuO). As already mentioned, the higher the oxidation rate of a metal, the higher the dissociation pressure of its compound with the oxidant. If the oxidation reaction takes place at an oxidant pressure lower than the dissociation pressure of the compound with a higher degree of metal oxidation, the scale is single-phase. In slightly different conditions, *e.g.* at the same temperature and higher pressure, a two- or a multi-phase scale can be formed. Fig. (**3**) presents an example of the formation of such a film, indicating that the external coating is formed by the oxide in which the metal ion has the highest degree of oxidation (Fe_2O_3). The thickness of each of these coatings depends on their growth rate. Since the oxide with the largest number of defects, and the largest deviation from the stoichiometric composition, grows the fastest; therefore, the wüstite phase (Fe_xO) is the thickest. At the same time, the chemical composition of individual phases and their mutual conditioning are affected by the temperature at which there is oxidation of the alloy (the mobility of lattice defects depends on the temperature).

The kinetics of alloy oxidation are therefore of a complex nature, linked to the mechanism of structural changes in the growing scale. For ferrous alloys with a chromium content lower than 14%, the oxidation rate follows the curve O-F (Fig. **1**). With this type of oxidation, a hetero-phase scale is formed, built from chromium-iron spinel ($Fe_{3-x}Cr_x$) O_4 and iron oxides (wüstite). By increasing the concentration of chromium or reducing the operating temperature of the casting, there is a transition from the hetero-phase to a single-phase scale [10]. For alloys with a content of 14 - 18% Cr, oxidation takes place according to the curve O--C-E or O-A-C-D. In the conditions corresponding to the line O-A, there is an incubation period preceding the scale puncture (A-C-E). The incubation period depends on the oxidation temperature and the presence of water vapour in the atmosphere. The occurrence of the scale puncture is due to the formation on the coating of Cr_2O_3 spinel, which can be removed by the re-creation of a pure coating of Cr_2O_3 (segment C-D).

Alloys containing 25-30% Cr form (single-phase) scale, corresponding to Cr_2O_3, with a trace amount of haematite - up to 2%. These films are more stable and have better protective properties than those containing chromium spinel. The oxidation process of these alloys is characterised by the O-A-B curve (Fig. **1**).

As is clear from numerous studies [9, 11], the mechanism of scale formation is based on the occurrence of crystallographic lattice defects within the scale and ions contained in the interstitial spaces. If the oxide crystal lattice has excessive cationic vacancies, the compound has an excess of metal. However, in the presence of anionic vacancies, the compound has an excess of oxidant. If cations or anions are present in the interstitial spaces, the compound has an excess of metal or an excess of oxidant, respectively.

Oxides and sulphides of most metals have cationic lattice defects (Fe_3O_4, FeO, Cu_2O, NiO), and in high-melting metal oxides, defects occur in the anionic lattice (ZrO_2, WO_3, NbO_2). Since, in accordance with the theory of metal oxidation at high temperatures, diffusion of the components is performed by shifting ions (see Wagner [9]), along with electrons, through ion and electron defects; therefore, in scales with a defective cationic lattice, the oxidation process takes place as a result of the outward diffusion of ions and electrons (Fig. **2a**), and, in scales with mostly anionic defects, by the inward diffusion of the ions and electrons of the oxidant (Fig. **2b**). If there are cationic and anionic defects, oxidation may take place by inward diffusion of the oxidant and outward diffusion of metal (Fig. **2c**).

The oxidation mechanism of iron alloys is more complex, due to the influence of admixtures, grain boundaries, presence of other phases, *etc*. At a temperature above 570ºC, wüstite scale is formed, while haematite forms only at about 1300ºC. In the binary system Fe-O at temperatures above 570ºC, the following scale is then formed: $Fe\text{-}FeO\text{-}Fe_3O_4\text{-}Fe_2O_3\text{-}O_2$. At the Fe - FeO boundary, there is a process of iron passing from the matrix into the wüstite phase. This is performed by filling cationic vacancies in the wüstite with bivalent iron ions and filling electron holes with electrons. Through diffusion of Fe ions and electrons to the second film (Fig. **3**), the wüstite phase grows, and magnetite decomposes to FeO. Further diffusion of Fe ions and electrons is performed through the second film to the third film, where the magnetite phase grows. In turn, at the external boundary IV, there is the incorporation of oxygen ions in the crystal lattice of haematite and transport of Fe^{+3} ions.

In Fe-O alloys with alloy additions, *e.g.* Fe-Cr-O, the presence of wüstite is a confirmation of the low oxidation resistance of the alloy. According to [12], in the metallic core, only thermodynamically stable phases can be present. In the two-component system Fe-O, the simultaneous presence of several phases, iron, wūstite and magnetite, occurs at a given temperature and pressure. Introduction of a third component (*e.g.* Cr) causes a shift of the quadruple point to higher temperatures. This means that the wūstite causes an increase in the oxidation rate to occur at a higher temperature than for the pure composition Fe-O. This has been confirmed experimentally [12].

This mechanism of metal oxidation refers to those conditions in which the scale is a solid solution of oxides (for low-alloy steels). Deviation from these conditions is also caused by small quantities of other elements, such as carbon.

2.3. The Rate of Oxidation

The oxidation process of metals and alloys is dependent on the products formed by the metal-oxidant reaction. At high temperatures, the oxidation rate grows due to increasing free energy of oxidation (Fig. **1**), according to the reaction:

$$M_s + O_g = MO_s + \text{energy g- gas; s- solid} \quad \textbf{(3)}$$

The oxide film formed protects the metal against further oxidation. To maintain the continuity of oxidation reaction, oxygen must diffuse through the film to the metal. The course of this reaction may be:

Linear (Fig. 4)

If the product is volatile, liquid (low-melting eutectics) or solid (with numerous pores, cracks), the oxidation rate is conducted according to the linear law (Fig. **4**):

$$\frac{dM}{dt} = k \qquad \textbf{(4)}$$

giving $\Delta M = k \cdot t$,

where: M – weight loss,
t - reaction time,
k - constant

In the first two cases (gas or liquid products), the reaction product is formed by oxidation on the surface of the oxidised metal, and in the latter one the flow of oxidant through the pores to the oxidised surface is so rapid that the reaction at the metal - scale boundary, rather than the diffusion process, is the slowest oxidation process.

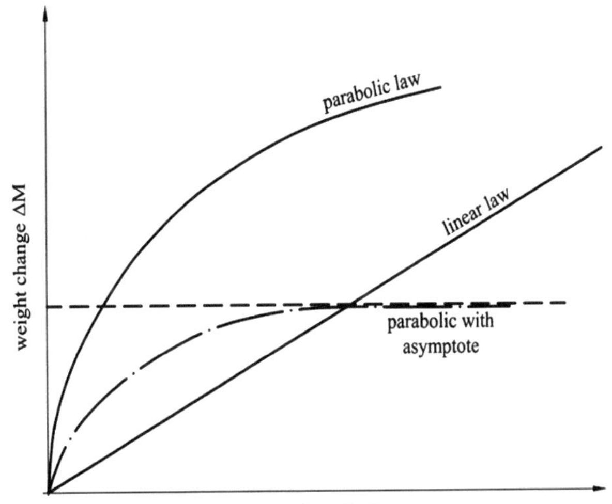

Fig. (4). Different forms of kinetic oxidation in metals.

Under other conditions, the most common in practice for compact scale, a layer of scale separating the oxidant from the metal is formed on the reaction surface. In early stages of oxidation, when the protective layer is still thin, the diffusion rate of oxidant may be still high. However, over reaction time, the transport route grows and diffusion becomes the slowest partial process, depending on the degree of scale defect and temperature.

Parabolic

If the scale is compact, strongly associated with the metal matrix, weight gain per unit is of a parabolic nature (Fig. **4**), called the parabolic oxidation law:

$$\frac{dM}{dt} = \frac{k_p}{M} \quad \textbf{(5)}$$

where: k_p - parabolic constant of oxidation rate [$kg^2/m^4\ s$].

After formation of oxide film on the metal surface, the oxidation rate is determined by the diffusion rate (of oxygen atoms) through the scale to the metal surface and the metal atoms to the surface. Under certain conditions of temperature and alloy chemical composition, the oxidation process takes place according to the law:

$$\frac{dM}{dt} = \frac{k'_c}{x^2} \quad \textbf{(6)}$$

where: x - scale thickness,
k'_c - exponential constant of the oxidation rate

Depending on the oxidation conditions, the exponent can take various values (even fractions).

At low and medium temperatures (up to 700°C), the oxidation process may take place according to exponential dependencies (*e.g.* following a logarithmic law). Under such conditions, the corrosion rate decreases over time due to the very good protective properties of the scale. However, these types of oxidation conditions are rare, especially at lower temperatures.

The most stable oxide films are those whose diffusion coefficients are low: Cr_2O_3 on steels or SiO_2 on silicon cast iron [11]. Characteristics of scale stability should always take into account service conditions. For example, in steels, a film of Al_2O_3 is twice as stable as a film of Cr_2O_3 [11], but alloys of Fe-Cr-Al cannot be used at temperatures above 500°C, due to their poor creep resistance [13]. To ensure good creep resistance, good strength and oxidation resistance at temperatures above 900°C, steels or cast steel such as: Fe-(20-25%) Ni-(12-15%) Cr-(3-4%) Al-(1-3%) Nb (%), must be used [13].

The kinetics of alloy oxidation depend significantly on the temperature. In a range of temperatures corresponding to the kinetic law, the oxidation rate is characterised by the Arrhenius equation:

$$k = A\exp(-\frac{E}{RT}) \quad \textbf{(7)}$$

where: k - constant of the oxidation rate,
R - gas constant,
A - constant,
E - process activation energy.

The value of activation energy E refers to the slowest process determining the scale growth rate. It is not a measure of the oxidation rate and most often does not describe the mechanism of oxidation, because scale formation is subject to a number of other processes, particularly with multi-component alloys. In contrast, it describes the oxidation mechanism well in the temperature range in which the kinetic law is important.

In various temperature ranges, the same metal or alloy is subject to various kinetic laws. For example, iron is subject to the linear law up to 200°C, and later, up to temperatures of 1100°C, to the parabolic law. On the other hand, nickel is oxidised to 300°C by the logarithmic law, at 300 - 400°C by the law of exponent 3 and above 400°C by the parabolic law.

In the case of obtaining solid oxidation products on the metal surface, the oxidation rate is not characterised by the scale thickness, but by the weight gain, Δm per unit area q (Δm/q). For example, for Fe-Cr alloys, changes in the weight gain may take place in accordance with Fig. (**1**). In Fe-Cr alloys containing up to 14% Cr, oxidation occurs according to the curve OF, and in alloys containing 20-25% Cr, according to the curve OAB. Such a small weight gain indicates the good protective properties of the film of Cr_2O_3. The OACE curve indicates that after the initial incubation period (OA), there is a puncture of scale and the oxidant reaches the metal layer freely. The oxidation is then performed in a cumulative manner. Under certain conditions, there may be a stable film of Cr_2O_3 formed again across the puncture site, and oxidation will occur according to the curve OACD.

In oxidation processes, weight loss can also occur, (Fig. **4**), since some metals form gaseous oxides. This applies to tungsten, which at temperatures above 1300°C forms the volatile oxides WO_3 and WO_2.

Data mentioned here characterise the oxidation of metals and alloys in general terms. More detailed information can be found in [7, 9, 12], and in [1 - 3] for steel castings.

3. HEAT-RESISTANT CAST STEELS

Division of this group of cast steel is made either with respect to the chemical composition, or in accordance with the matrix microstructure. According to the first division, castings intended for work at high temperatures can be divided into four groups:

- chromium cast steel (Fe-Cr) used in the temperature range 873 - 1473K at low static loads and uniform heating at the above mentioned temperatures;
- chromium - nickel cast steel (Fe-Cr-Ni) having not only good heat resistance, but also adequate creep resistance, necessary when there are mechanical loads on the castings. The presence of nickel is the basis for classifying the cast steel as a heat resistant alloy. As austenitic cast steels, they have higher strength and plasticity than the previous group of cast steel. They work under oxidising and reducing conditions;
- nickel-chromium cast steel (with predominance of nickel over chromium), has a high nickel content (over 25%), used under carburising and nitriding conditions. Also resistant to changes in temperature and to an atmosphere containing more than 0.05% H_2S [14];
- Ni-Fe-Cr alloys, used particularly under carburising conditions.

The most common heat-resistant cast steels grades are presented in Table **2**. It shows that the primary function of the heat-resistance is fulfilled by chromium, silicon, nickel and molybdenum. It is chromium that is the most important element with regard to improving heat-resistance, and its function consists in the formation of an oxide film of Cr_2O_3, closely adjacent to the alloy matrix. In pure iron heated above 570°C, an FeO film (wüstite) is first formed, followed by magnetite Fe_3O_4 and, on the outside, haematite Fe_2O_3. In Fe - Cr steels containing more than 12%Cr, the Cr_2O_3 oxide film is formed under the outer film of Fe_2O_3 [6].

Nucleation of Cr_2O_3 is favoured due to the higher chromium content in the alloy and the increase in temperature. The beneficial effect of chromium is revealed by the oxidation rate of Fe-Cr alloy, which is best in the temperature range 900 - 1200°C (Fig. **5**). Fe-Cr alloys containing about 25% Cr have the lowest oxidation rate. Characteristically, the minimum is independent of the temperature.

Table 2. Chemical compositions and characteristics of heat-resistant cast steels [3, 5].

Desig. ASTM	max. wt.%							Temperature Limits [°C]
	C	Si	Mn	Cr	Ni	Mo	Service Conditions	
HA	0.20	1.0	0.35-0.65	8-10	-	0.9-1.2	oxidation resistant.	650
HC	0.50	2.0	1.0	26 - 30	4.0	0.5	sulphur and oxidation. res. low UTS YS;	1095
HD	0.50	2.0	1.50	26 - 30	4-7	0.5		
HE	0.2-0.5	2.0	2.0	26 -30	8-11	0.5		
HF	0.2-0.4	2.0	2.0	18 - 23	8-12	0.5		815
HH	0.2 - 0.5	2.0	2.0	24-28	11-14	0.5	oxid. res.; good strength	1090
HI	0.2-0.5	2.0	2.0	26-30	14 - 18	0.5	Improved oxid. compared to HH	
HK	0.2 - 0.6	2.0	2.0	24 - 28	18 - 22	0.5	good strength up to 1150°C, good res in oxid. & reduc. conditions	1150
HL	0.2 - 0.6	2.0	2.0	28 - 32	18 - 22	0.5	improved sulphur res. compared to HK	
HN	0.2 - 0.5	2.0	2.0	19 -23	23 - 27	0.5		
HP	0.35-0.75	2.5	2.0	24 - 28	33 - 37	0.5	resistant to oxid. and carburising atm.	1090
HT	0.35–0.75	2.5	2.0	15 - 19	33 - 37	0.5	used in thermal shock condit.; oxid. & red. flue gases; salts	1150
HU	0.35–0.75	2.5	2.0	17 - 21	37- 41	0.5	better strength than HT	
HW	0.35–0.75	2.5	2.0	10 - 14	58 - 62	0.5		1120 in oxid. atm. and 1040 oxid & red. products of combustion
HX	0.35–0.75	2.5	2.0	15 - 19	64 - 68	0.5	hot gas corr (in air, carburising gases, combustion gases, flue gases	1150

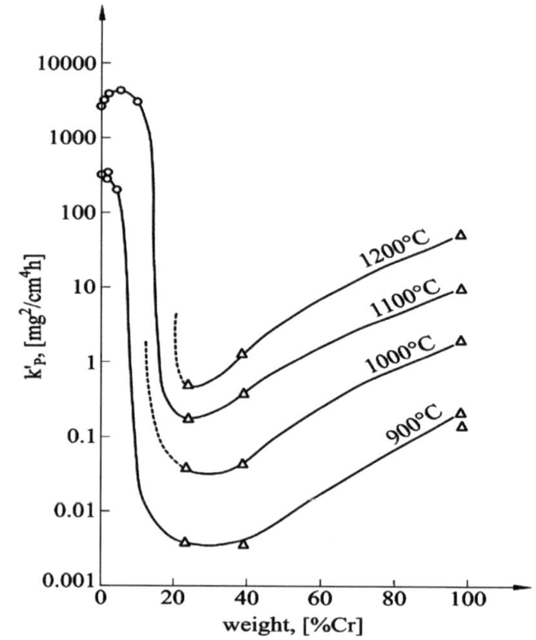

Fig. (5). Chromium and niobium carbides along a grain boundary [15].

The effect of **nickel** is revealed more in creep resistance than in heat resistance. A large addition is used in two cases: when a pure austenite structure is required or to ensure high temperature strength. The application area of high-nickel cast steel is however restricted by the presence of H_2S in the working environment of castings, since such a reducing atmosphere leads to the creation of low melting eutectics Ni-S ($T_t = 1060K$) or Ni-NiS ($T_t = 917K$).

Carbon acts similarly to nickel. Increasing its content raises the share of austenite and creep resistance. The upper limit of carbon concentration should not exceed 0.4% [2], due to the precipitation of carbides at grain boundaries, especially when the casting stays at high temperatures for an increased period of time (Fig. **6**).

A higher carbon content (0.6-0.8% C) is, however, necessary in case of additional requirements, namely, high abrasion resistance for heat-resistant cast steel, *e.g.* conveyor rollers in a catalytic device or a tunnel furnace.

Silicon has similar effects to chromium. Heat-resistant cast steel includes 2.5% of Si, which leads to a reduction of its plasticity at high temperatures (due to its brittleness, possibility of a σ phase forming). An increase in the concentration of silicon above 2% does not provide an increase in oxidation resistance, either.

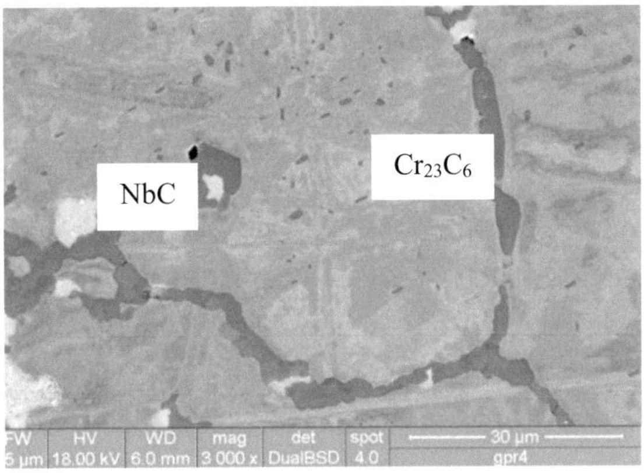

Fig. (6). Chromium and niobium carbides along a grain boundary [15].

Manganese does not affect heat resistance, and its addition promotes the formation of the austenitic structure. Nitrogen, niobium, and tantalum act similarly, stabilising the austenite.

Molybdenum plays an important role in heat-resistant alloys, because even a small addition increases the high temperature strength. Under alkaline conditions in the presence of molybdenum, low-melting compounds are formed, causing deep pits on the casting surface.

Aluminium has a positive effect as an alloying addition, since it creates a film of spinel containing Al_2O_3, which adheres well to the alloy matrix (especially in Fe-Cr-Al cast steel). Aluminium oxidises selectively. In the presence of 23.5% Cr, 1.8% Al and 1.3% Si [10, 12] at a temperature of 1473K, the scale formed contains 49.7% Al, and 3% Cr (without the presence of iron and silicon). The influence of aluminium on the oxidation rate of Fe-Cr alloys is presented in Fig. (7). To ensure a constant oxidation rate, higher amounts of aluminium at lower concentrations of chromium (at 17% Cr, at least 3.5% of Al at a temperature of 1100°C) should be added into the steel. These data are also the basis for intensively developed studies on new alloys of the formula Fe-Cr17-Al 6 [5, 14]. Their chemical composition is optimised by additions of Mo, V and Nb.

Tungsten and cobalt increase the creep resistance of chromium-nickel cast steel. Cast steels with the addition of these elements are used on heavy-duty castings, operating at temperatures above 1273 K (*e.g.* pipe clamps of cast steel with a chemical composition as follows: 0.35%C, 0.5%Mn, 1.5%Si, 26%Cr, 35% Ni, 15%Co and 5%W) [2].

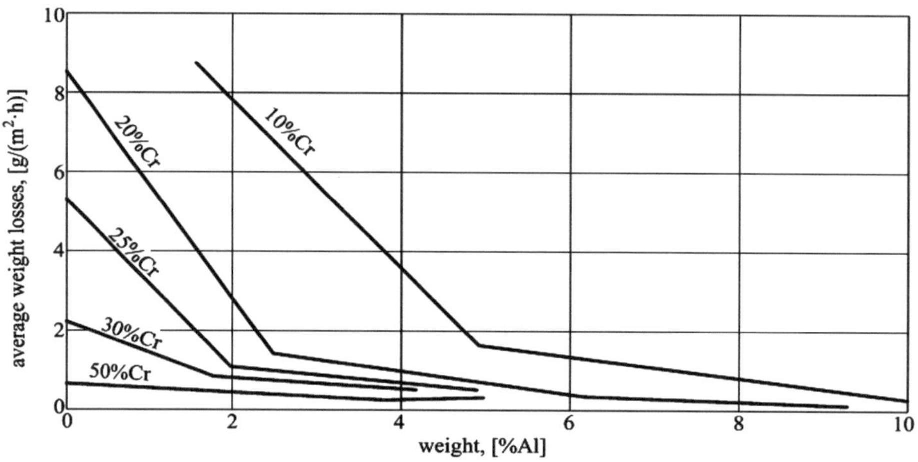

Fig. (7). Effect of aluminium on the oxidation of Fe-Cr alloys [3, 16].

The addition of **yttrium** is of particular interest. A small concentration of this element in chromium steel (0.02-0.10% Y) results in a slower oxidation rate of alloys at temperatures above 1473 K [13, 14].

4. MICROSTRUCTURE OF HEAT-RESISTANT CAST STEELS

Steels for heat-resistant castings are modelled on previously developed heat-resistant steels. In terms of the microstructure formed, these steels are divided into:

1. martensitic steels
   - containing 5 - 6% Cr and additions of Mo and Al; oxidation resistant to 873 - 1073K, applied in the petrochemical industry,
   - at 7 - 10% Cr, and in the presence of Mo and Si, applied on valves,
   - at 13% Cr - used on turbines and valves.
2. ferritic steels
   - containing 17% Cr with the addition of Si; resistant to a temperature of 1173K and applied on burners and parts of annealing furnaces,
   - at 27 - 30% Cr, and often 3 - 5% of Al, applied in a sulphur atmosphere at temperatures up to 1473K.
3. austenitic steels
   - containing 18% Cr and 9% Ni, resistant to a temperature of 1173K,
   - containing 25% Cr and 12% Ni, resistant to temperatures of 1373 - 1423K,
   - containing 25% Cr and 20% Ni with the addition of silicon, resistant to 1423K.

The group of martensitic steels is used in those industries where there is an effect of hydrogen or hydrocarbons (especially at temperatures up to 773K and under pressure). The stability of the mechanical properties of the cast steel grades up to 773K is very often used. Their most common uses are in pipes for the oil and petrochemical industries, in equipment for chemical synthesis and elements of steam boiler heaters. Due to the increased resistance of these steels to the action of sulphur (sulphur compounds) compared to cast steel of the 18 - 9 type, they are used for castings in the oil industry.

Cast steels containing 13% Cr also belong to the group of martensitic cast steels. Depending on the carbon content, three subgroups can be distinguished here [16, 17]: below 0.15, from 0.2-0.4 and from 0.6-1.0% C. Nickel (cast steel 13 - 4) is usually introduced as an alloying addition. Castings of this steel are widely used for massive elements and operating at low temperatures, due to their low critical quenching rate. In thick-walled castings, and with various wall thicknesses, a uniform structure and properties are obtained.

Ferritic steels of both groups (<0.12% C and 16-18% Cr, and <0.35% C and 24-30% Cr) have chromium ferrite and varying amounts of chromium carbides in their structures. This structure is not subject to change when heated up to temperatures of 1273K. In the upper range of these temperatures, the carbides are dissolved in the ferritic matrix and the grain grows. With the increase in the chromium content from 13 to 18%, the impact toughness decreases as a result of shifting the brittleness threshold to higher temperatures.

Due to the lack of any α-γ transformation, heat treatment of ferritic cast steel consists in normalising annealing at a temperature of 1073 - 1173K. After this treatment, the microstructure consists of ferrite and carbides. Cast steel containing 17% Cr is resistant to oxidation up to 1123K. It also has another valuable advantage: the thermal expansion coefficients of the scale and the matrix are virtually identical. As a result, the scale does not come off during the cyclic heating and cooling of castings (as opposed to Cr-Ni cast steel). An increase in the heating temperature, *e.g.* to 1273K, leads to the dissolution of carbides and precipitation of austenite. During rapid cooling in water, martensite is then obtained besides ferrite.

Cast steels containing 24-30% Cr are subjected to the same heat treatment (normalising 1123K/air). Cooling in air eliminates the formation of brittleness '475', which high-chromium cast steel is frequently subject to. The structure of cast steel containing 26% Cr is composed of ferrite and carbides precipitated both at the grain boundaries and in the alloy matrix. Very often, the grain boundaries reveal precipitation of another phase. According to L. Colombier and J. Hochmann [5], the presence of carbon, nitrogen and oxygen can lead to the formation of austenite after long-term heating at high temperatures.

Very often, cast steels containing 25-28% Cr have higher concentrations of carbon (0.2-0.4%). The structure of this cast steel is then composed of ferrite and carbides, precipitated in inter-dendritic spaces Fig. (**8**).

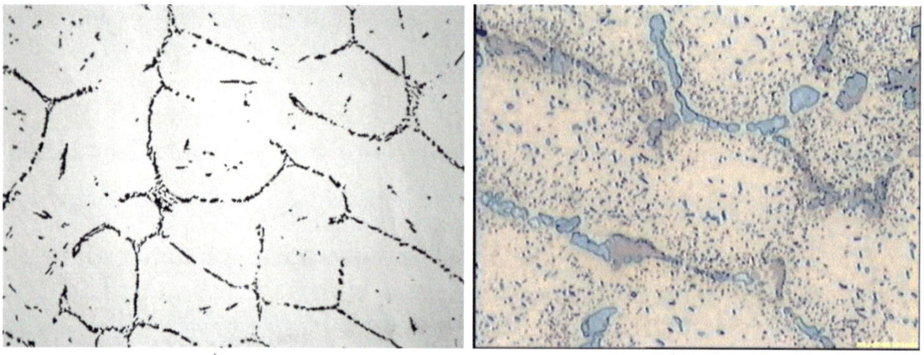

a) b)

Fig. (8). Chromium carbides on the prior ferrite grains in Fe-Cr26 cast steel (a- Magn. 400X) and in Fe-Cr 20-Ni32Nb (b – Magn. 2,000X) [15].

The increase in carbon concentration to 0.35% does not lead to deterioration of the oxidation resistance, and significantly extends the range of application of castings with such a chemical composition.

Fe-25%Ni-20%Cr cast steel has a more stable structure than Fe-Cr-Ni cast steel and is less sensitive to the presence of brittle phases (*e.g.* σ, χ). With the increase in the nickel content, the tendency to carburise and the high temperature strength decrease. Along with the increase in the nickel content, the resistance to thermal shock and thermal fatigue also increases. The increase in chromium causes an increase in the oxidation resistance and in the resistance to gas corrosion (SO_2).

5. THE EFFECT OF TECHNOLOGICAL FACTORS ON THE IMPROVEMENT OF HEAT RESISTANCE

In the first place, the effect of fragmentation of the structure must be mentioned here. Oxidation of the alloy matrix takes place by the diffusion of reagents at grain boundaries. This was practically confirmed by increasing the heat resistance of alloys by preventing the growth of grains at the operating temperatures of these alloys. Introduction of calcium, strontium, barium and rare earth elements into alloys significantly increased their heat resistance. For example, calcium (0.2%) increased the stability of the alloy Ni80-Cr20 four times and that of misch-metal ten times [18].

Both in Ni-Cr type 80-20 and 65-18 alloys, as well as in Fe-Ni35-Cr20 and Fe-Cr30 alloys, the strongest influence is made by cerium, calcium, boron and molybdenum. Some of them are modifiers and lower the grain size significantly, but not all; *e.g.* molybdenum does not change the grain size. That is why the theory on the beneficial effect of a fragmented structure cannot be applied to all additions.

Therefore, the theory that the beneficial effect on heat resistance results from those elements that have a large atomic volume should be considered more promising. Thus, these large atoms fill all the gaps in the crystal lattice of chromium oxides and inhibit the diffusion of metal atoms inside the alloy surface.

Another technological parameter is the type of the matrix and the form of the carbide occurrence. Rapid cooling from temperatures close to the solidus temperature removes all signs of segregation. However, subsequent heating at the operating temperatures of casting causes precipitation of carbides from supersaturated solid solution areas. They reduce the plasticity of the alloy, creep strength and corrosion resistance. Carbides are most unwanted when precipitated at the grain boundaries. Such a network of carbides facilitates selective attack (of oxygen, sulphur or H_2S) from the atmosphere in which the casting operates.

Fig. (9). Effect of modification by Fe – Ti on the primary grain size [19].

There have been studies carried out [18] on the modification of the structure and carbides in cast steel (0.4%C-26%Cr) during its solidification. An example of significant fragmentation of the structure in this cast steel is presented in Fig. (**9**). Introduction of Fe-Ti into the ladle during the taping of the steel removes brittle fractures in castings, giving a very fine structure [19]. Modification of the microstructure of 28%Cr chromium cast steel was used to eliminate cracks in casting grates, working at 900°C in tunnel furnaces. The effect of Ti, Nb or Fe-CaSi is based on the formation of heterogeneous crystallisation nuclei in liquid steel. Their action results in an increase in the amount of grains, and the process can be controlled and confirmed before tapping into moulds, which is important for engineers in practice [18, 19].

An introduction of niobium, titanium or calcium-silicon into the cast steel also increases concentrations of these elements in the inter-dendritic spaces. Due to the higher affinity of Nb for carbon than chromium, fine precipitations of NbC are also formed, which reduce the carbon concentration in the matrix and increase the concentration of chromium. Change in the form of occurrence of carbides (from $M_{23}C_6$ into NbC) and significant fragmentation of the structure increase the plasticity and creep resistance of the alloy. Thereby, the working time of castings is extended and the costs lowered by reducing cracks and tears in the castings.

The network of carbides at primary grain boundaries is generally undesirable, because it causes an increase in brittleness, and during long-term operation of the casting, it facilitates the formation of brittle phases.

6. SULPHUR CORROSION

The main factor leading to destruction of the oxide film or oxidation of the alloy matrix is the atmosphere in which the casting operates. The protective oxide film is formed only in an oxidising atmosphere, *e.g.* with an excess of air or oxygen, and when sulphur is in the form of SO_2. In contrast, a reducing atmosphere (sulphur in the gaseous form, H_2S) destroys the protective film and leads to changes in the structure of the alloy. The presence of sulphur in the environment (both in the form of SO_2 and H_2S) significantly increases the oxidation rate. This is visible in Fig. (**10**), presenting the effect of sulphur on the oxidation of an alloy with the composition 0.15% C, 9.5% Cr and 12% Ni. At a temperature of 1173K, the presence of 0.5% SO_2 in the coke oven gas surrounding the casting does not lead to oxidation of the alloy, while 0.5% S increases losses due to the reaction of sulphur with the alloy. Change in the environment from an oxidising into a reducing one increases the mass losses of the casting significantly.

A similar detrimental effect is caused by particles of ashes (V_2O_5 and Na_2SO_4) and cement in the presence of alkaline components, since chromium oxide reacts with bases forming low-melting chromates.

The formation rate of sulphide scales is definitely higher than the formation rate of oxide scales, because sulphide scales have more defects in their crystal lattices. Also, due to the low melting temperatures of sulphides, there is a tendency to disrupt the protective role of such a coating. It was found [20] that below 1000°C in the liquid state, the following eutectics can already exist: Ni-Ni_3S_2 (T_m = 635°C); Fe-FeS (T_m = 988°C); Co-Co_4S_3 (T_m = 877°C); FeO-FeS (T_m = 940°C);

CrS-Cr_2S_3 (T_m = 1350°C) [17, 21]. In addition, the sulphides NiS and Ni_3S_2 melt at temperatures of 810 and 787°C respectively. Therefore, the presence of nickel in cast steel working in a sulphur atmosphere is in some cases detrimental.

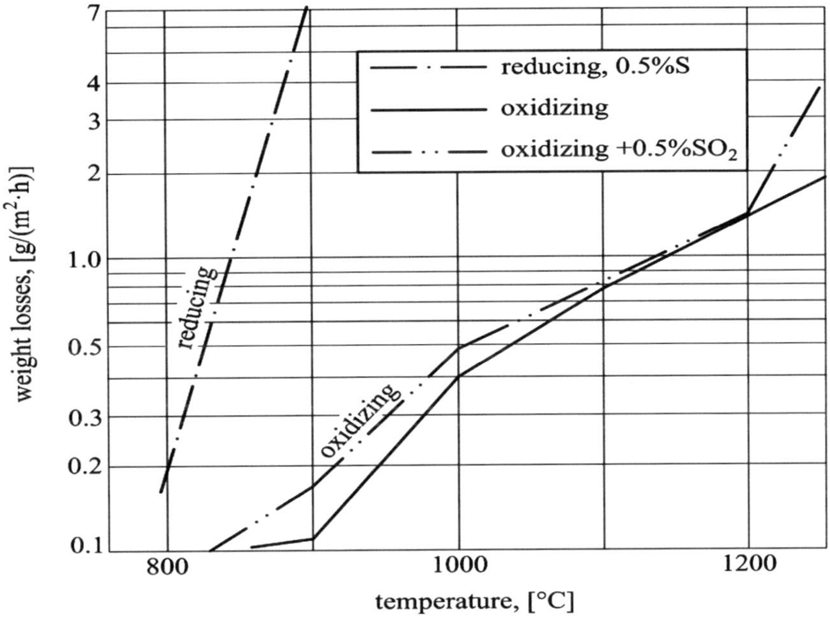

Fig. (10). Oxidation of steel containing 0.15% C-19.5% Cr-2%Si-12% Ni in an atmosphere with equal sulphur content [2, 22, 23].

It should be noted that scales can be composed of oxides and sulphides, because under certain conditions, oxidation occurs in gases containing air and sulphur dioxide. For these reasons, a distinction between oxidising and reducing conditions can be difficult. It is assumed [6] that reducing conditions occur when there are no chromium oxides in the scale. Typical environmental conditions, in which there is sulphur corrosion or sulphur corrosion with oxidation, are primarily these working conditions, in which there are mixtures of air and SO_2 present. In contrast, atmospheres containing hydrogen - H_2S, or flue gases including sulphur and carbon gasification products, create reducing conditions. If in a casting or another element of a machine, there are precipitates containing sulphur, corrosion occurs intensively, even when there is oxygen in the ambient atmosphere. At too low a concentration of oxygen in the environment or too high a concentration of sulphur, chromium sulphides, rather than oxides, are formed on the surface. There are data [12, 13] suggesting that in order to maintain good resistance to sulphur corrosion, the content of chromium should be above 25% and that of nickel less than 20%.

The presence of sulphur dioxide in the corrosive environment causes an increase in the scale growth rate (compared to pure oxidation), due to the formation of chromium sulphides under the protective layer of Cr_2O_3. The mechanism of formation of sulphide and oxide scales is based on a temporary increase in the activity of sulphur on the outer surface of the scale. Areas of increased activity are responsible for nucleation and growth of sulphides, formation of discontinuities in the scale, through which there is transport of SO_2 to the metal-scale surface.

The exact mechanism of sulphide scale formation is the subject of ongoing discussions [6, 7], since it has been stated that sulphide scales depend on such environmental conditions as: temperature (a change in temperature of 50°C leads to significant modifications in scale morphology [20]), the presence of free sulphur reacting with the chromium of the alloy matrix [6], and the presence of nickel, which may form films of Ni_3S_2 in the scale [7]. An important factor in corrosion in air - SO_2 atmospheres is thc introduction of water vapour particles. Under these conditions, the corrosion rate increases, but its mechanism is not known. There are suggestions [6] that water vapour reduces the scale plasticity followed by its separation from the matrix.

The maximum operating temperatures of the selected alloy grades in atmospheres of SO_2 are given by [6] in Table **2**. In H_2 - H_2S reducing atmospheres (catalytic pipes, desulphurisation devices), there are sulphide scales with various chemical compositions. If the alloy contains less than 20% Cr, sulphides of iron and chromium are formed. The parameter controlling this process is the transport of iron and chromium ions by the scale, subject to the parabolic kinetic law.

Under reducing conditions, Fe-Cr alloys with a high chromium content (more than 18% Cr) and additions of aluminium and silicon are used, which increases the stability of the scale. In the presence of nickel in the alloy, low-melting eutectics Ni - Ni_3S_2 are formed, which lead to rapid corrosion. For this reason, many conditions affecting the behaviour of various alloys in atmospheres of hydrogen sulphide were developed and the corrosion rates of these alloys under such conditions, at variable temperatures and concentrations of H_2S, were calculated. It is generally accepted [6] that austenitic steel (cast steel) (0.25% C - 25% Cr - 20% Ni) has beneficial properties at high temperatures and an H_2S atmosphere. The corrosion rate of austenitic steel under these conditions is given in Fig. (**11**). Good resistance to H_2S environments is revealed by the following grades of steel (and cast steel): 18Cr-9Ni, 25Cr-12Ni and 25Cr-20Ni. They owe their resistance to their stable scale film not coming off from the matrix surface. In type 25%Cr-32% Ni-Nb austenitic cast steel, the sulphurisation rate can be reduced by the following additions: Nb, Ti, Zr, V and Ce [24]. In catalytic pipes cast centrifugally from this cast steel, fragmentation of the structure (elimination of the dendritic structure in favour of equiaxed grains) reduces the sulphurisation rate about ten times [25]. Introduction of micro-additions of Ti+Zr+Ce (less than 0.1% each) refines not only the cast steel matrix, but also the sulphide scales, and improves scale adherence [25].

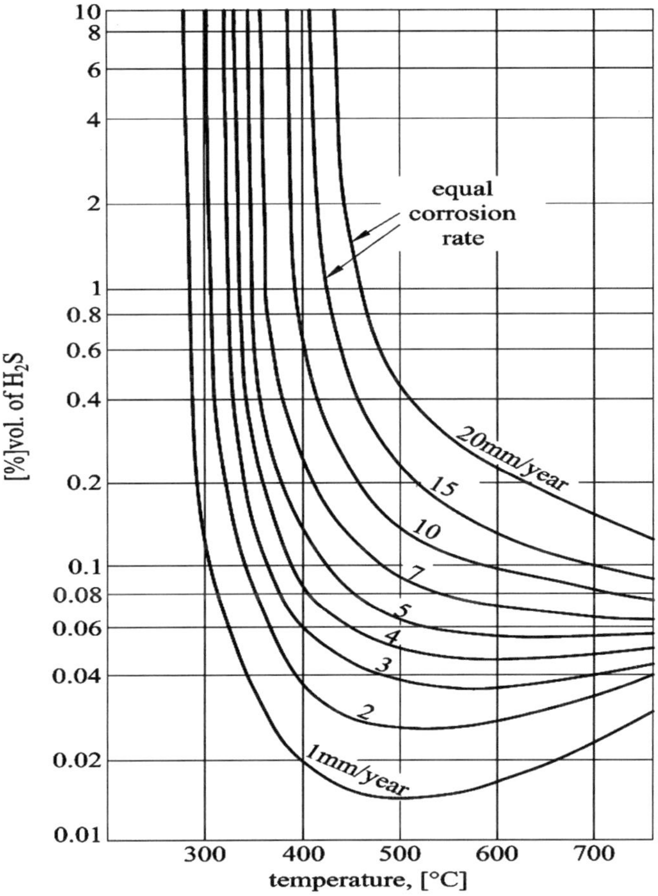

Fig. (11). Effect of temperature and content of H_2S on the corrosion rate of Cr18-Ni-9 steel [16].

However, Co-Cr alloys are more resistant. The process of their sulphurisation proceeds in accordance with the parabolic law, while their scale has a multi-layer structure: an outer layer of cobalt sulphide and an inner solid solution of cobalt and chromium sulphides [26].

7. CASTINGS WORKING UNDER CARBURISING CONDITIONS

Castings and machine elements exposed to the effect of CO, CO_2-CO and hydrocarbons at high temperatures are subject to a carburising process, and chromium carbides are formed on their surfaces (at the grain boundaries), (Figs. **6** and **8**). They cause cracks resulting from the increase in their volume. The temperature, the duration of the environmental impact and the forms of contact (gas or liquid centres) affect the corrosion rate. The carburising potential of the environment and the alloy composition are also important parameters.

The kinetics of the carburising process are subject to the parabolic law, in which the carburisation depth (x) depends on the time of carburising (t) and is described by the equation [27]:

$$x^2 = 2k \cdot t \tag{8}$$

where: k – constant.

The process of destroying the alloy under carburising conditions consists of simultaneous internal oxidation and a carburisation reaction. For example, in Ni-Cr alloys working in an atmosphere of CO - CO_2, a hetero-phase layer of Cr_2O_3 and chrome-nickel spinel is formed on the surface, (Fig. **12**). Deeper into the metal, there are precipitates of Cr_2O_3 in the alloy matrix, and then a carburised zone composed of chromium carbides and the matrix depleted in chromium (chapter 9) [12, 28]. The carburised layer consists of a mixture of Cr_2O_3, SiO_2 and Al_2O_3.

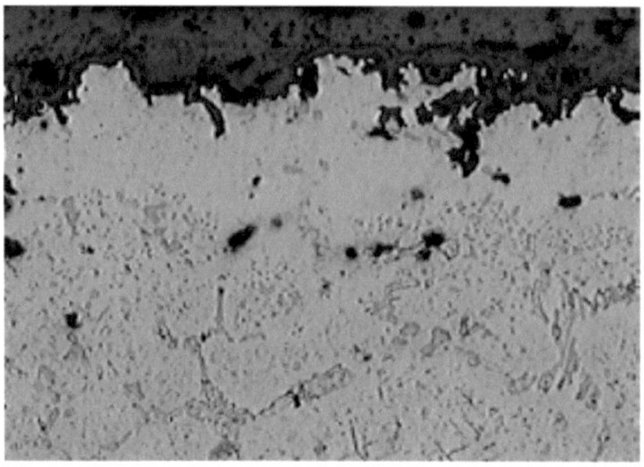

Fig. (12). Oxide scale on a centrifugal casting; Fe-Cr24-Ni32Nb cast steel, service conditions: 800° C - 5 years. Magn.800X [15].

The choice of alloy for operation under carburising conditions is based on the development of the appropriate chemical composition. Silicon (along with Cr, Ni and Mo) is one of the most important elements in heat-resistant cast steels. The increase in its content in cast steel leads to a reduction of the carbon content in the surface layer of the casting. However, the higher the concentration of silicon, the worse the weldability, creep and rupture strength, and there is also a tendency for cracks to form in castings. In the grades that are intended for operation under carburising conditions, the silicon content should be kept above 1.3%, while in creep-resistant working grades, it should be kept below this value [1].

The effect of nickel and chromium is varied; in HC, HD and HE grades, the carburisation resistance decreases with the increase in nickel content, and in grades with high nickel content (HW, HX), it increases. Nickel decreases the solubility of carbon in austenite, contributing to a reduction in carburisation compared to grades with a lower concentration of nickel.

In heat-resistant cast steel castings, an important role is played by all the surface defects: cracks, micro-shrinkage, porosity and non-metallic inclusions. Carburisation products, such as soot and combustion products, are accumulated in these places. Such products pass into the alloy matrix, accelerating the diffusion of carbon. There are a large number of casting alloys intended for work under carburising conditions, see Table **2**. Some examples of such alloy grades are given in this table; their broader range of application and the diversity of introduced additions (B, Zr, Y and V) can be found in references [16, 29].

CONFLICT OF INTEREST

The author (editor) declares no conflict of interest, financial or otherwise.

REFERENCES

[1] M. Blair, *Steel Castings Handbook.* Chapter 22. 6th ed. ASM Intern, 1995.

[2] K. Roesch, H. Zeuner, and K. Zimmermann, *Stahlguss.* Verlag Stahleisen GmbH: Düsseldorf, 1982.

[3] D.J. Young, *High Temperature Oxidation and Corrosion of Metals.* Elsevier: Amsterdam, 2008.

[4] M.P. Brady, Y. Yamamoto, M.L. Santella, P.J. Maziasz, B.A. Pint, C.T. Liu, Z.P. Lu, and H. Bei, "The development of alumina-forming austenitic stainless steels for high-temperature structural use", *JOM,* vol. 60, pp. 12-18, 2008. [http://dx.doi.org/10.1007/s11837-008-0083-2]

[5] L. Colombier, and J. Hochmann, *Stale odporne na korozję i stale żaroodporne,* Katowice Śląsk, Ed., , 1964. (in Polish)

[6] S. Mrowec, and T. Werber, *Nowoczesne materiały żaroodporne.* WNT: Warszawa, 1982. (in Polish)

[7] E. Brauns, A. Rahmel, and H. Christmann, "Die Verschiebung des Nonvarianzpunktes zwischen Eisen,Wüstit,Magnetit und Sauerstoff im System Eisen-Sauerstoff durch Legierungselemente oder fremde Oxyde, Auswirkungen auf das Verhalten von Eisenlegierungen beim Verzundern", In: *Arch. Eis.* vol. 30. , 1959, pp. 553-564.

[8] G.Y. Lai, *High temperature corrosion of engineering alloys.* ASM Materials Park: Ohio, 1990.

[9] C. Wagner, "Beitrag zur Theorie des Anlaufvorgang", *Z. Phys. Chem.,* no. 22, p. 25, 1933.

[10] M.P. Brady, Y. Yamamoto, M.L. Santella, P.J. Maziasz, B.A. Pint, C.T. Liu, Z.P. Lu, and H. Bei, "The

development of alumina-forming austenitic stainless steels for high-temperature structural use", *JOM,* vol. 60, pp. 12-18, 2008.
[http://dx.doi.org/10.1007/s11837-008-0083-2]

[11] M. Blair, *Cast Stainless Steels, Metals Handbook.* vol. 1. 10th ed. ASM Intern, 1990, pp. 908-929.

[12] J. Benard, *L'Oxydation des Métaux, t.2.* Gauthiers–Villars: Paris, 1967, p. 141.

[13] N. Birks, *Proc. Intern. Symp: Properties of high temperature alloy,* 1976

[14] G. Furgał, *The effect of Ce and Yt additions on the oxidation of single crystals Fe-Cr26, PhD, Univ.Sc.a.* Technology: Cracow, Poland, 2002.

[15] B. Mikulowski, J. Glownia, G. Boczkal, and G. Tecza, *Investigation in centrifugal tubes of Cr25-Ni35NbTi.* Industry Project: Refinery Jedlicze, Poland, 2009.

[16] C. Houska, *Castings – Stainless steel and Nickel-base, Ref.Book Series/11022*, 2012.

[17] J. Kelly, and A. Thekdi, "Materials selection considerations for thermal processing equipment", In: *Heat Treating Progress*, 2003, pp. 51-55.

[18] M. Brzeziński, and J. Głownia, "Anwendung der Thermischen Analyse zur Impfbeurteilung von ferritischen Stahlguss", *Giessereitechnik,* vol. 36, no. 6, pp. 168-172, 1990.

[19] J. Głownia, *Technology Production of casting grates, Industry project, Univ.of Sc.a.Techn., no 5.170.183, Cracow,* 1995. (in Polish)

[20] "Materials selection considerations for thermal process equipment", *US.Department of Energy, Industrial Technologies Program,* 2004.

[21] E.B. Backensto, and J.W. Sjoberg, "Iso-corrosion rate curves—for high temperature hydrogen-hydrogen sulfide—a contribution to the work of NACE Technical Group Committee T-8", *Corrosion,* vol. 15, no. 3, pp. 41-43, 1959.
[http://dx.doi.org/10.5006/0010-9312-15.3.41]

[22] E. Houdremont, and G. Bandel, "Vorgänge beim Angriff hitzebeständiger Stähle in heissen Gasen", *Arch.Eis,* vol. 11, pp. 131-138, 1937/38.

[23] K. Tjokro, *High temperature corrosion of heat resistant alloys; ACOM Rep. Nr 3 – 1991.* Avesta Sweden, 1991.

[24] R. Zapala, Z. Zurek, and J. Glownia, "High temperature sulphidation of cast austenitic alloy Cr25-Ni32-Nb with microadditives Ti, Zr and Ce", *High-Temp. Mater. Process.,* vol. 27, no. 2, pp. 127-134, 2008.
[http://dx.doi.org/10.1515/HTMP.2008.27.2.127]

[25] F. Kröger, *The chemistry of imperfect crystals.* 2nd ed. North-Holland: Amsterdam, 1974.

[26] N.S. Stoloff, *Cast Stainless Steels, Metals Handbook.* vol. 1. 10th ed. ASM Intern, 1990, pp. 951-980.

[27] A.J. Sedriks, *Corrosion of Stainless Steels.* 2nd ed. J. Wiley and Sons, Inc.: NY, 1996.

[28] N.S. Basu, and G.J. Yurek, "Effect of alloy grain size and silicon content on the oxidation of austenitic Fe-Cr-Ni-Mn-Si alloys in pure O2", *Oxid. Met.,* vol. 36, pp. 281-315, 1991.
[http://dx.doi.org/10.1007/BF00662967]

[29] K. Nishino, H. Takamiya, Y. Awano, Y. Genma, and M. Suzuki, "Effect of alloying elements on properties of 16Cr ferritic heat-resistant cast steel", *AFS Trans.,* vol. 104, no. 96 – 87, pp. 245-250, 1996.

CHAPTER 11

Wear – Resistant Cast Steels

Abstract: Steel castings are chosen for several wear conditions as elements of machines working in mining, transport and military industries. They work mainly in contact with ceramic materials (quartz sands, different ores, water-solid particles). In general, castings work with abrasion condition with high stresses. This chapter shows the types of wear, effect of hardness, load, and temperatures on weight losses. In the present work, the microstructure of used grades, their heat treatment and mechanical properties are shown.

Keywords: Abrasion, Cast steels, Hardness, High-chromium, High manganese cast steels, Types of wear.

1. INTRODUCTION

Selection of the proper cast steel grade to operate under abrasive conditions (and even more broadly wear resistant - working under the conditions of adhesive, erosive or erosive - corrosive wear) is important in such industries as machine building, mining, transport, power industry and processing of minerals. All castings, as machine parts working under these conditions, are subjected to strong wear and are eliminated from the operation. For these reasons, hundreds of thousands tons of steel (metallic materials) are used every year, putting each country at large losses. Any extension of the work of the casting as a part of the whole device brings measurable economic effects.

2. DEFINITION AND GENERAL CHARACTERISTICS OF WEAR

The concept of the material wear[1] is understood variously (depending on the service conditions of considered materials) [1, 3]. In this chapter, material wear is defined as alloy loss due to the long-term friction of adjoining materials. Unequal approach to the issue is related to the impact of a lot of factors on the wear process. For example, on the adjoining surfaces, there may be corrosion, chemical reaction of one phase (carbides, nitrides) with the matrix or a local change in the hardness as a result of the temperature increase.

When considering the wear, the following conditions of wear should be considered:

Jan Głownia

1. friction of metal against metal, *e.g.* wear at sliding friction with lubrication (*e.g.* roller in bearings) and dry friction (brake blocks in the drum), wear during rolling (also with lubrication, *e.g.* in bearings) and dry - crane wheels working on rails,
2. friction of metal against non-metal; as previously, wear during sliding friction (ploughshares) and rolling (ball mills) and additionally by throwing on (cleaning of castings with shot, pumps transferring sludge, sand, *etc.*) can be distinguished here,
3. friction of metals against liquid or gas (steam and gas turbines),

Generally, there is resistance to movement (friction) or resistance to destruction of the surface (wear). Both mechanisms may occur in a dry system or in a system involving the environment {material - lubricant (water, oil), material-alloy}. For such characteristics, there are also parameters, such as the type of movement: rolling, sliding, abrasion and load, temperature, stress or corrosion. In practice, there are usually several of these conditions at the same time, it is said about mixed or complex wear. This is due to the fact that the wear mechanism should take into account whether hard particles (*e.g.* carbides, nitrides) participate in the wear, and whether the material is plastic, and whether wear of passing surfaces does not occur by their breaking (fracture). In the case of occurrence of several wear types, the leading process should be distinguished [2].

In the case of wear without lubricant, the presence of carbides or other non-metallic particles causes the occurrence of not only the wear of plastic, but also local stresses (particularly high pressure, *e.g.* in mills at breaking stones, in a cement plant - clinker, *etc.*). Similar phenomena occur during hitting of liquid or gas drops in turbines (cavitation). The reduction of wear in this case may occur at obtaining high plasticity of the material (without cracks or fractures).

Due to its nature, the wear resistance is then difficult for the qualitative characteristics. In practice, there are a lot of types of wear; therefore, plenty of alloys recommended for use in industrial conditions. In every tribology system, there is a combination of factors that best specify the nature and rate of wear. These factors can be unified in the following groups:

- structural (chemical composition, hardness, heat treatment, other physical, chemical, technological properties),
- working conditions (load, temperature, duration, environment) characterising the mechanism of interaction (the type of movement of both elements, the type of contact and lubricant).

2.1. Structural Factors

The analysis of these factors should begin with recognition of the tribology system and its co-operating components. Such considerations include the output circuit, possible laboratory tests and final cooperation of interacting elements. These are usually two abrading alloys, lubricant (optionally pollution, dust or other solid particles) and the surrounding environment (air, corrosive solution) [4]. Considering the wear process, it should be taken into account whether the set of factors is in the open system (the materials are supplied to the system and released from the system on a continuous basis), or in the closed system.

The results of the wear measurement and the wear rate depend on a number of material parameters affecting the materials. These include: chemical composition of abraded materials, hardness, their thermal conductivity, modulus of plasticity, geometry and topography of the abrasive surfaces, grain size and chemical composition, viscosity, temperature and basicity of (the environment) corrosive solution.

There is a common belief that among the structural factors, wear is most affected by hardness. This is illustrated well in Fig. (**1**), in which four various types of cast steel are compared. This type of dependence is true when considering a large amount of metals and alloys, and comparable under similar conditions (pure metals, steel, cast steel, carbides *etc.*) [5].

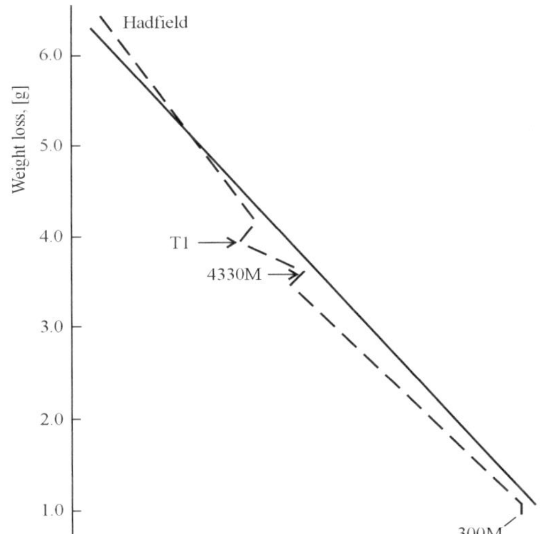

Fig. (1). General dependence of wear on hardness [1 - 4, 6].

For iron alloys, such dependence is limited due to the significant influence of other factors: chemical composition (the content of carbon), heat treatment (martensite, pearlite) and the type of the alloy matrix (presence of carbides). A good example is Fig. (**2**), which indicates comparable wear of alloys with a different structure (matrix) and varied hardness. The close dependence of wear resistance (of the specific type) on hardness occurs only when one selected alloy (*e.g.* 0.35%C-Mn-Si-Mo, 2.0%C-Cr12) of variable hardness obtained by the diverse heat treatment (*e.g.* tempering temperature after quenching and tempering) is considered. As this figure shows, hardness is only a general indicator in the evaluation of wear resistance of the alloy. Characteristic of the wear resistance of the alloy consists in therefore on how to measure, or even better in which conditions there is wear rate measurement.

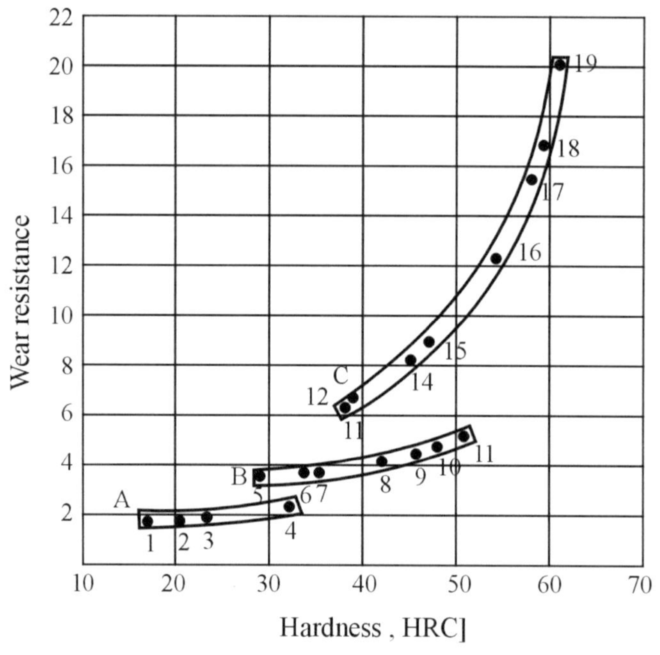

Fig. (2). Comparison of wear resistance of various alloy grades, A, B – quenched and tempered steels, C – high chromium cast steels [4, 7].
1. 0.35%C-1.2%Mn
3. 0.75%C-1.6%Cr
4. 0.5%C-3%Cr-0.5%Mn
6. 1%C-1.6%Cr-0.3%Mo
8-11. As[4] other heat treatment
12. 3%C-12%Cr annealed
13. 15%Cr-3%Mo
18. 20%Cr-2%Mo-1%Ni (Q,T)

It is known that for that reason, in individual laboratories, there are differences in the evaluation of the measurements of two various types of wear, or in industrial practice – due to a different mapping of alloy working conditions.

Some laboratories take into account the rate of the applied pressure and temperature, some - the rolling conditions of metal on metal (crane wheels), and some - abrasion of metal against the rocks, of course, so in winter as in summer. This calculation still does not include the wear of alloys by abrasion occurring in addition to the effect of the corrosion centres (pumps). The effect of other centres or working conditions, such as wear + creep, wear + oxidation, wear + fatigue, can be also mentioned here.

The choice of one of the alloys presented in Fig. (**2**) should therefore be made not only on the basis of one basic parameter - wear resistance. Very often, other reasons decide on the suitability of a given alloy; these are its plasticity, weldability, possibility of easy machining, and finally - all combined characteristics of the costs of the casting made of this alloy. If castings are subject to high requirements for weldability, avoiding brittleness and necessity of mechanical processing, an introduction of cast steel should be considered. No requirements for elongation or impact toughness allow for use of high chrome cast iron.

The issue of wear resistance is associated with so many parameters (properties) of the environment [4], that is, in practice, significant changes in abrasion can be observed at introduction of small deviations in the wear process. Most often, this involves not only material properties of the worn alloy (*e.g.* chemical composition, hardness, matrix structure, surface geometry or properties of the counter-specimen (another alloy, stone, sinter *etc.*), but also characteristics of the medium (air, water solutions, water, sand) in which the wear takes place. Tribology parameters also include: the type of stress, load, deformation rate and temperature - as operating parameters [4]. It has been found, for example, that the type of medium in which abrasion occurs can change the wear loss by the factor from 1 to 200 [8]. In this case, hardness, size and shape of the grains, their brittleness, the ability to stick and presence of other admixtures are decisive.

Generally, the known principle - the higher the hardness, the higher the wear resistance - is important only when considering a wide range of operating conditions of castings (*e.g.* a large number of metals or alloys), (Fig. **2**). In iron alloys, such dependence is less significant because of high importance of the chemical composition, matrix structure and heat treatment [3]. Very large areas of variable wear resistance for one alloy confirm this conclusion.

2.2. Operating Conditions

They describe all the factors affecting operating conditions of the tribology system. The basic factors include: the type of motion (sliding, rolling, impact and their combination), the load (acting perpendicularly or diagonally to the contact surface of both materials), the speed of both materials (total or component), temperature (initial and its increase during abrasion) and duration of wear. In many cases, it is necessary to know the disturbances affecting the operating conditions, these include vibrations and thermal radiation.

2.3. Factors Characterising the Mechanism of Interaction

Operating conditions of the whole system affect structural factors over time. They change the type of abrasive contact of the materials and the type of environment because a part of abraded alloys is involved in the further abrasion process. The impact of these materials and the stress distribution are characterised on a microscopic scale, creating a set of features that describe the wear mechanism. Thus, changes in the microstructure and impact of stresses lead to variable wear rates attributed to various types of wear, such as adhesive, abrasive wear *etc.*

2.4. Types of Tribology Wear

As a result of interaction of two plastics, tribology wear occurs as a result of removing the materials from their surface. This wear is resulted from the occurrence of friction or friction with emission of heat, or friction with chemical reaction. In alloys, pure mechanical impact is rare; it is usually connected with temperature and corrosion. The main causes of tribology wear are:

- plastic and elastic deformations in the tops of the surface irregularities,
- pressing a part of the surface layer of one alloy (material) into the layer of the second material,
- adhesive connection of adjoining surface layers,
- formation of friction of brittle oxides on the surface,
- enrichment of the surface layer with hydrogen [9].

Due to the complexity of tribology wear processes, there are a lot of types of wear [1, 3, 6, 10,]. For practical and teaching purposes, only basic types of wear have been presented. And so, the most common division distinguishes [1, 11]:

A. **Abrasive wear** - It represents about 80-90% [9] of all types of wear. Scratching, ploughing, cutting, grinding or gouging are the leading processes here. It consists in the separation of small parts of the material surface layer under conditions of sliding friction caused by the occurrence of harder abrasive particles between friction surfaces. During the relative movement of the friction surfaces, there is piling, moving, pulling out and shearing of surface elements, causing: micro-cutting, tearing or shearing of both surface irregularities (Fig. **3**) [9].

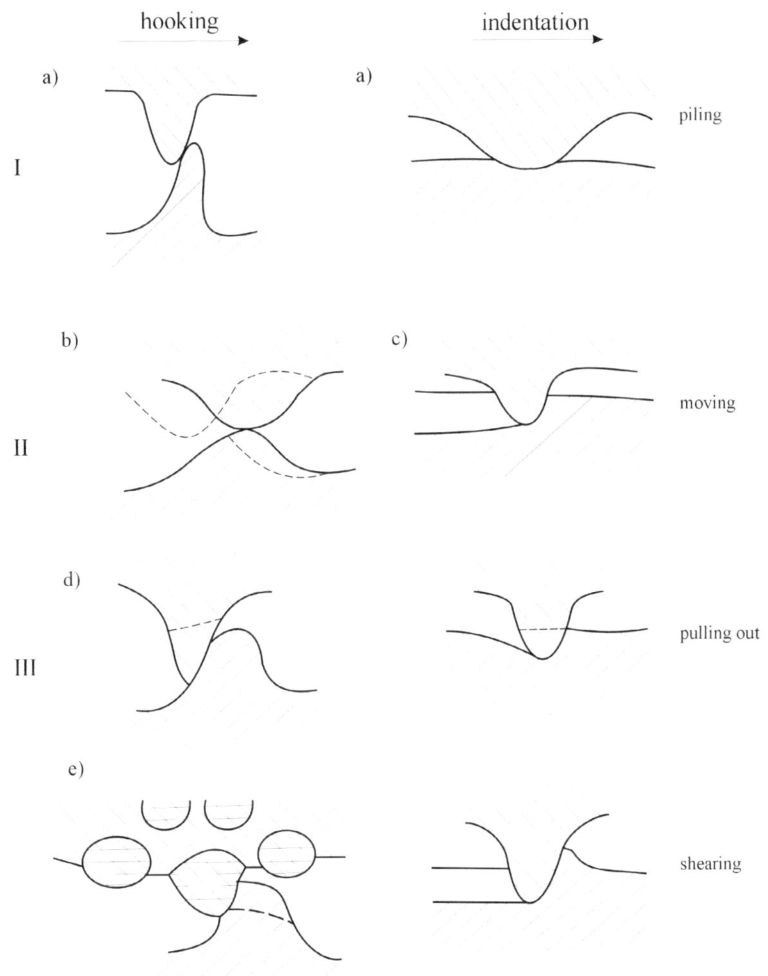

Fig. (3). Different mechanisms of abrasion wear [9].

Depending on the working conditions of both elements, the abrasive role is fulfilled by oxides, sand, coal, ore, dust or chipping. There is a division of this kind of wear for three various operating conditions: at low-stress abrasion, high-stress abrasion and impact-abrasion [12].

Wear under conditions of low stress (I) occurs when the abrasive is not subject to cracking or impact stress. Movement of the ore (of cement, coal, sand) on the conveyor belt, without the occurrence of impact, can be presented as an example here. Wear occurs due to scratching and micro-cutting of the metal surface. Abrasive of sharp shape increases the wear rate (II). For the abrasive conditions, cast steel grades are suitable that are characterised by high hardness.

Wear under high stress (III) occurs when the abrasive is fragmented and penetrates deep into the surface of the compressive elements (jaw crushers, ball mills, rock drills). These conditions occur when the abrasive works between two loaded surfaces. The abrasion surface is subject to wear due to the combined effect of scratching and plastic deformation. The reduction in the wear rate can be achieved after application of alloys with high strength and hardness for castings.

A lot of castings are subject to impact conditions. The wear mechanism is characterised by local occurrence of stress on the surface of the abraded casting. Such wear conditions can be met in pulverising mills, chutes and shovels. Cast steel is a typical material used under these conditions.

The basic parameters characterising abrasion resistance or wear rate are: hardness, crystal structure (and grain orientation), microstructure, alloying additions, heat treatment, presence and morphology of other phases (carbides, nitrides).

Hardness - it has been shown that the higher the hardness, the lower the wear, For pure elements, there is a linear dependence of wear on hardness. Richardson, however, showed that for elements crystallising in the BCC network, this dependence occurs at a smaller angle of inclination.

In steels, hardness has a strong influence on the abrasive wear resistance only at higher hardness values (above 550-600 HV) (Fig. **2**). Effect of carbon and chromium content is beneficial to increase the abrasion resistance. However, high hardness, despite the increased wear resistance, leads to a reduction in the plasticity of the material, especially in the impact toughness. Under the conditions of sudden loads, a compromise must be applied: reduce the hardness and maintain a certain plasticity of cast steel (elongation, impact toughness).

Microstructure - to obtain proper abrasion resistance, it is necessary to obtain a bainite, sorbite or martensite structure. Wear resistance defines their degree of carbide phase dispersion. There is also a case for the use of steel for castings of austenitic structure whose hardness after solution annealing equals 280 - 340 HB. Its characteristic feature is strengthening in the service conditions. As a result of the occurrence of dynamic stress, the surface of the casting is strengthened by an increase of up to 500 HB [13 - 15]. It is characterised by unusual strain hardening through deformation twinning [4, 13, 16 - 18] and through formation of martensite in the direct transformation $\gamma \rightarrow \alpha$ martensite, or in the indirect conver-sion $\gamma \rightarrow \varepsilon \rightarrow \alpha$ martensite [19].

Chemical composition - Introduction of alloying elements is a method used very widely to increase the wear resistance. Introduction of Cr, Mo, V, Mn into cast steel improves the performance of the cast steel by formation of the second phase in the matrix (carbides). Heat treatment also reinforces material properties as a result of modification of carbides formed.

B. **Adhesive wear** - It occurs under high pressure and at low friction velocity. Then, adhesive connections are formed (Fig. **3**) in which the coupled material is often harder than the matrix of both friction materials. If this type of wear is sometimes called joins type I, joins type II are also met [2] occurring under conditions of high pressure, high friction rate and at no lubricant layer. Under these conditions, the material is heated to temperatures in which plasticity is increased and both abraded surfaces become bonded due to rubbing. A similar chemical composition of abraded surface, similar crystal lattice and no lubrication increase the adhesive wear rate. In steels, the significant tendency to join is revealed by chromium, copper, aluminium, zinc, nickel, chromium and titanium.

The leading processes under the conditions of adhesive wear are: formation of joins, their tearing off, seizing, scratching and transferring particles to another material (alloy). If the adhesive join has a strength higher than the alloys forming them, shearing (pulling out) occurs within the weaker material and increases quickly (often rapidly), leading to the strong wear of machine components. Total abrasive and adhesive wear (scuffing) is often found in operating conditions of crushers for grinding molybdenum and chromium ores [12, 17, 20].

C. **Oxidative wear** - It occurs under friction conditions by adsorption and diffusion of oxygen into the surface layer. As a result, in steels (cast steels), there are iron oxides formed $Fe_2O_3 \rightarrow Fe_3O_4 \rightarrow FeO$ - the alloy matrix causing the increase in hardness of these films in relation to the hardness of the alloy matrix. Formation and removal of these films are the leading processes of this wear. Oxidative wear occurs when the rate of formation of oxide films is higher than the abrasion rate of the surface. Such a process takes place at sliding and rolling friction,

D. **Fatigue wear** - It occurs under cyclic strength - plastic stresses. This wear is caused by wear of the surface layer in these elements, which have curvilinear friction surfaces (protrusions, alloy discontinuities, oxides - non-metallic inclusions).

The mechanism of fatigue wear assumes overlapping of (residual and external) stresses in places where there are irregularities in the surface, their accumulation and leading to material strain. As a result, there are micro-cracks formed, tearing off particles of the surface layer, and finally – macro-cracks. There are two types of fatigue wear:

- spalling,
- pitting.

The former occurs under conditions of unit pressure and a large number of load cycles. The intensity of wear by spalling depends on the depth of plastic deformation of the surface structure and the mechanical properties of materials subject to friction under fatigue conditions.

However, pitting wear occurs most often on the surfaces of rolling bearings and gears (under conditions of friction with lubrication). The role of the lubricant is to protect both friction surfaces before their adhesion and material stress. The lubricant reduces the unit pressure and retards fatigue wear and formation of micro-cracks. In the later stages of fatigue, it enters micro-cracks, causing them to increase and tear off. This type of wear is several times slower than wear by spalling [21]. It occurs on the surfaces of die-casting dies and hot-working dies. To reduce fatigue wear, one should seek to reduce thermal cycling and to improve the quality of materials (by elimination of inclusions, gases and others flaws which initiate the cracks).

E. **Erosion wear -** occurs under conditions of destroying the surface layer by mechanical and corrosive effect of loose abrasive (solid or liquid) particles. In practice [22], solid particle erosion and slurry erosion are distinguished. The former mechanism occurs in the transport of sand and moulding sand (conveyor, transport of waste sand), and the other - mostly pumps in the mining industry (transport of salt mine water with coal and sand, transport of sediments, pipelines that transport in slurry rocks to the processing plant). Since, in a lot of operating conditions, a stream of particles with high kinetic energy has a double effect (erosion, corrosion) on certain surfaces, significantly accelerating their wear (pumps, compressors, turbines, pipelines).

Under conditions of solid particle erosion, the most important parameters of the process are: particle shape and size, particle hardness, particle flux, embedding of erodent fragments and erosion rate E are:

$$E = k \cdot v^n \tag{1}$$

where: k – constant,

v – particle velocity, [m/s]
n - velocity exponent (2-2.5 for metals)
E – [g/g], mass (or volume) of removed material per mass (or volume) of erodent.

The value of E depends on the angle of incidence α, which changes the wear rate: in plastic materials, the wear rate is the highest at small angles α and in brittle materials - at angles of α ≈ 60-90°. This dependence brings various effects in steels. The size and velocity of particles, heat treatment (ferrite, pearlite or martensite) and also the type of abrasive material used are important for individual grades of steel. Practically obtained values of E should refer to individual steel and service conditions of wear.

Erosion-corrosion wear, consisting in the destruction of the surface layer, as a result of a cyclic effect of the mechanical loads of the corrosion medium, is a particular case of this kind of wear (Fig. **4**). Metal oxides formed on the mechanically exposed friction surface and formed from the passive film on the surface of the worn element are a characteristic symptom of such wear (Fig. **4** Ident.2). This kind of impact can take place in the stream of solid particles (pneumatic transport in the foundry), liquid particles (flow pumps - sea water, acids, solutions) or most often - in the mixture of liquid and solid particles (sand, sludge, minerals - power plants, mines, contaminated industrial water).

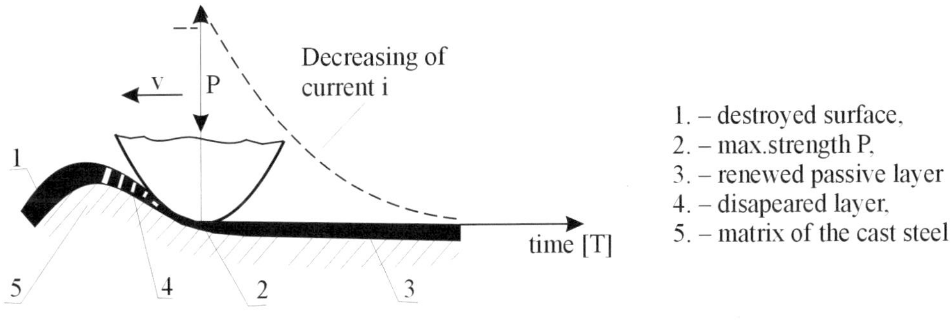

Fig. (4). Mechanism of corrosion-erosion wear [9, 11].

Erosion - corrosion wear is the case often encountered in the industry, therefore, there are a lot of wear models for these wear conditions. For one of them, characterising the flow of the mixture of water (saturated with oxygen) and sand in the pipeline with the diameter of 100 mm, it is indicated that the highest erosive wear occurs (summary effect of erosion and corrosion) on the brinks and edges of grooves in the pipelines. These observations were confirmed under practical conditions [12, 23].

This type of wear is also defined as slurry erosion. It is characterised as the transport of the mixture of solid particles in the water (also oil, water-coal, salt water sand) of such a consistency that it can be readily pumped. Abrasion increases with the concentration of the solid particles. At the concentration of the solid phase above 10%, the wear rate begins to be equal.

This indicator is based on the comparison of wear in relation to weight losses of Fe-27%Cr sample. The higher the indicator value, the faster the wear of the alloy.

F. cavitation wear - It occurs under the conditions of liquid flow in which there are discontinuities (gas bubbles, bubbles being a mixture of gas and liquid). This type of wear relates primarily to manufacturers and users of turbines, propellers and turbine engines. A more detailed description of cavitation wear can be found in works [2, 9, 11].

3. WEAR-RESISTANT CAST STEELS

In literature [5, 8, 17], there are four groups of wear-resistant cast steel. This classification has been made for utility reasons, indicating their operating conditions and a range of application.

Carbon Cast Steel - Group 1

This is an alloy used rarely under wear conditions. Castings of wagon wheels, which operate under the conditions of metal - metal friction, are the exception. Comparison of wear of various grades of cast steel (under conditions of abrasion test at the surface pressure) is presented in Fig. (**5**) [1]. The best results are revealed by cast steel with the addition of chromium. These data are used in practice in the castings of gears, drive and crane wheels, in which the matrix (carbon cast steel) is plastic and the area is carburized and quenched on the surface. As an example of the use of high carbon cast steel, castings of rolls to the rolling mill are given [24]. Large mechanical loads and work in the dust are conditions in which these steel grades are applied.

Low-Alloy Cast Steel - Group 2

This group of alloys is important, owing to its wide range of applications. In this alloy group, there is cast steel of pearlitic and martensite structure. Cast steel of pearlitic structure is used for armour plates for mills (for grinding ores), walls of screen and wedge shaped plates. These castings are operated under conditions of high dynamic loads. In relation to castings made of high manganese cast steel, they have higher plasticity, therefore, do not reveal "plastic flow". Cast steel marked from 1-P to 3-P (Table **1**) is intended for castings of large impact loads and wall thickness of 50 - 150 mm. Cast steel 3-P is used for balls with a diameter of 50 - 125 mm.

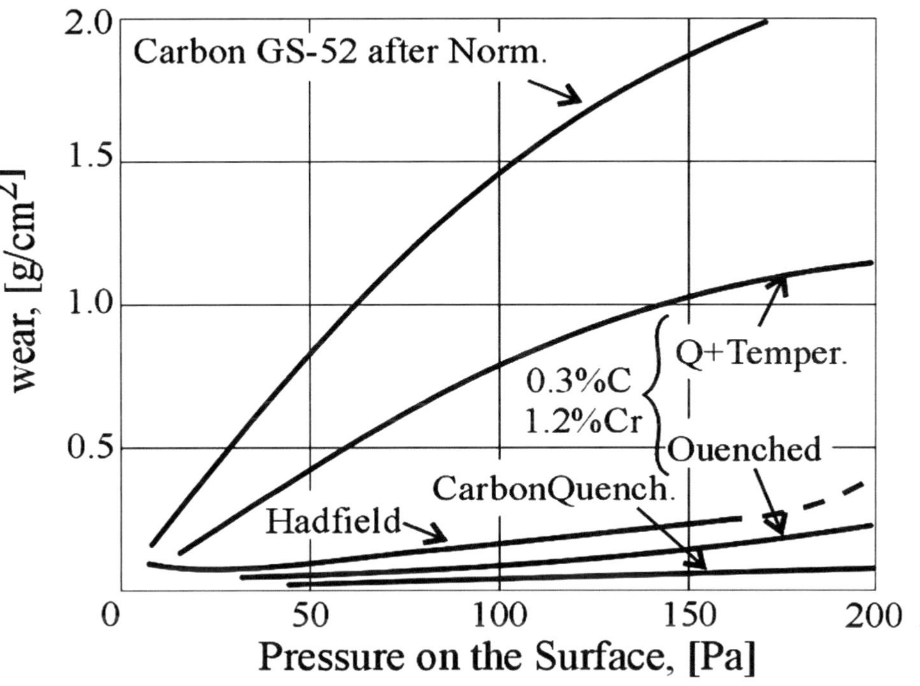

Fig. (5). Rate wear of carbon cast steels [1].

After quenching and tempering, they have a pearlitic structure and hardness of 300-420HB, and are used for castings of liner and screen castings for ball mills, with a wall thickness of 50-150 mm. The heat treatment is chosen (Table **1**, Ident. P) so as to obtain uniform properties of the wall thickness and proper hardness as compared to the abrasive material. Comparison of the abrasion resistance of various grades of cast steel [4] shows that pearlite cast steels have less wear than carbon cast steel, but they are inferior to low alloy cast steel grades of martensitic structure.

The latter comprise the largest part of quenched and tempered alloys, intended for operation under abrasion conditions. The choice of chemical composition and heat treatment is based on such a combination of strength and plastic properties, which provide the highest abrasion resistance. Hence, the lowest possible temperature of tempering at a relatively high carbon content (more than 0.35% C) is important. An increase in the carbon concentration enhances hardness, wear resistance and hardenability. However, above 0.4% C, the tendency to quench cracks increases; therefore, castings of such a carbon content are not quenched in water, but in oil or in air.

On the other hand, cast steel with lower content of carbon (0.25-0.30%) has a lower wear resistance, but better plasticity. Castings from such grades have better weldability, lower costs of repair (and removal of risers - elimination of heating before welding) and there are no quench cracks.

With proper selection of the carbon content, the concentration of other alloying additions should provide the required hardenability, suitable for wall thickness of the castings made. Most preferred properties are obtained when the casting structure is fully martensitic after quenching and before tempering. This cast steel grade includes additions of Cr, Ni, Mo, Si, and vanadium and boron as modifiers.

Chromium is maintained in the range up to 1.5%. This results from the tendency of chromium to form quench cracks, temper brittleness and segregation of carbides, usually at the austenite primary grain boundaries.

Molybdenum has a strong effect on hardenability, but it reduces the temperature of the martensitic transformation. In thick-walled castings, it also forms carbides at grain boundaries, impairing the plasticity of cast steel. Significant effect of molybdenum is based on elimination of the temper brittleness (in castings which were tempered in the temperature range of 450- 520°C). In wear-resistant cast steel, Mo content does not usually exceed 0.8% (above this content it significantly reduces hardenability and increases the tendency to form carbides at grain boundaries).

Nickel enhances the effect of molybdenum, chromium and manganese. Its concentration in abrasion resistant cast steel does not exceed 3% because it lowers the temperature of the martensitic transformation, especially at high concentrations of carbon. An improvement of the plasticity, particularly at low temperatures, is an unquestionable benefit of the introduction of nickel into wear-resistant cast steel.

The effect of phosphorus, sulphur, hydrogen and nitrogen is reflected in decreased plasticity, which is of particular importance in cast steel of high hardness. Concentrations of these elements should be as low as possible. Concentration of aluminium is also essential. After deoxidation conducted properly in the ladle, its final content should vary in the range of 0.02 ÷ 0.04%. At higher concentrations, aluminium reduces plasticity of cast steel (Fig. **13** in chapter 8).

On the basis of these premises, a number of cast steel grades has been developed, which are used in practice. Their diversity is so large that it goes beyond the compilation presentation. Hence, Table **1** includes only those grades that are representatives of a lot of cast steel groups. For example, T1 cast steel is applied on responsible castings with the wall thickness of 50-80 mm. After quenching in water from the temperature of 1173K and tempering at 610÷890K, UTS=820 MPa, YS=700MPa, El=15% and RA=35% are obtained [28]. When higher resistance to abrasion (at the expense of plasticity) is required, castings of this grade should be tempered to the hardness over 320 HB (at 380HB the tempering temperature is usually 673- 700 K). Due to the low carbon content, the abrasion resistance of cast steel is lower than that of other grades of cast steel (Fig. **5**). It is used for castings of bulldozer parts, dipper teeth, track shoes, crawler wheels, parts of fittings and cast-weld pumps. Two other cast steel grades: HY80, 8620M have a similar field of application. HY80 cast steel is used for castings with a wall thickness up to 200 mm.

Cast steels from the group Cr - Mo and Ni - Cr - Mo (Ident. GS-25CrMo4 and Ni-Cr-Mo) are used for machine parts, which require high abrasion resistance and good strength (YS > 700MPa). Cast steels of these alloy groups, but containing less carbon (*e.g.* GS-25CrMo4), are used for work at low temperatures. To obtain good plasticity, homogenisation is recommended quenching from 1173K in water and tempering 700K. Ni - Cr - Mo cast steel has a higher concentration of carbon; therefore, it is more resistant to abrasion than cast steels T1, HY-80 and GS-25CrMo4. When castings operate at minus temperatures, it is recommended to conduct smelting so as to obtain significantly lower concentrations of sulphur and phosphorus (especially for cast steel GS-25CrMo4) [1, 25]. Deoxidation should be performed so as to shift sulphides in the least detrimental form (type III or IV). Quenched and tempered Cr-Mo cast steels are used for castings of shredders. For small and medium-sized castings, usually cast steel matching the chemical composition of cast steel GS-42CrMo4 is used, and for thick-walled castings - cast steel 4135, with the addition of about 1.5% of nickel and modified with vanadium is used.

Other cast steel grades contain concentration of carbon from 0.24 to 0.35%. For this reason, they are cast steel grades most commonly used for abrasion resistant castings. They are very common in use in highly industrialised countries (*e.g.* USA [25, 26]). They are characterised by good combination of abrasion resistance and plasticity [27]. Wearpact cast steel, which contains boron to increase its hardenability, is the example here. Besides, it requires deoxidation with rare earth metals to obtain sulphides with the most beneficial (spherical) morphology and low concentrations of sulphur. Only when these conditions are fulfilled, its better properties (abrasion resistance at low temperatures) can be used properly. After normalisation at 1230K, quenching from 1170K and tempering at 590K (HB=450

- 490), UTS=1600MPa, YS=1350 MPa, El=10% and RA=20% can be obtained in this cast steel [25]. It is used for parts of construction machinery working at low temperatures, such as road construction machinery, parts of continental pipelines operating at the temperature up to "- 60°C".

300-M cast steel has a higher silicon content (compared to 0.5% of Si in other grades). This avoids brittleness in the range of 530 - 600K (the first brittleness range) because silicon increases the temperature at the beginning of the martensitic transformation. At double tempering at 590 K, the cast steel reveals UTS=1400MPa. At the hardness of 470 to 500 HB, castings of this cast steel are used for track shoes and cement mill plates.

GS-44CrSiMo12-4 cast steel has a higher content of chromium and silicon. It is characterised by good resistance to tearing, impact toughness and wear resistance. A typical field of application are castings of pump housings and impellers, coats and ball mill walls, dipper teeth and parts. Also, due to the increased carbon content, the proper heat treatment depends on the quality of cast steel (quantity of the inclusions and gas content).

Chromosil LC cast steel (also Chromosil HC with a higher carbon content - Table **1**) is quenched in air (from 1173 - 1223K). For castings with a wall thickness above 100 mm, compressed air should be used (due to the low rate of quenching) and reduced chromium content of up to 0.9% and increased nickel content of up to 1.5% (Fig. **6**).

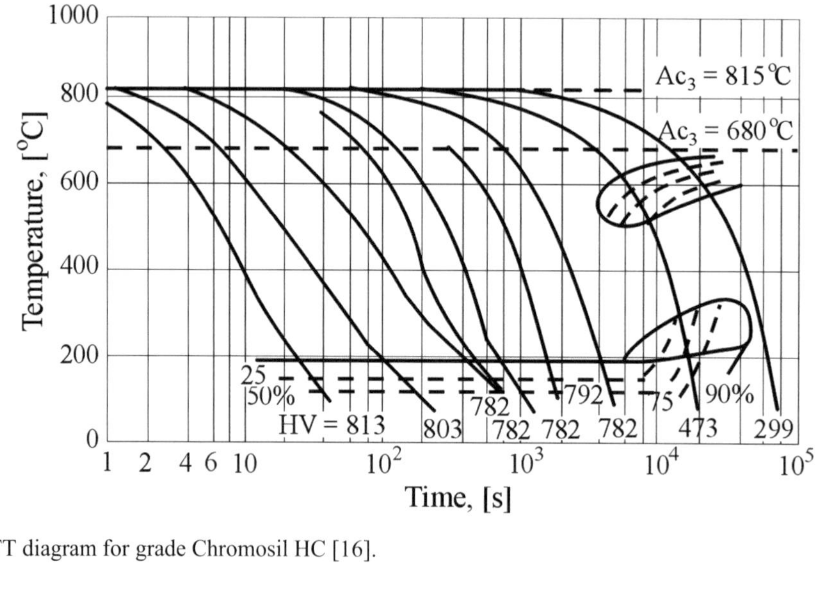

Fig. (6). TTT diagram for grade Chromosil HC [16].

Table 1. Example of wear resistant cast steels [1, 25].

Designation	Wt.% of elements							
	C	Mn	Si	Cr	Ni			
Carbon cast steels								
GS–52=ISO270-480	≤ 0,25	≤ 1,2	≤ 0,6	≤ 0,35	≤ 0,4	≤ 0,15	≤ 0,05	
GS-60=ISO 340-550	≤ 0,25	≤ 1,5	≤ 0,6	≤ 0,35	≤ 0,4	≤ 0,15	≤ 0,05	
1030AISI	0,28÷0,33	0,6÷0,9	0,4÷0,6	–	–	–	–	
Low alloyed cast steels								
1–P	0,90	0,7	0,50	2,20	–	0,4	–	Q/ air; T 230
2–P	0,80	0,80	0,50	1,70	0,80	0,30	–	Q/air
3–P	0,50	0,60	0,30	1,00		0,2	–	Q/air
P	0,50÷1,10	0,60÷0,90	0,30÷0,80	1,50÷2,50	do 1,00	0,30÷0,50	–	Q 920-980/air;T 420÷450
L40GM	0,35÷0,45	1,40÷1,80	0,40÷0,80	–	–	0,15÷0,30	–	–
L40HM	0,32÷0,40	0,60÷0,90	0,30÷0,40	1,00÷1,30	do 0,60	0,40÷0,60	–	–
L35GSM	0,32÷0,40	1,20÷1,40	0,60÷0,80	< 0,30	<0,30	0,30÷0,40	–	–
GS–30Mn 5	0,27÷0,34						–	–
Mn–Mo	0,28÷0,33	1,20÷1,55	^0,50	–	–	0,25÷0,40	–	N 900; Q 860/ water; T 590÷700
AISI4330	0,30	0,8	0,40÷1,70	0,82	1,85	0,40	0,10	2 xT 320
– << – 8620 M	0,18÷0,23	0,70÷0,90	0,20÷0,35	0,40÷0,60	0,40÷0,70	0,15÷0,40	–	–
– << – 8630	0,28÷0,33	0,70÷0,90	0,40÷0,60	0,40÷0,60	0,40÷0,70	0,15÷0,25	–	–
– << – 9530	0,28÷0,33	1,00÷1,30	0,40÷0,60	0,55÷0,75	0,40÷0,70	0,30÷0,40	–	–
T1	0,10÷0,20	0,60÷1,00	0,15÷0,70	0,40÷0,65	0,70÷1,00	0,40÷0,60	0,03÷0,08	Q900/ w; T 540÷620
GS–25CrMo 4	0,22÷0,29	0,50÷0,80	0,30÷0,50	0,80÷1,20	–	0,20÷0,30	–	Q880/ w,air; T 500÷650

(Table 1) contd.....

Designation	Wt.% of elements							
	C	Mn	Si	Cr	Ni			
GS–34CrMo 4	0,30÷0,37	0,50÷0,80	0,30÷0,50	0,80÷1,20	–	0,20÷0,30	–	
GS–42CrMo 4	0,38÷0,45	0,30÷0,50	0,50÷0,80	0,80÷1,20	–	0,20÷0,30	–	
AISI4135	0,33÷0,38	0,60÷0,90	0,40÷0,60	0,80÷1,10	–	0,15÷0,25	–	
HY80	≤ 0,20	0,55÷0,75	≤ 0,50	1,50	2,50÷3,25	–	–	
Ni–Cr–Mo	< 0,25	0,35	0,30	1,20	2,4	–	–	Q 900/ w; T 430
Wearpact	0,23÷0,33	1,30÷1,80	0,30÷0,60	0,40÷1,00	< 0,50	–	0,003÷0,006	N 950; Q 900/ w; T 320
300M	0,26÷0,33	0,60÷1,00	1,48÷1,80	0,65÷0.90	1,65÷2,00	0,05÷0,15	–	Solution 1010÷1200;Q 900/oil; T 320
Chromosil LC	0,40÷0,45	1,40÷1,60	0,90÷1,20	1,35÷1,60	–	–	–	–
HC	0,50÷0,60	1,40÷1,70	0,90÷1,20	0,65÷0.90	–	–	–	Q 900÷950/air: T 200÷700
L 100H6M	0,90÷1,10	0,20÷0,50	0.20÷0,70	5,50÷6,50	–	–	–	
Utaloy 700	0,66	1,10	1,00	1,50	–	–	–	
L 65HNM	0,60÷0,70	0,90÷1,20	0,20÷0,50	1,00÷1,50	0,90÷1,50			

The effect of the wall thickness is revealed in the method of performing quenching: castings with an average wall thickness are quenched in still air, thick-walled castings - even in a stream of compressed air. It is cast steel widely used in the United States [26] for elements of ball mills, replacing castings of pearlitic cast steel.

The last grade of wear-resistant cast steel (Utaloy 700) in the Table **1** was developed in the USA [4] on shell liners in ball mills. The carbon content higher than 0.5% gives high wear resistance, particularly in mills for grinding ore with large quantities of quartz. Castings of this cast steel grade often have increased chromium (up to 3%) and manganese (up to 1%) to increase hardenability and wear resistance. They are oil-quenched and tempered to the hardness of 420- 480 HB.

High carbon steels of martensitic class are used for operation under conditions of abrasive wear [22], which require better plasticity and wear resistance than steels of the pearlitic class. Special heat treatment in these steels leads to precipitation of secondary carbides in the martensitic matrix. Thus, for example [22], the use of spheroidizing annealing, followed by quenching in oil and tempering at 200°C, allowed to obtain up to 10% of secondary carbides in steels containing 0.8-1.1% C; 2-3% Cr and 0.5-0.6% Si and significant increase in abrasion resistance in the environment of silica grains.

Abrasion resistant cast steel containing more than 0.4 - 0.5% carbon comprise basic material for liner of mills for grinding quartz, ores and coal. This results from low material costs and sufficient hardenability. The main problems in the production processes are low plasticity and tendency to brittleness.

The increase in the hardenability of these cast steel grades is improved by optimising the content of chromium (1 - 3.5%), molybdenum (up to 0.3%) [1] and manganese (up to 1%) [1, 28].

Steels containing 0.8%C are used after quenching from the temperature of 900°C (in oil or air) and tempering most often at 220 - 250°C (in air). After performing such a heat treatment, depending on the wall thickness of the casting, a structure composed of: martensite, troostite and sorbite is obtained. Receiving pearlite or retained austenite is also possible in thick-walled castings (200-400 mm), (Fig. **9**). Such reduced concentrations of molybdenum, chromium and manganese (within the required limits) as well as the increase in segregation in hot spots shift the place of occurring pearlite to areas solidifying slower [17].

High chromium alloys - here is a number of wear conditions in which it is required to obtain high resistance to abrasion, with minor requirements of plasticity (*e.g.* transport of bulk materials, coal in heat and power plants, ores, clinker and sand in mines). In the case of small requirements for impact toughness of the alloy, the absence of rapid stresses and dynamic loads, a range of alloys with a higher content of chromium is applied (Table **1** - L100H6M).

From the system Fe -Cr (Fig. **7**), it results that up to the content of about 12%, there is a complete transformation of γ - α (pearlite - martensitic steels and cast steels), in the range of 12 - 14% - partial transformation of γ - α, and above 14 - 15% - the structure of chromium ferrite is obtained.

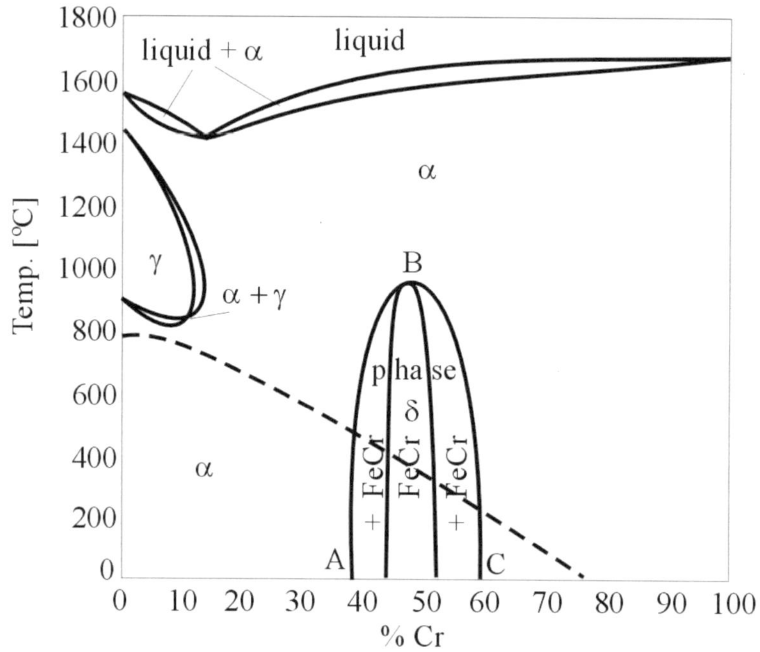

Fig. (7). Fe-Cr system [10].

However, in alloys of Fe - Cr - C, depending on the concentration of carbon and chromium, there are complex carbides, such as $(Fe, Cr)_3C$, $(Cr, Fe)_7C_3$ and $(Cr, Fe)_{23}C_6$, (Fig. **8a**). The increase in the carbon content (*e.g.* up to 1.2 - 1.3%) leads to solidification and precipitation of crystals γ Fe, but due to the segregation process, at primary grain boundaries, the eutectic from chromium carbides (Fig. **8b**), cementite and at higher concentrations of chromium and slower cooling of the casting - also ledeburite are released.

Steels with 0.7 - 1.2%C and 5 - 6.5%Cr are quenched in air (from 950 - 980°C) or in oil (from 870 - 900°C). The first method is possible only in castings with a wall thickness up to 70 mm [17], and in castings of thicker walls intermediate structures (in addition to martensite also bainite or bainite + pearlite) with a higher proportion of retained austenite are obtained.

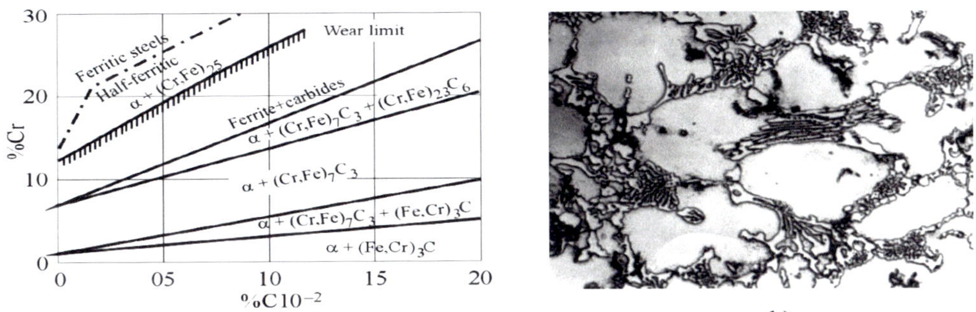

a) b)

Fig. (8). Fe-Cr-C system, a - [16]; b - eutectic chromium carbides in grade:1.4%C-12.5%Cr-0.6%-o-0.7%Mn, Magn.400X [29].

The problem of retained austenite in these alloys is often very troublesome in practice. Its amount depends on the carbon content, quenching temperature and the cooling rate. For each chemical composition, there is an optimal quenching temperature and, after exceeding it, the amount of retained austenite increases (Fig. **9**).

Fig. (9). Retained austenite in interdendritic areas and martensite needles, Magn.400X.

Also, the rate of cooling has a significant effect on the amount of retained austenite: slow cooling increases its amount (hot spots, casting walls with large thicknesses). The remaining amount of austenite in those parts of the casting is unstable and undergoes further transformation into martensite. This transformation is associated with the increase in volume (by 6% [5, 15]), leading to the increase in stress and very often to cracks in the castings during their storage. Therefore, castings should be tempered immediately after quenching. There are two temperature ranges in which the transformation or stabilisation of austenite occur: 150 - 300°C and 450 - 520°C. The first temperature range is used when there is certainty that in castings there is a small amount of retained austenite or when castings are required to be high abrasion resistant. If the castings are subject to dynamic loads, or have large amounts of retained austenite, double tempering must be performed at *e.g.* 440°C.

The degree of segregation of elements is an important parameter in this case. In the wall connections, hot spots and at primary grain boundaries, concentrations of carbon, chromium, molybdenum (and other elements) are significantly higher than in the areas of primary solidification. This changes the conditions of limit solubility, carbides are formed and hardenability is decreased. The increase in the concentration of carbon or manganese in these areas leads to a stabilisation of austenite and reduces the rate of austenite - martensite transformation. Thus, in these places of castings, there is the most retained austenite. Controlling the structure in the solid state (heat treatment, working conditions - temperature, stress, plastic deformations) aims at the increase in hardness by reducing the retained austenite (especially in the high chromium alloys).

When heating the casting to the austenitizing temperature (above 900- 950°C), there is the equilibrium established between the austenite and carbides. During the course of this reaction, supersaturated austenite loses (is impoverished in carbon and chromium). During cooling below 250°C, martensite is formed, which is further subject to tempering at temperatures 400 - 500°C. With such a performed heat treatment, only 5 ÷ 10% of the retained austenite is obtained.

Although in a lot of tribology systems in which there are impact loads, the beneficial effect of martensite structure with a small amount of retained austenite has been confirmed; however, during heat treatment, a more complex reaction takes place, which may lead to cracks in castings. Its course is as follows:

$$\text{austenite} + \text{carbides } M_7C_3 \rightarrow \text{martensite} + M_7C_3 + \text{austenite} + \text{secondary } M_7C_3$$

Secondary carbides may be nuclei of cracks and the retained austenite is unstable. During its transformation into martensite, the volume increases, leading to increased stress. This is important for thick-walled castings of, *e.g.* balls for grinding, rollers, where it is very difficult to select the optimal structure for the working conditions (frequency stress, abrasion intensity). Thus, knowledge of phase transformations and the proper selection of heat treatment parameters are very important.

Cast Steel Containing 12% Chromium

Due to the requirements of good abrasion resistance, they contain large amounts of carbon, usually above 0.8%. At these concentrations of carbon and chromium, according to the equilibrium system of Fe - Cr - C (Fig. **8a**), a certain amount of eutectic is formed, consisting of complex carbides $(Fe,Cr)_xC_y$. During solidification, it is precipitated in the interdendritic areas (Fig. **8b**) and leads to lowering plastic properties of the alloy. After further cooling (or during heat treatment), these carbides undergo transformation into others with more complex chemical composition.

Steels containing 12- 15% Cr are quenched in oil or in air. At 2.1% C recommended quenching temperatures equal 930 ÷ 960°C, at 1.65% C – 980-1010°C. Maximally obtained hardness, depends on the quenching temperature.

Quenching steel containing 1 - 2% C and 12 - 17% Cr in air results from shifting the transformation temperature by carbon and dissolution of carbides and their reaction with austenite. The mechanism of this process is as follows: above the narrow area of γFe, a part of ferrite δ can be transformed into austenite with the participation of carbides. This reaction proceeds according to the scheme: $\delta \rightarrow \gamma +$ carbides. It leads to the formation of structures corresponding to the pearlite transformation. For example, after heating steel containing 17% Cr to the temperature of 1100°C, the austenite is transformed into martensite during the final cooling.

Dissolution of carbides in austenite requires higher temperatures and longer heating. The increase in carbon and chromium content after dissolution of carbides in the matrix causes the increase in hardenability and wear resistance. Consequently, the amount of retained austenite is also increased. In steels containing 12% Cr and more than 0.5-1.0% C, this process takes place already at temperatures of 950 - 1000°C. Strong dissolution of carbides with increasing temperature of austenite decreases the martensitic transformation point and increases the amount of retained austenite.

Steel containing 12% Cr and 1% C is a typical example of chrome steel quenched in air. After increasing the temperature to 1150 - 1200°C, an entirely austenitic structure with spread carbides is obtained (Fig. **10**). This austenite is not stable and is easily decomposed during tempering at temperatures between 400 and 600°C.

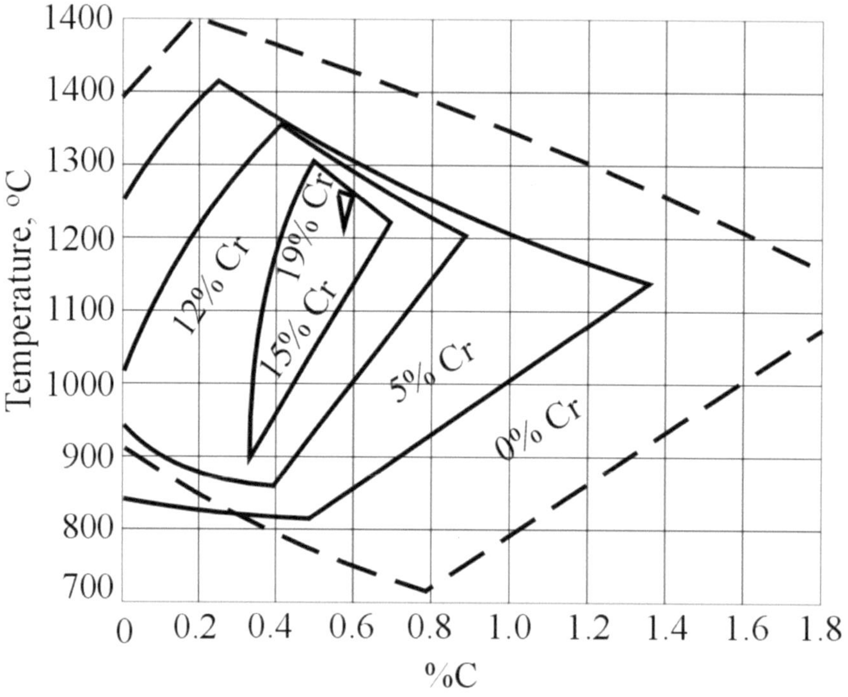

Fig. (10). Effect of carbon and chromium on the area of austenite in Fe-Cr-C alloys [3, 5].

Formation of retained austenite in cast steel from 13% Cr is similar to its formation in cast steel from 6% Cr (at ambient temperatures there is transformation of austenite and the casting cracking; at the extended cooling there is thermal stabilisation of the retained austenite). In cast steel with 12% Cr, austenite is more stable, although there is smaller amount of it. In steels 1.5% C and 12% Cr, the complete transformation of retained austenite takes already place during tempering at temperatures of 200 - 350°C [15].

Increasing the carbon concentration results in an increase of the wear resistance, amount of retained austenite and worse hardenability (carbon binds chromium into carbides). Therefore, to improve hardenability (and to avoid pearlite in castings with a wall thickness even up to 25 mm), other elements, such as Mo, Mn, and Ni, should be introduced. Therefore, these steels generally contain from 0.5 - 1.0% Mo because each 0.25% Mo compensates for the increase in carbon concentration by 0.3% [15]. Molybdenum does not increase the amount of retained austenite and does not reduce hardness, either (as *e.g.* Mn). The addition of vanadium also reduces the amount of retained austenite (at 1.2% from 20% to 10%) and grinds grain size significantly.

Mechanical properties of steels with 12% Cr are dependent on the heat treatment applied. After quenching, they are very brittle and should be tempered. Optimal impact toughness occurs during tempering at 200 - 300°C, and interestingly, does not depend on the hardness obtained.

Cast steel with 13% Cr is widely used in these devices, which work for wear in the mineral processing industry (shell liners in cement mills and for crushing ores - low wear but high impact toughness required). Further examples of application include hammers and other parts for crushers, blades for blast machines, impact plates for furnaces, parts of devices for loading charges, *etc*.

Chromium steels containing 0.8 - 1.1% C and 12 - 15% Cr have increased abrasion resistance (in relation to steel 0.8 - 0.9% C and 14% Cr) as a result of increased amount of carbides. The addition of 3% cobalt improves abrasion even more. Presence of cobalt, tungsten and molybdenum allows for air quenching of products made of steel containing up to 2.2% C and 12 - 18% Cr. This is a typical abrasion-resistant steel, applied on tools, plates, broaching rings and castings of machine parts. These are, however, steels (cast steels) with a specific field of application because they require complying with heat treatment technology very strictly. They belong to a group of ledeburite steels.

4. HIGH MANGANESE CAST STEELS

The known grades of manganese cast steel can be divided into several basic groups of these alloys [1, 30]:

1. 6Mn – 1Mo (0.75-0.95% C, 6-7% Mn,1% Mo,0.5% Si, max.0.04% S or P),
2. 8Mn – 1Mo (0.65-0.95% C, 7-9% Mn, 1% Mo, 0.5%Si, max.0.04%S or P),
3. 12Mn - 1Mo (0.6-1.1% C, 11-14% Mn,1% Mo, 0.5%Si, max.0.04% S or P),
4. 12Mn - 14Mn (1.1-1.4% C),
5. 17Mn – 2-3Cr (0.9%C, 16-18%Mn, 2-3%Cr),
6. 13Mn – 9-16V (2.6-3.3%C, 12-13.5%Mn, 9-16%V),
7. High manganese matrix with additions of Ti, Nb, V or nitrides or ceramic gluts.

The fourth grade of cast steel (invented in 1988 in England by R. Hadfield and hence often called Hadfield cast steel) has a chemical composition so well chosen that the ratio of manganese to carbon equals 10. After solution annealing from 1050°C and cooling in water it reveals a completely austenitic structure. Completely stable austenite is obtained when: $\%Mn + 13 \cdot \%C > 17$ [31]. At large surface loads, high manganese austenite shows large strengthening at ambient (and lower) temperatures. The strengthening is much stronger than for other grades of austenitic cast steel (*e.g.* Cr-Ni grades). Due to the fact that the strengthening includes surface areas, the casting does not lose plasticity on the wall thickness and hence use of the cast steel in the machine construction is so popular (bending resistance and impact toughness). Typical examples of the use of the cast steel include castings of hammers and jaws for crushers, crusher mantles, grinding balls, dipper teeth, track shoes and rail crossings (frog castings).

Under conditions of rapid wear, the surface strengthening is insufficient and then abrasive wear takes place. In these cases, it is advisable to introduce alloy elements (Cr up to 3%, Mo up to 2%) that form carbides after heat treatment.

Obtaining a fully austenitic structure in castings with a wall thickness over 70 mm (at 1.2% C and 12 - 14% Mn) is possible only at the introduction of 2 - 4% Ni into steel. Then, a homogenous structure through the whole wall thickness of the casting is obtained, but nickel reduces the abrasion resistance. It should be noted that in addition to the mentioned properties (UTS, YS, El), it is necessary to take into account impact toughness and resistance to wear (abrasion).

Concentration of phosphorus is important in high manganese cast steel. Its content is typically much higher because when filling up large amount of manganese during the smelting of steel, the application of Fe - Mn introduces considerable amounts of phosphorus. Its presence increases the hardness of high manganese cast steel, but reduces its other parameters, such as impact toughness [23]. Characteristic reduction in plasticity after exceeding the phosphorus concentration over 0.09% is obtained in two other cases:

1. when using high solution temperatures,
2. during possible local overheating of the casting when removing risers (*e.g.* with a burner) or further repairing defects (by welding).

In the first case, the plasticity changes are caused by a coarse structure due to overheating the alloy, in the other one - by another precipitation of carbides within the range of 500 - 600°C [1].

The production of castings of high manganese steel has not been reduced for several years. Properties of this alloy make it difficult to replace with other materials in certain industries. These properties include good wear resistance and excellent plasticity at dynamic loads, due to high rate of strengthening at the impact. These properties are mainly due to the chemical composition of Hadfield cast steel. Castings of this material are used in such operating conditions, which require high impact toughness and wear resistance (construction machinery, excavators, coal and transport industries). However, with low mechanical loads, they are low-resistant to abrasion (transport of sand or moulding sand, elements of track drive, *etc.*). Because these castings are usually of large weight, therefore, they are thick-walled and of diverse wall thickness. This means that in various parts of the casting (*e.g.* with a wall thickness of 60 mm and 150 mm), very diverse structures are obtained in the as-cast condition; these structures are characterised by the presence of carbides in hot spots. Under industrial conditions, this may lead to uneven properties and in the case of further inaccurate supersaturation - to cracks in castings.

An important advantage of this material, which is used in practice, is the ability to its surface strengthening due to the sudden increase in pressure or load (impact).

The basic technological disadvantage of this material is the tendency to crack at various moments in the manufacturing process. Most often, this is due to coarse crystalline structure [32] segregation of phosphorus in the interdendritic areas and stresses caused by low coefficient of thermal conductivity. The number of cracks can be reduced by rapid cooling of castings in the range of 600 - 650°C. Therefore, they must be knocked out in the temperature range of 800 - 850°C and quickly cooled or moulds should be knocked out when they are at the temperature of 400 - 450°C [33].

The structure of the alloy depends on the content of carbon and manganese and the heat treatment applied. In practice, it is chosen so that, after solution heat treatment in water, it obtains pure austenite, free of undissolved carbides (Fig. **5**, chapter 13). In practice, the structure is also dependent on the cooling rate of castings. For castings, whose wall thickness does not exceed 30-40 mm, a purely austenitic structure is obtained. In castings with the majority of wall thickness 50-70 mm, there are carbides at grain boundaries and phosphorus eutectic, Fig. **5** in chapter 13). Elimination of the effect of wall thickness (reduction in the effect of the cooling rate) is brought by the correction of the chemical composition (reduction of carbon concentration below 1% and introduction of 2 - 4% Ni). This type of stabilisation of the structure is obtained, however, only for castings with a wall thickness up to 80 mm.

Mechanical properties of Hadfield cast steel usually obtain the following values: UTS above 600 MPa, YS above 380 MPa, El = 18-20%, RA = 20-30%, K_U = 100 - 250 J/cm^2. These values are dependent on the technological parameters, such as wall thickness of castings, pouring temperature, phosphorus content or, especially, the diameter of grains [34]. Due to the high content of carbon and manganese, the pouring temperature varies in the range of 1430-1450°C. Exceeding it is dangerous especially for thin-walled castings because the grain grows and plastic properties of cast steel are reduced. For these reasons, in practice, the optimal pouring temperature, adapted to the wall thickness of the casting, must be kept.

NOTES

[1]according to [1] it is an undesirable separation of the solid body surface as a result of

mechanical load acting on it for a longer period of time, according to [2] tribological wear is a type of wear caused by friction processes.

Tribology (Greek trybos - friction, logos - science) is a field of science dealing with phenomena occurring on the friction surface of machine parts. The tasks of tribology cover two major problems: reduction of friction resistance and increase in wear resistance of these machine parts which are subject to friction. The first issue is considered mainly by mechanics, the second comprises a discipline of founders or specialists in material engineering. The last issue is also a subject of discussion in this chapter.

CONFLICT OF INTEREST

The author (editor) declares no conflict of interest, financial or otherwise.

REFERENCES

[1] K. Roesch, and K. Zimmermann, *Stahlguss* Verlag Stahleisen m.b.H.: Düsseldorf, 1966.

[2] J.H. Tylczak, Abrasive Wear. *Friction.* vol. 18. Lubrication and Wear Technology, 1995.ASM Handbook

[3] D. Olson, and C.E. Cross, *Friction, Lubrication and Wear Techn., Metals Handbook.* vol. 4. ASM Intern, 1991.

[4] T. Burakowski, and T. Wierzchoń, *Surface Eng. of Metals: Principles, Equipment, Technologies.* CRC Press Boca Raton: London, 1999.

[5] S. Nesic, J. Postlethwaite, and S. Olsen, "An Electrochemical Model for Prediction of Corrosion of Mild Steel in Aqueous Carbon Dioxide Solutions", *Corros. Sci.,* vol. 52, no. 4, 1996.

[6] R.H. Sailors, and J. Owens, "Cast High Chromium Media in Wet Grinding", *Proceedings of Intermountain Mineral Symposium, Climax Molybdenum Company, AMAX,* 1982pp. 53-62

[7] K. Röhrig, "Verschleissbeständige Cr-Mo-Gusseisen", In: *Konstr. u. Giessen* vol. 3. , 1976, p. 21.

[8] K-H. Zum Gahr, *Microstructure and Wear of Materials.* Elsevier, 1987.

[9] J. Janecki, and M. Hebda, *Tarcie, smarowanie i zużycie części maszyn* WNT, Warszawa, 1969. (in Polish)

[10] J. Dodd, *Trends in the Production, Utilization and Development of Abrasion resistant Alloys in America* Climax Molybdenum Co.

[11] J. Wallace, *Toughness of Cast Steels as Influenced by Heat Treatment and Other Processing Variables, Journal of steel Castings Research, no.74.* SFSA, 1976, pp. 1-15.

[12] T. Burakowski, and T. Wierzchoń, *Inżynieria powierzchni metali* WNT, Warszawa, 1995. (in Polish).

[13] M. Hebda, and A. Wąchala, *Trybologia* WNT, Warszawa, 1990. (in Polish).

[14] F. Henke, *Taschenbuch der Giesserei – Praxis*, 1973, pp. 181-292.

[15] K. Röhrig, "Gefüge und Beständigkeit gegen Mineralverschleiss von carbidischen Gusseisen", *Giesserei,* vol. 58, p. 697, 1971.

[16] G. Colette, C. Crussard, A. Kohn, J. Plateau, G. Pomey, and M. Weiz, "Contribution to the study of the transformations of the austenites to 12% Mn", *Rev. Metall.,* vol. 54, no. 6, pp. 433-486, 1957. [http://dx.doi.org/10.1051/metal/195754060433]

[17] T.E. Norman, "A Review of Materials for Grinding Mill Lines", *Proc. of Conf. for the Mining Industry,* 1974

[18] A.V. Levy, and S. Jahanmir, "The Effects of the Microstructure of Ductile Alloys on Solid Particle Erosion Behaviour", In: *Corrosion-erosion Behaviour of Materials* TMS-AIME, 1980, pp. 177-189.

[19] H. Lin, and C-J. Zhang, "An investigation of the role of secondary carbide in martensitic steel during three-body abrasion wear", *Wear,* vol. 176, pp. 103-109, 1994. [http://dx.doi.org/10.1016/0043-1648(94)90203-8]

[20] K.C. Ludema, Sliding and Adhesive Wear, *Friction, Lubrication and Wear Technology, ASM Handbook*, 1995, pp. 236-241.

[21] Designing with High – Strength Steel Castings, *Symp.* Materials for the Mining Industry, Climax Molybd. Comp, 1974.

[22] S. Hansen, *Relative Erosion Resistance of several Metals, Erosion, Prevention and Useful Application, STP 664.* ASTM, 1979, pp. 148-162. [http://dx.doi.org/10.1520/STP35800S]

[23] I. Finnie, *The Mechanisms of Erosive Wear in Ductile Metals, Corrosion-Erosion Behaviour of Materials,* K. Natesan, Ed., TMS-AIME, 1980.

[24] H. Lüling, Verschleissfeste Werstoffe für die Hartzerkleinerung*Symp. Zerkleinern Frankfurt am Main*, 1962, pp. 631-645.

[25] F. Henke, Niedrig- und hochlegierter verschleissfester vergütungsstahlguss*Giesserei – Praxis* vol. 23/24. , 1975, pp. 377-407.

[26] E. Albertin, and A. Sinatora, "Effect of carbide fraction and matrix microstructure on the wear of cast iron balls tested in a laboratory ball mill", *Wear,* vol. 250, pp. 492-501, 2001. [http://dx.doi.org/10.1016/S0043-1648(01)00664-0]

[27] K. Bungardt, E. Kunze, and E. Horn, "Untersuchungen über den Aufbau des Systems Eisen-Chro--Kohlenstoff Arch", *Eis.,* vol. 29, pp. 147-203, 1958.

[28] L.A. Vermeulen, and D.D. Howat, "Abrasive and impactive wear of grinding balls in rotary mills, J.S", *Afr.Inst.Mining Metal.,* vol. 83, pp. 189-197, 1983.

[29] "Microstructure of industry balls for cement plant", In: *Project of Univ.of Sc.a.Techn.*Cracow, 1993no. no. 6.170.26, .

[30] S. Sobula, and G. Tęcza, "Stabilization of the Hadfield cast steel microstructure at increased temperatures in the presence of chromium", In: *Przegl.Odl.*, 2009, no. 3, pp. 132-137.

[31] O. Bouaziz, S. Allain, C.P. Scott, P. Cugy, and D. Barbier, "High manganese austenitic twinning induced plasticity steels: A review of the microstructure properties relationships", *Curr. Opin. Solid State Mater. Sci.,* vol. 15, pp. 141-168, 2011. [http://dx.doi.org/10.1016/j.cossms.2011.04.002]

[32] E. Bayraktar, C. Levaillant, and S. Altintas, "Strain rate and temperature effect on the deformation behavior of the original Hadfield steel", *J. Phys. IV,* vol. 3, pp. 61-66, 1993. [http://dx.doi.org/10.1051/jp4:1993706]

[33] F. Maratray, High Carbon Manganese Austenitic Steels, The Intern, *Manganese Inst.*Paris, 1995.

[34] E.G. Moghaddam, N. Varahram, and P. Davami, "On the comparison of microstructural characteristics and mechanical properties of high-vanadium austenitic manganese steels with the Hadfield steel", *Mater. Sci. Eng. A,* vol. 532, pp. 260-266, 2012. [http://dx.doi.org/10.1016/j.msea.2011.10.089]

CHAPTER 12

Pouring Systems and Rising

Abstract: The quality of steel castings depends upon how the liquid metal enters into the mould and solidifies. The technology of pouring molten steel to the ladle and then to the mould must be controlled if no defect is to be obtained. These factors must be accounted for in designing the gating system, the feeding and risering of castings. Gating system ought to control a directional solidification of each part of the casting in such a way, that no wall thickness is isolated from the feeding during the freezing time. Position and size of the riser ensure elimination of shrinkage which takes place when a molten steel changes from the liquid to solid state.

Keywords: Bottom pour ladle, Calculation of their size, Gating systems, Positioning of risers, Pouring temperatures, Types of risers.

1. INTRODUCTION

Preparation of liquid steel is the first step in obtaining steel castings. For proper quality of castings, steel melted in any furnace (EAF, induction furnace or converter) must be very well deoxidized, heated to a temperature above the pouring temperature and degassed (low contents of nitrogen and hydrogen). Obtaining the intended chemical composition is not sufficient. After tapping from the furnace into the ladle, molten steel is transferred to the mould pouring place. It must therefore be overheated by the amount of heat lost in the ladle during tapping and the amount of heat lost during transportation. In the ladle, steel is not only deoxidised, but very often modified. After tapping, steel is protected against the loss of heat and an increase in the concentration of oxygen, nitrogen and hydrogen with the slag layer. The steel castings are poured with bottom ladles (Fig. **1**). The pouring rate is controlled by opening and closing a nozzle with the stopper. These ladles and method of pouring have the advantage of pouring because the molten steel is separated from the slag and in addition, inclusions flowing upwards leave pure metal at the bottom of the ladle.

Filling moulds with molten steel are subjected to important principles:

- mould filling velocity should ensure smooth flow of metal, without turbulent filling because:

Jan Głownia

-high flow velocity can lead to erosion of the mould or core (mould or core erosion),
-rapid mould filling may result in: premature freezing in thin sections, increase in thermal stresses between thin and thick sections or promotes the entrainment of gases,

- pouring steel should protect against slags, ceramic materials from the ladle liner or gating system flowing into,
- conditions of pouring additional moulds should be similar to the previous ones.

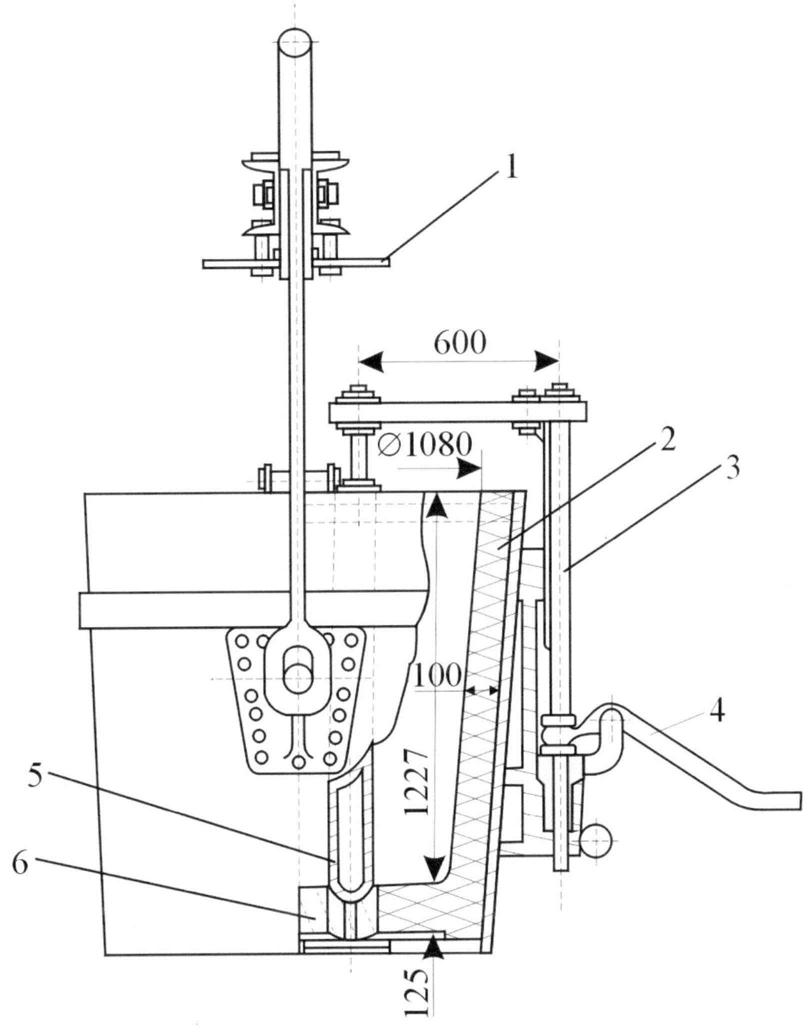

Fig. (1). Bottom ladle used in steel foundry: 1-lifting sling, 2,6-lining of the ladle, 3,4-stopper, 5-plug

To meet these requirements, the gating system consists of the following components, (Fig. **2**): pouring cup, sprue, runner and ingates. The runner is filled with liquid steel before ingates, which allows for a reduction in turbulences and stabilises the full-flow conditions of filling. There is also a principle that the total cross-sectional area of the ingates should be smaller than the cross-sectional area of the runner. Placing the ladle with liquid metal over the gating system generates a high metallostatic pressure. To reduce it in the mould, two gating systems have been developed:

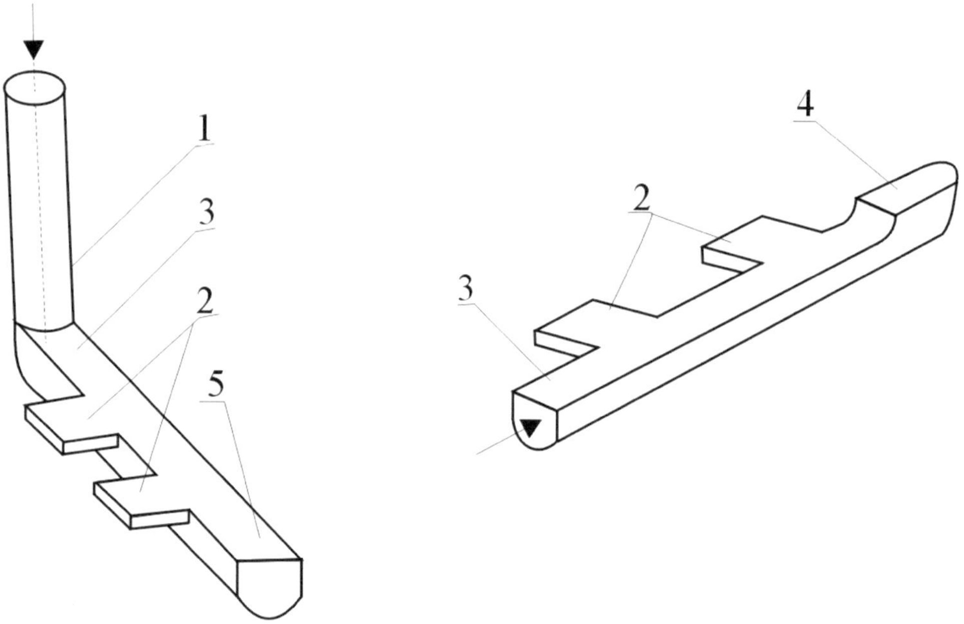

Fig. (2). General view of gating system: 1-sprue; 2-ingates; 3,4,5- runner.

unpressurised, in which the ratio of the area of sprue: runner: ingates is increasing in area *e.g*:

1:2:2 or 1:2:4 or 1:4:4 [1]
1:1.3:1.5 or 1:1.5:2 [2, 3]

This system reduces the velocity of molten steel in the gating system, which reduces turbulent flow. In this system, all elements of the gating system are filled with liquid metal, which facilitates the outflow of inclusions and moulding sand particulates.

pressurised: 4:8:3 [1]

1.1:1.05:1 or 1.4:1.2:1 [2]
4:3:2 [4]

leads to high velocity just as the molten steel enters the casting and finally - to the erosion of the mould and cores.

Construction of Gating Systems

There are a lot of solutions to bring metal to the casting, (Fig. **3**). Generally, there are two basic systems: horizontal and vertical gating systems [1, 3, 5, 6]. In practice, often there is bottom gating applied because the supply of liquid steel to the lowest place of the casting allows for a smooth filling of the mould cavity.

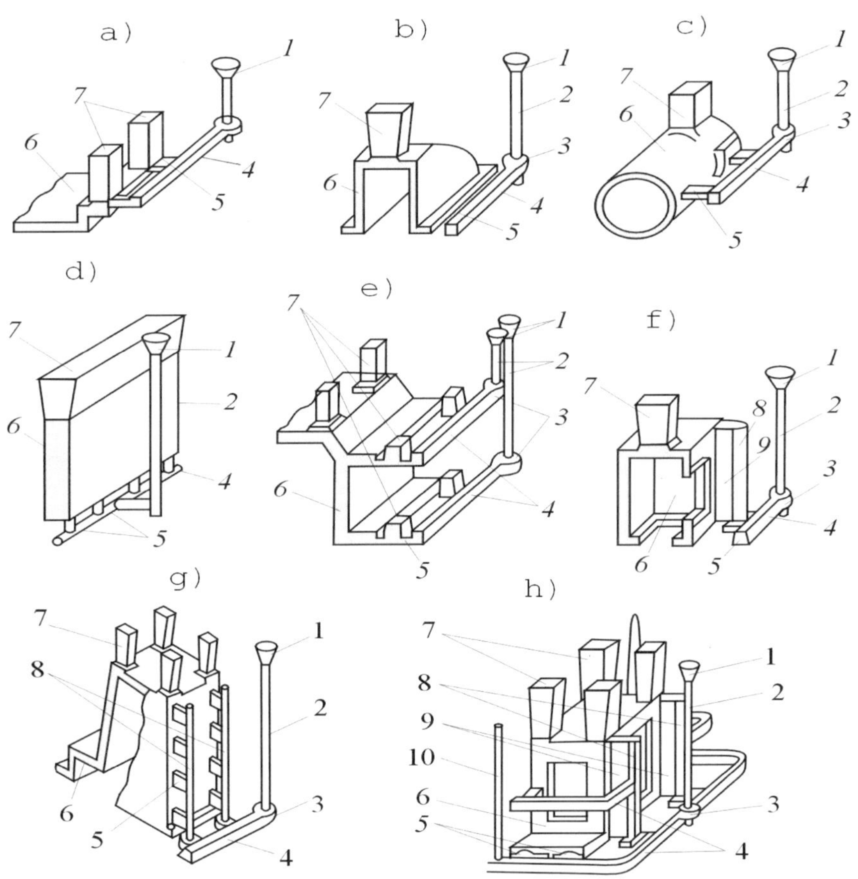

Fig. (3). Different constructions of gating systems: 1,2 – sprue; 3,4,5 – runner; 6 – casting; 7 – riser,8,9,10 – additional sprue [2].

Vertical gating system: Bringing the liquid steel from the top ensures directional solidification (the highest temperature of steel is under the riser). The disadvantage of this solution is the washing of moulding sand out of cores and moulds. During falling of the metal stream on the core, it is broken and the metal is heavily oxidised. Therefore, this system is rarely used, and rather in combination with other bringing of metal (when the metal stream may be supplied from the top onto the liquid metal in the mould cavity).

Horizontal gating system: The liquid steel is brought to the lowest part of the casting from the side. This ensures smooth filling, but lower elements of the casting are overheated, and the upper elements of the casting are filled with cooler metal. Therefore, this system is applied in small, less massive castings.

Bringing metal from the side, on several levels is a kind of this system, (Fig. **3**) *e.g*. In this solution, more portions of the casting arc filled with metal of a higher temperature, which is favourable to directional solidification and a decreasing rate of the metal stream.

Bottom gating systems: This system ensures the most peaceful filling of the mould. However, a high metallostatic pressure may cause splashing of the metal stream, especially in high moulds, such as ingot moulds, rolls and massive crusher mantles.

Combined gating systems: In dependence on the weight and construction of the casting (complex construction, wall thickness, number of cores), in practice there are a lot of solutions of gating systems, which allow to obtain castings without defects. These include: fissured gating system (Fig. **3g**) and a combination of two or three types of the gating system, (Fig. **3h**). Fissured gating system is applied to high castings (H/S > 50) [2]. It allows for uniform introduction of liquid steel into the mould cavity and ensures directional solidification of the casting.

2. CALCULATION OF GATING SYSTEM

In contrast to filing of molten steel from the teapot ladle, filling of molten steel from the bottom-pour ladles has some advantages:

1. separates the slag from the liquid steel,
2. the velocity of metal flow is regulated by the diameter of the hole in the bottom ladle and metallostatic pressure in the ladle,
3. movement of liquid steel in the gating system has a turbulent character,
4. sprue diameter is always larger than the diameter of the hole in the ladle, which ensures continuous flow of steel out of the ladle,
5. during the outflow of steel from a ladle, induction of the air into the gating

system occurs (thus possibility to increase the content of oxygen, nitrogen and hydrogen in the casting),
6. high metallostatic pressure can wash ceramic material out of tap holes, gating sleeves and ingate.

During the filling of the mould from the bottom ladle, there is a permanent change in the metallostatic pressure, both in the ladle and the mould cavity (it is the highest at the beginning of pouring, and gets lower after each subsequent pouring of the mould). Control of the pouring rate of metal from the bottom ladle is very difficult. Practically, the size of the ladle and the diameter of the tap hole decide upon the pouring rate and the time of filling the mould, according to the formula:

$$t = 4D\cdot(\sqrt{G_{init}} - \sqrt{G_{fin}}) / d^2 \sqrt{\pi\cdot\rho\cdot g} \qquad \textbf{(1)}$$

where: t – pouring time [s]

D – ladle diameter, [cm]

d – diameter of tap hole, [cm]

G_{init} and G_{fin} – weight of liquid steel in the ladle at the beginning and at the end of pouring, [kg]

ρ – density of molten steel, [7.8g/ml]

g – acceleration of gravity, [981cm/s^2]

Calculation of the gating system proceeds as follows:

1. determination of the optimal time of pouring,
2. checking the lift velocity of the molten steel surface in the mould cavity,
3. calculation of cross sections of: sprue, runner and ingates.

Optimal pouring time t is calculated by the formula:

$$t = k\cdot\sqrt[3]{G}\cdot\delta_o \qquad \textbf{(2)}$$

where: t – pouring time, [s]

G - weight of casting, including risers and gating system, [kg]

δ_o – average wall thickness, [mm]

k - coefficient, dependent on the wall thickness and weight of the casting, (from 1.75 - 0.75 [2, 3]). Higher values are assumed for castings poured to dry sand moulds and smaller - for green sand moulds.

In casting practice, there are nomograms to calculate the time of pouring (Fig. **4**).

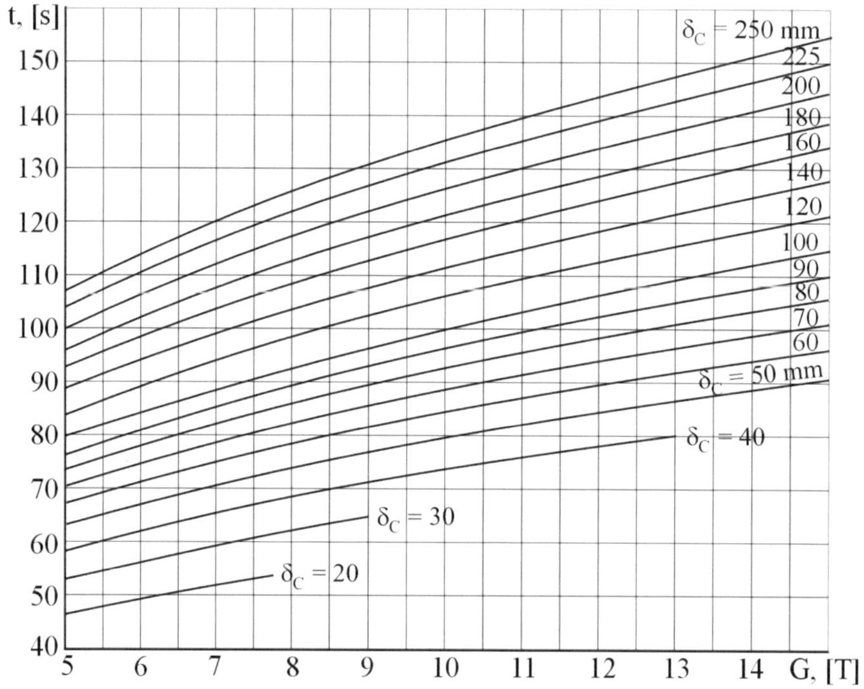

Fig. (4). Nomogram for calculation of pouring time [5].

Deviations from these values occur when making massive castings weighing more than 7-12 tons. They result from differences in the wall thicknesses [5], the use of self-hardening sands and the method of pouring (with two or three ladles).

Pouring time calculated by the formula (2) should be combined with the lift velocity v_{lift} [m/s] of the molten metal surface in the mould cavity. This velocity is determined by the formula:

$$v_{lift} = H/t \qquad \textbf{(3)}$$

where: H – height of mould cavity, [m]

t – pouring time, [s]

With the slow filling of the mould, the surface of the liquid steel oxidises and forms high-melting oxides, and quickly loses heat. The result is the occurrence of defects, such as cold shut, misrun and discontinuity in castings. Recommended lift velocity of the metal bath is given in Table **1**.

Table 1. Filling rates v_{lift} for castings with different wall thicknesses g [2].

Cast steel	**g = max. 4 mm**	**4 -10**	**10 – 40**	**40 - 70**	**>70 mm**
Carbon	100 - 20	20 - 16	16 - 12	12 - 10	10 - 8
high-alloyed	-				
0.1%C-18%Cr-9%Ni	-	36 - 29	29 - 22	22 - 18	18 - 14
1.1%C- 13%Mn	-	25 - 20	20 - 15	15 - 13	13 - 10
0.4%C-24%Cr-12%Ni	-	40 - 32	32 - 24	24 - 20	20 - 16

Calculation of the Diameter of the Tap Hole in the Ladle

The outflow rate of steel from the ladle depends on the metallostatic pressure H\` (the level of molten steel in the ladle) and the tap hole diameter d_o, Fig. (**5**). The diameter d_o is calculated by the formula:

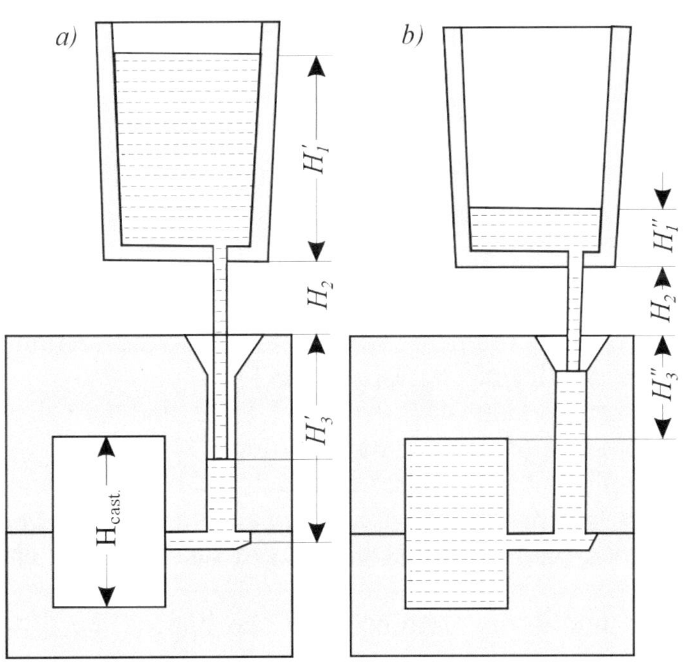

Fig. (5). Initial (H\`) and final (H\`\`) level of molten steel in the ladle during pouring [6].

$$G/t = \mu \cdot F_o \cdot \rho \cdot \sqrt{2g \cdot h_m} \quad (4)$$

where:

v_{ladle} = G/t outflow rate of liquid steel from the ladle, [kg/s]

F_o area of tap hole [$10^{-4}m^2$]

ρ- density of molten steel, [kg/m^3]

μ resistance coefficient of outflow of molten steel from the ladle in general 0.8 [6], 0.17- 0.9 [3]

g acceleration of gravity, [m/s^2]

h_m metallostatic pressure [10^{-3} m].

In this formula, one should consider increasing the hole d_o because the steel flows out of the ladle and the variable value H\` at the beginning and at the end of pouring. In practice, average values are taken. In the case of massive castings (several tons), changes in the values $d_o$ and H\` can be significant. Most often, this applies to making massive castings made of high-manganese steel (12-14% Mn, 17% Mn + 3%Cr).

Calculation of Area of Ingates F_{ingate}

The complexity of the calculation of the area of ingates is presented by the assumptions regarding casting steel from the bottom ladles. They can be collected in the following observations:

1. the set of all elements of the gating system should be expansive, *i.e.* cross-sectional area of sprue F_s should be smaller than the cross-section of the runner F_r and the basal area of ingates F_{ing} larger than $F_{r,}$
2. it is often taken that the value of F_s = (1.15-1.25) d_o and the value of F_r = (1.2-1.5) F_{ing}, for keeping the proper casting conditions [5].
3. gating systems applied in the foundry are developed by the experienced staff.
4. the friction factor is dependent on the chemical composition of materials used in the ladle stopper, (carbon cast steel, alloyed steels and the capacity of the ladle).
5. when filling the mould, the ladle operator can impede the flow of steel by closing a tap hole.

These two examples of the construction of the gating system confirm the essence of the issue. Usually, the practice of technical staff and casting conditions developed (mould sands, pouring temperatures, construction, wall thickness and weight of the casting) decide on the construction of the gating system.

3. TYPES OF RISERS AND THEIR CALCULATION

The primary objective of the use of risers is the elimination of shrinkage cavities, formed during the casting solidification. Differences in liquid and solidification contraction of steel are the direct cause of the formation of shrinkage porosity, and consequently - the use of risers. To calculate risers, the contractions in liquid (I) and the mushy zone (during solidification) (II) must be taken into account (Fig. **6**). Value of the liquid contraction equals 1.6%/100°C [6] or 1.6-1.8%/100°C [7], and the total volume contraction – from 11.14% (for cast steel containing 0.1%C) to 12.08% (for cast steel containing 0.8%C) [8].

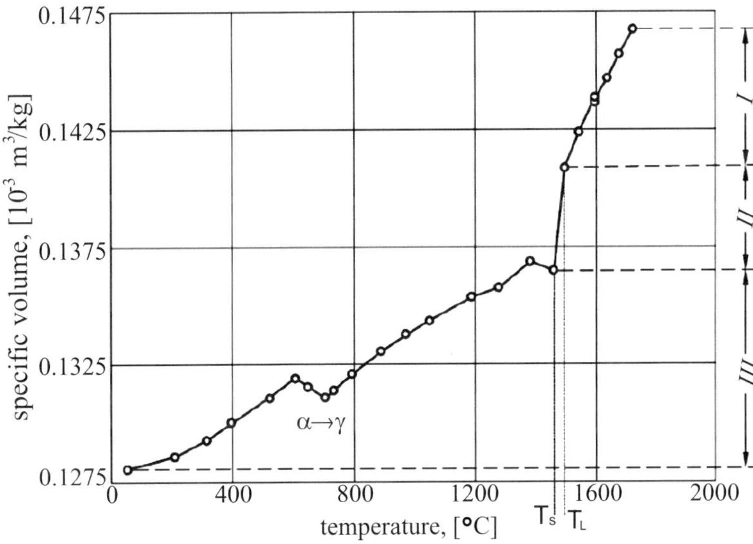

Fig. (6). Three types contraction of carbon cast steel (0.35%C) [6].

The value of the volume contraction is affected not only by the carbon content, but also by the temperature and gas content in the liquid steel. In the alloy cast steel, contraction is also changed by the presence not only of carbon (increase in the volume shrinkage 4.11% per each %C), but also Al (1.7%/1%Al), silicon (1.03%/1%Si) and chromium (0.12%/1%Cr) [2]. Changes in the contraction during solidification of the casting result in the formation of shrinkage cavities, microporosity and dishing cavities. These defects are eliminated by the proper positioning of risers and calculation of their volumes in relation to the size of occurring shrinkage cavities.

3.1. Feeding of Hot Spots

The effective riser ensures bringing of liquid steel to every place in the casting before it solidifies. This means that the riser solidifies longer than the casting. These conditions shall be determined by selecting the proper shape of the riser and a calculation of the riser's volume.

The first rule, which should be fulfilled by the riser, is the determination of its area, which it can feed into the casting, so called feeding distance FD (Fig. **7**). The value FD determines the number of risers, which must be set on the casting surface in order to avoid defects resulting from shrinkage.

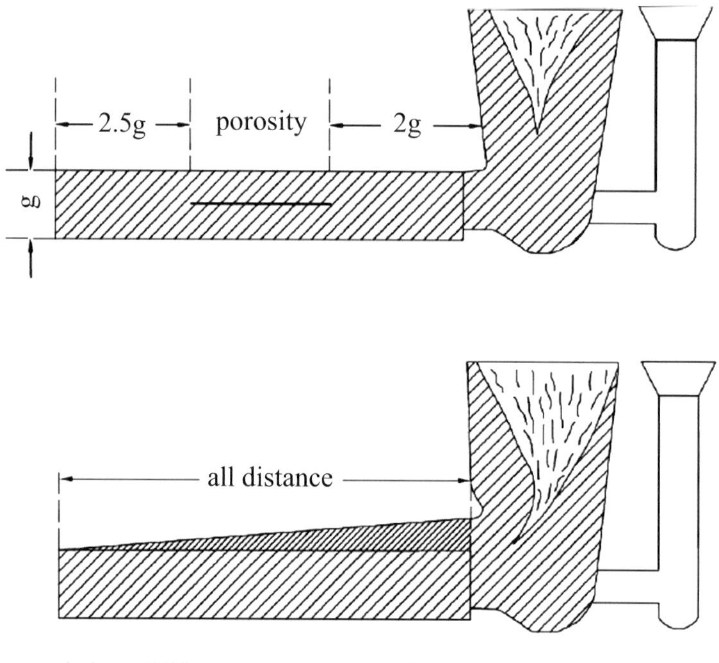

Fig. (7). Feeding distance during solidification of steel castings [6].

It depends on the method of bringing the liquid steel, its overheating, chemical composition and construction of the casting (intersections of two or more walls thicknesses - hot spots). In older solutions, there is the value FD as 2g (g - wall thickness) (Fig. **8a**), but when considering the impact of two risers (Fig. **8b**), or application of chills (Fig. **8c**), the scope of the riser activity is expanded. From the research of W.S. Pellini [8], it can be observed that maximum FD for a plate is 4.5 g.

The problem of feeding hot spots is essential in order to obtain castings without shrinkage defects. This issue is subject to continuous practical researches, which take account of additional technological parameters. For example, W.D. Spielberg [10] introduced the term of solidification gradient as a criterion of formation of shrinkage porosity, and E. Niyama *et al.* so-called Niyama criterion N_V

$$N_v = G/\sqrt{\dot{T}}, \qquad [\sqrt{K \cdot s}/mm] \tag{5}$$

where: G – temperature gradient [K/mm]

$\dot{T}$ - cooling rate [K/s]

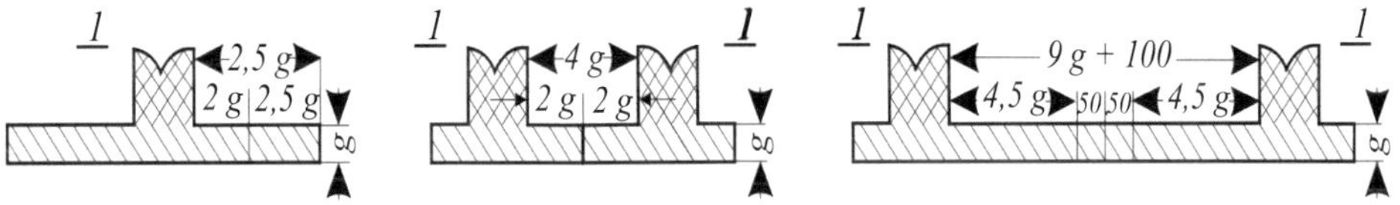

Fig. (8). Effects of risers positioning on the feeding distance, [5, 6, 9].

This criterion was verified in X-ray examinations ASTM shrinkage x-ray level [9] and also in numerical modelling of the feeding process. The results of these tests are shown in Fig. (**9**). They relate to low alloy steel castings and indicate possibilities for obtaining higher FD than before [11].

4. RISER CALCULATIONS

In the technological process of making castings, one of the main tasks is to determine the spot of placing risers, and to calculate their volume, weight, dimensions, shape and type. The riser fulfils its function if:

1. liquid steel in the riser solidifies later than in the hot spot,
2. the shape of the riser does not hinder the process of solidification and moulding,
3. the level of liquid steel in the riser is always higher than the level of metal in the fed hot spot,
4. volume of liquid steel in the riser must compensate for the volume of the shrinkage cavity,
5. mass of liquid steel intended for risers is the minimum to ensure the total feeding of the casting.
6. it can be easily separated from the casting, and the cost of usage is economically viable.

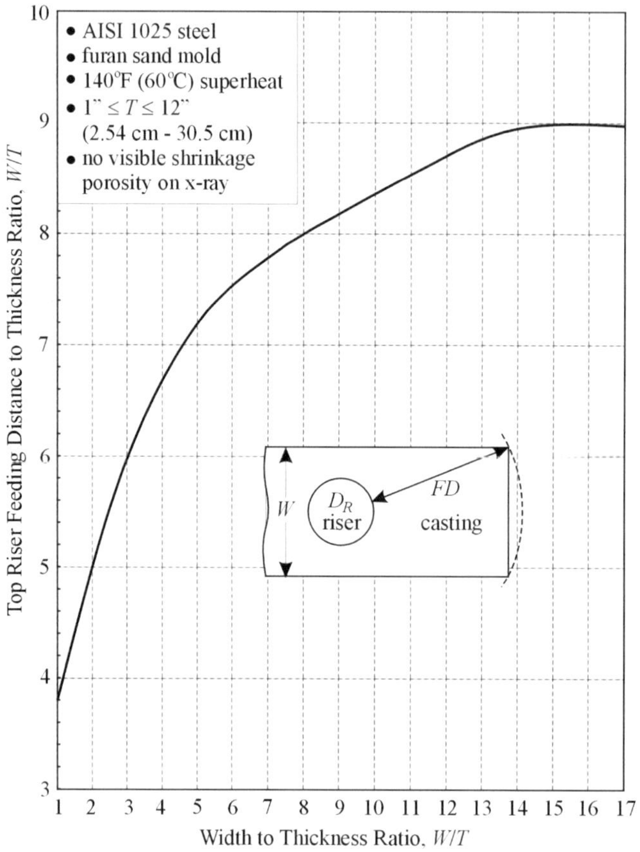

Fig. (9). Feeding distance as a function of ratio W/T [9].

In order to meet these requirements in practice, there are a lot of types [12] of risers classified according to: activity - open riser (e, f), blind riser (a), atmospheric riser, exothermic riser (b,c), pressure head riser, side riser (a), shape - elliptical riser (f), cylindrical riser, conical riser, spherical riser (d), moulding method - isolated riser, necked-down riser, exothermic mini-riser. Some of these types are shown on Fig. (**10**).

The proper course of action for calculation of risers includes [5, 10, 13, 14]:

1. dividing the casting as a set of simple areas of:
   hot spots (the places of risers),
   side-riser section with end-effect,
2. determining the feeding zones, feeding distance FD, with knowledge of the wall thickness of the casting T and casting section width W,

3. calculation of riser sizes V_R,

$$V_R = 2.51\ V_C\ (SF)^{-74}$$

V_{C} - casting volume = L · W · T (fed by the riser)

For cylinder riser with height $H = D_R$

$$D_R = \sqrt[3]{3.2 V_C\ (SF)^{-0.74}}$$

4. compute the distance from the edge of the riser to the furthest point of the casting section (fed by the riser),
5. check that feeding distance is greater than to the distance found above.
   - feeding distance FD, [cm]
     is given graphically as a function: FD/T against W/T [9, 11]
   - riser diameter D_R [cm]
   - length section L, [cm]
   - width W, [cm]
   - shape factor SF=(L+W)/T

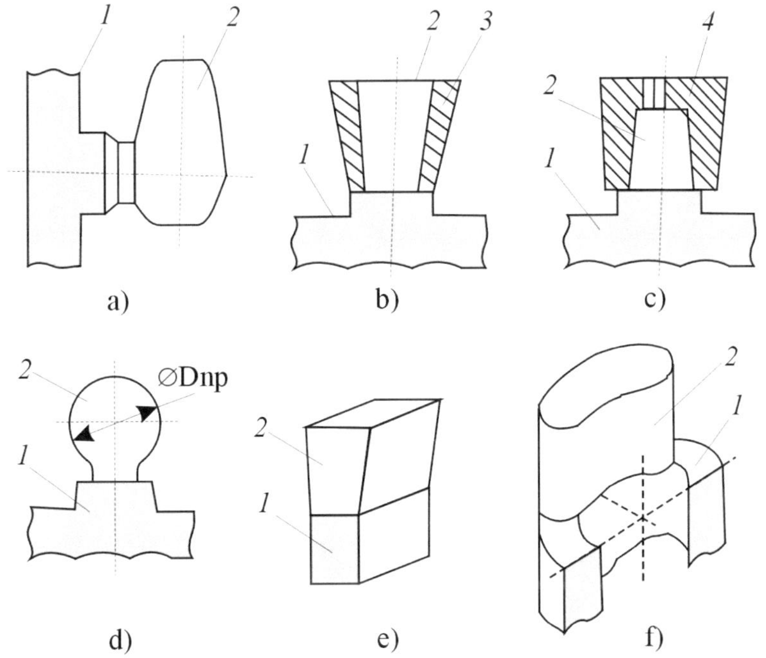

Fig. (10). Examples of different types of risers: 1 – casting; 2 – riser; 3,4 – exothermic sleeve [12].

This method of calculation of risers is also presented graphically [14, 15], (Fig. **11**).

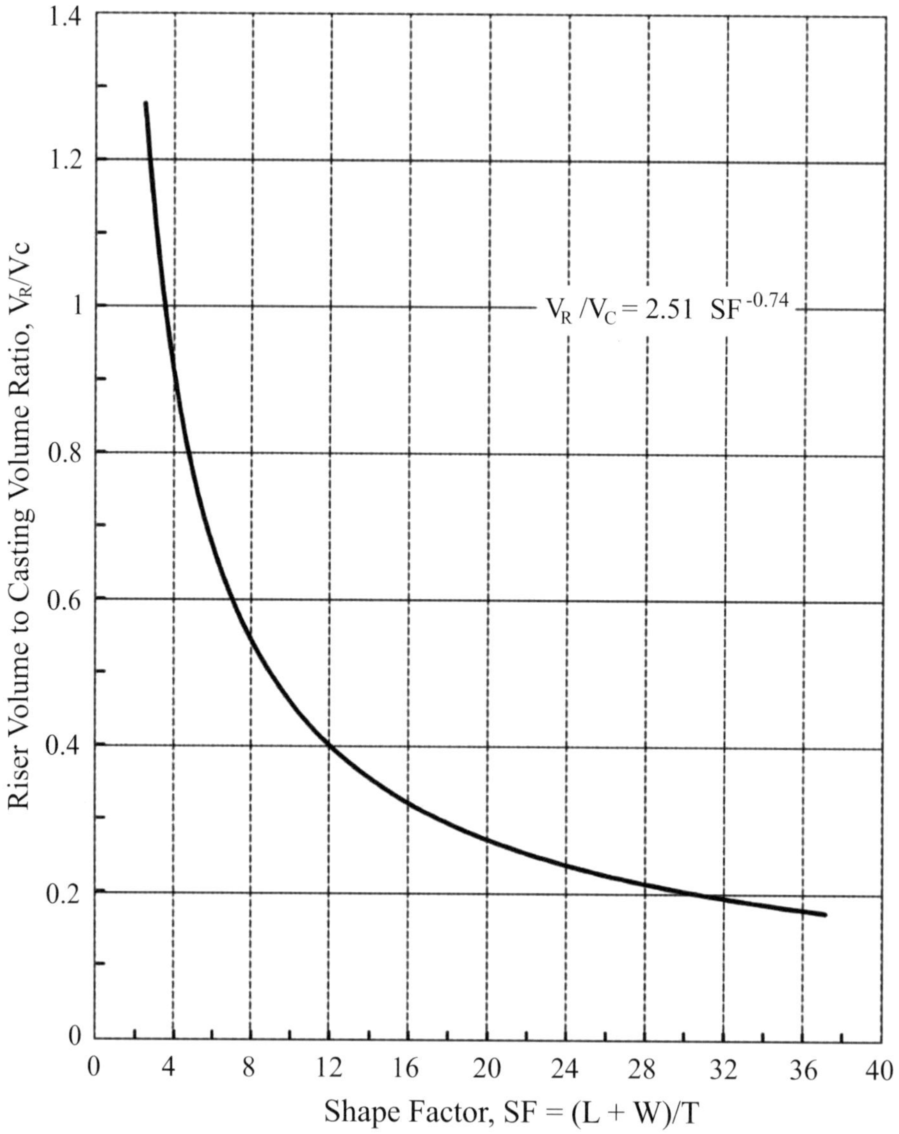

Fig. (11). Ratio V_R/V_C as a function of the shape factor [14].

The Modulus Method

This method is based on the assumption that the freezing time of casting and riser is proportional to their modules, defined as:

$$M_C = V_C/A_C \qquad M_R = V_R/A_R \tag{7}$$

where: M_C, M_R are modulus of casting and riser, respectively [cm]

A_C, A_R are surface area of casting and riser, respectively [cm^2],

V_C, V_R are volume of casting and riser, respectively [cm^3].

These calculations are based on Chvorinov`s rule, which says that the time of solidification of the casting or its part "t" is equal to:

$$t = k^2 \cdot (V_C / A_C)^2 = k^2 M_C^2 \qquad [s], \tag{8}$$

k – constant governed by grade of steel and mould

R. Wlodawer [16] simplified this formula and assumed that instead of calculating the time of solidification of individual hot spots, this rule can be accepted to the solidification time of the whole casting.

Thus $t \approx V_C / A_C = M_C$. In this method, it is also assumed that good feeding of the casting occurs when the value of $M_R = 1.2\ M_C$.

Simplification of the module calculation is of significant practical importance of R. Wlodawer's method [16]. This method assumes that multiple connections of the casting walls correspond to simple geometric shapes, and thus the module M_C can be easily calculated without the need to count the area and the volume of the casting element. This condition allowed for the presentation of calculations of risers in the form of practical charts [16]. Such a solution is very advantageous in engineering practice.

5. COMPUTER METHODS

Recent solutions of calculations of risers are based on counting heat transfer in the solidification process of the casting-riser- hot spot [17 - 19]. The basis for calculations is to predict the places with last drops of liquid steel. This way, it is known where to place a riser or use a chill. An example of such a solution is shown in Fig. (**12**). It shows that the method of calculation the place of last liquid drops of steel, has a significant impact on the location of theoretical shrinkage cavities and consequently - risers.

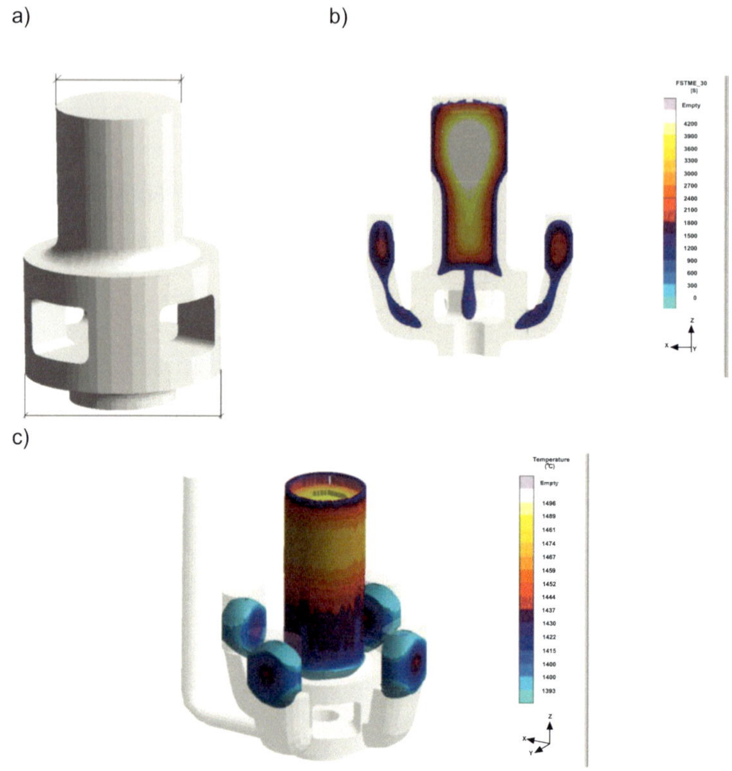

Fig. (12). Example of yoke casting (weight 700 kg) - a, hot spot distribution after 18 minutes of solidification – b and after 82% of solidification time – c [18].

CONFLICT OF INTEREST

The author (editor) declares no conflict of interest, financial or otherwise.

REFERENCES

[1] A.L. Sushil, "Lee.A", *Plutshack, ASM Casting, ASM International,* vol. 15, pp. 589-597, 1988.

[2] A. Tabor, and J. Rączka, *Projektowanie odlewów i technologii form* Fotobit: Cracow, 1998. (in Polish)

[3] Ju. A. Niechendzi, and J. Rączka, *Stalnoje litjo* Mietalurgizdat: Moskwa, 1948.

[4] K. Roesch, H. Zeuner, and K. Zimmermann, *Stahlguss.* Verlag Stahleisen: Düsseldorf, 1982.

[5] P.F. Wasilewskij, *Technologija Stalnowo Litja, Maszinostrojenie.* Moskwa, 1974. (in Russian)

[6] G. Kniaginin Katowice, "Śląsk, Staliwo,1972 (in polish)",

[7] W.S. Pellini, "Factors which determine riser adequacy and feeding range", *Trans.AFS,* vol. 61, pp. 61-80, 1953.

[8] B. Matuschka, "Problems with solidification of steel casting", *26th Int. Foundry Congress,* 1959 Madrid

[9] Feeding and Risering Guidelines for Steel Castings, *Steel Founders' of America.* Illinoi: Barrington,

2001, pp. 1-44.

[10] W.D. Spielberg, *"Computation of Solidification Gradients in Cast Steel Sections"*, PhD. Thesis, Case Western Reserve University, 1970.

[11] Shouzhu Ou, K.D. Carlson, R.A. Hardin, and Ch. Beckermann, Steel Founders Soc. of America, *Feeding and Risering Guidelines for Steel Casting Attachment D* Illinoi: Barrington, 2001.

[12] L. Kozlov, W. Kolokolcev, E. Ten, K. Wdowin, L. Dolgopolova, and A. Filipienkov, *Proizwodstwo Stalnych Otliwok* MISIS: Moskva, 2005.

[13] V.E. Niyama, T. Uchida, M. Morikawa, and S. Saito, "A Method of shrinkage prediction and its application to steel casting practice, Americ.Foundrymen's Soc", *Intern.Cast Met.J.,* vol. 7, no. 3, pp. 52-73, 1982.

[14] R.C. Creese, and Y. Xia, "Tapered riser design optimization", In: *AFS Transaction* vol. 91-67. pp. 717-727.

[15] K.D. Carlson, S. Ou, R.A. Hardin, and Ch. Beckermann, "Development of new Feeding Distance Rules Using Casting Simulation, Part I:Methodology, Attachment C", In: *Feeding and Risering Guidelines for Steel Castings*, 2001.

[16] R. Wlodawer, *Directional Solidification of steel castings.* Pergamon Press Inc.: Long Island City, N.Y., 1966.

[17] R. Hardin, Shouzhu Ou, and Ch. Beckermann, "Yield Improvement Case Study: Computational Study Comparing Stacking and Riser Sizing, Attechment C", In: *Feeding and Risering Guidelines for Steel Castings, Steel Founders of America.*

[18] J. Głownia, D. Koshal, R. Skoczylas, and S. Sobula, "Simulation of solidification conditions in high quality castings for mining industry", *Arch. Metall. Mater.,* vol. 58, no. 3, pp. 791-794, 2013. [http://dx.doi.org/10.2478/amm-2013-0073]

[19] Ch. Beckermann, K. Carlson, and R. Hardin, "Yield Improvement and Defect Reduction in Steel Casting", In: *The University of Iowa, Award Number DE-FC36-02ID 14225*, 2004.

CHAPTER 13

Heat Treatment of Steel Castings

Abstract: This chapter highlights the fact that every steel casting should be heat treated. The structure of cast steel in as-cast conditions requires changes – this applies to virtually all types of cast steel grades. As a result of heating and cooling (according to various technological options), beneficial changes in mechanical and also other properties (*e.g.* hardness, corrosion resistance and wear resistance) are obtained. This chapter includes the types of heat treatment of cast steel, possible practical effects after each treatment and the structural changes caused during its course. There are examples of the microstructure in as-cast conditions, after quenching and after tempering at various temperatures.

Keywords: Hardness, Mechanical properties, Normalisation, Quenching, Tempering, Type of the heat treatment.

1. INTRODUCTION

Iron allotropic transformations play a primary role in heat treatment. In the solid state, depending on the temperature, there are three phases: αFe (BCC), γFe (FCC) and δFe (BCC). Each of these phases occurs in a specified temperature range and has various properties. Control of a cast steel structure consists of choosing appropriate phases and components of the microstructure, in order to obtain beneficial properties for the working conditions of castings. The basis for creating the required structures by the choice of basic procedures of heat treatment are the charts of:

- equilibrium system of Fe - Fe_3C, (Fig. **1**),
- austenite transformations under TTTi isothermal conditions,
- austenite transformations under continuous cooling TTTc.

The Fe - Fe_3C system includes structures obtained in the equilibrium state (at very slow heating and cooling). They are a reference to the structures obtained by practice and heat treatment terminology, but they do not include the time factor. A detailed discussion on the conditions of formation of individual phases, definitions of phases, *e.g.* ferrite or pearlite, and phase transitions can be found in

Jan Głownia

several textbooks on metal physics [1 - 6]. These data were used here for a detailed presentation of the types of heat treatment used for steel castings.

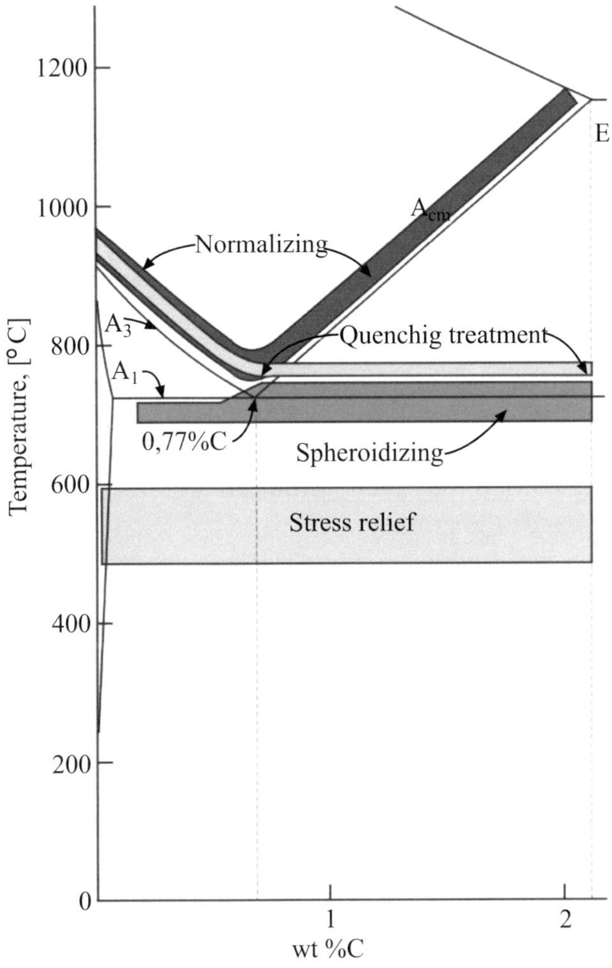

Fig. (1). The basic processes of heat treatment for steel castings.

2. MICROSTRUCTURES OF CARBON CAST STEELS IN AS-CAST CONDITIONS

In as-cast conditions, cast steel has some disadvantages. Steel that is subjected to plastic deformation (wrought, rolling and forging), and in which the macrostructure of the ingot or the slab is subject to complete transformation, has dendrites and primary grains that become fragmented, and the morphology of non-metallic inclusions is fully changed. After the plastic deformation, the primary inclusions are fragmented, and directed according to the rolling direction.

Therefore, the properties of steel are different in samples cut along the direction of rolling, compared to the properties of samples cut in the direction perpendicular to the direction of plastic work.

In steel castings, the properties are virtually uniform in all directions. Thus, the primary inclusions formed in the liquid steel, and the secondary inclusions formed in the range of liquidus – solidus, have the same size, shape and distribution in the steel matrix and in castings (in as-cast conditions). Thus, cast steel reveals different microstructure characteristics (and properties), as compared to forged steel. Therefore, in as-cast conditions, cast steel has some disadvantages:

- it has a dendritic structure, Fig. (**2**); the dendritic growth direction is opposite to the direction of heat transport. Compared to forged steel, cast steel also has a coarser structure,
- it is characterised by chemical heterogeneity. Due to segregation of elements during solidification (in the range of liquidus - solidus) at the primary grain boundaries, the concentration of elements with low partition coefficients (sulphur, oxygen, carbon, but also chromium and manganese) is higher at these boundaries than inside the grains,
- in hypo-eutectoid and hyper-eutectoid carbon, as well as low-alloy cast steel, there is an acicular structure providing low plastic parameters (especially impact toughness). This so-called Widmannstätten structure consists of ferrite and pearlite needles (in hypo-eutectoid cast steel) as well as cementite and pearlite needles (in hyper-eutectoid cast steel) (Fig. **3**).

Fig. (2). Dendritic structure in the shrinkage cavity of steel casting, SEM.

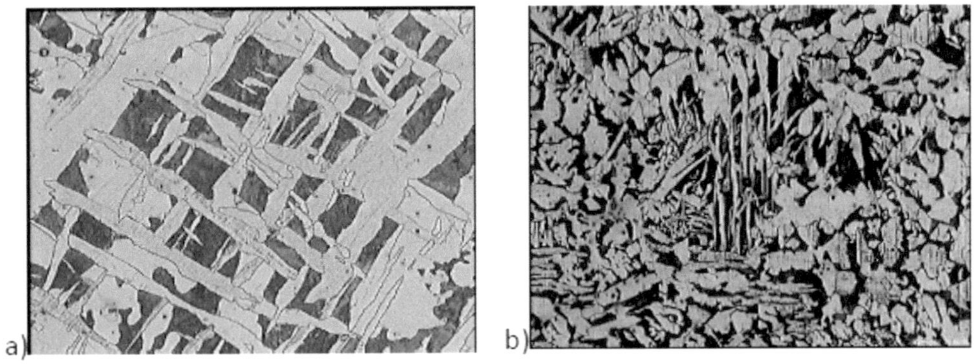

Fig. (3). Widmannstätten structure in carbon cast steels: **a**- in as cast conditions, Magn.400X **b**- after too short a time of holding at the austenite temperature, Magn.100X.

Normalisation

Normalisation is the primary heat treatment allowing for the removal of the Widmannstätten structure. This consists in heating the castings up to a temperature of 40 – 50 °C above Ac_3 or Ac_{cm}, holding at this temperature and then cooling the castings in air. From fine-grained austenite above Ac_3, as a result of its transformation during cooling with air, fine-grained pearlite and ferrite are formed. The results of this processing are shown in Fig. (**3b**). After normalisation, ferrite needles disappear, and the structure is more homogenous and fine-grained. It is also important to reduce stresses in castings and to improve mechanical properties, especially plastic parameters such as elongation and impact toughness. In low-carbon cast steels with a greater participation of ferrite, an impact toughness several times higher than the as-cast condition is obtained after normalisation (Table **1**).

Table 1. Effect of heat treatment on the mechanical properties of carbon cast steel (0.35%C) [7].

Heat treatment	UTS [MPa]	YS [MPa]	RA [%]	El [%]	KCU [J/cm²]	HB
As-cast	608	310	16.3	14.5	1.9	179
Normalisation 930°C/air	678	430	18.5	15.0	4.0	183
Normalisation 830°C/air	650	415	21	18.8	53	175
Normalisation 930°C/air Tempering 650°C/air	588	385	29	25.3	58	175
Quenching 900°C/water Tempering 650°C/air	660	553	32.8	21.8	62	202

Industrial problems with normalising annealing are associated with casting wall thickness and uneven heating in thick and thin walls; the heating time at the austenite temperature and heating rate have a significant impact on the homogeneity of the structure. The influence of wall thickness in castings and the carbon content increase the tendency for the Widmannstätten structure to occur, (Fig. **4**). In the case of an improper selection of heating time (unmatched to the thickest walls of the casting or hot spots), the microstructure reveals a presence of residual amounts of ferrite needles, Fig. (**3b**). This testifies to an incomplete process of normalising.

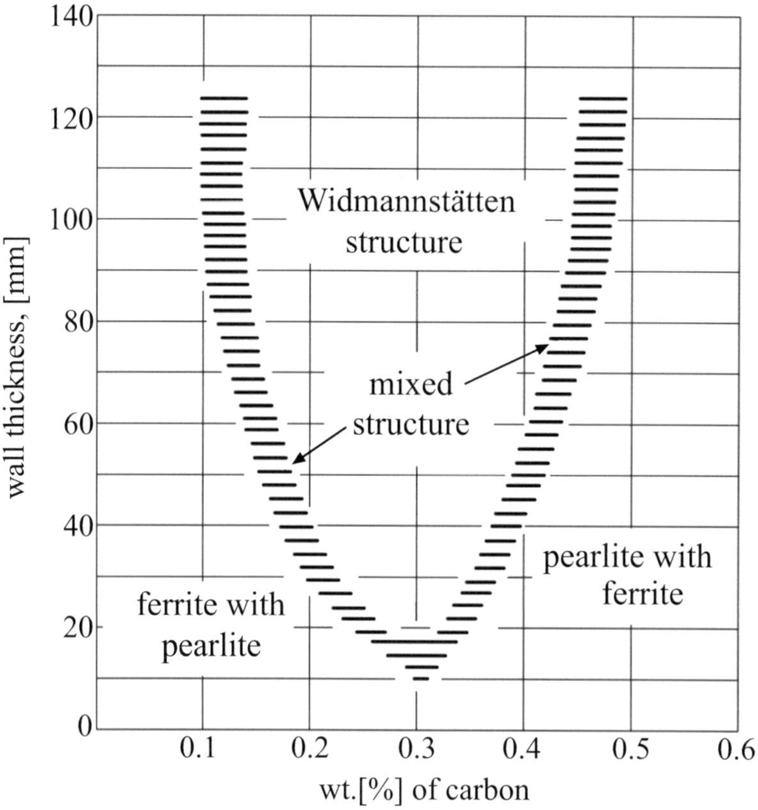

Fig. (4). Effect of wall thickness on the formation of Widmannstätten structure [5].

Annealing

The primary purpose of annealing is to soften the steel in order to reduce the stresses. In steel castings, such conditions occur under the following conditions: at solidification, rapid cooling of the castings with various wall thicknesses, quenching, while applying external chills, after welding, and during the removal of risers.

Stress Relief Annealing

This process aims at the removal of internal stresses, such as residual and transformation stresses, so as not to lead to structural changes in the casting. Castings are heated up to a temperature below Ac_1, held at this temperature for a time depending on the wall thickness of the castings and then cooled slowly within the furnace. Usually, this temperature is lower than 600°C, (Fig. **1**). The effectiveness of stress relief annealing is the greater, the higher the temperature and the longer the heating time. However, when the temperature exceeds ≈550°C, the yield strength, YS, decreases, which is unfavourable to the cast steel's properties.

Spheroidisation Annealing

This type of annealing is used for castings in order to allow their easier mechanical processing and to improve machinability. After performing the annealing, ferrite and spheroidal cementite are obtained. If the structure is pearlite, the cementite plates spheroidise. Spheroidisation annealing consists of heating castings to a temperature of 10 - 15°C above Ac_1 (for hyper-eutectoid cast steel) or 10 - 15°C below Ac_1 (for hypo-eutectoid cast steel), heating at this temperature for several hours and cooling at a rate of 10 - 20°C/h to a temperature of 600 - 650°C. Further cooling may be performed in air (Fig. **1**).

Solution Annealing (Treatment)

Most often, this concerns cast steel in as-cast conditions, after welding or heating in a temperature range of about 425 - 870°C [3]. Complex chromium carbides (Cr, Fe, $Mo_{23}C_6$) are formed at grain boundaries when castings are slowly cooled during the solidification process. The same process is observed in heat-affected zones (HAZ), if the castings are repaired by welding. At the austenite – carbide interphase, the matrix is depleted in chromium (below 12-14%) and the corrosion resistance of the alloy is diminished.

An improvement in the corrosion resistance of castings occurs by dissolving these carbides at an austenite temperature and rapidly cooling the castings in water. A typical course of solution annealing consists of heating cast steel to temperatures of 1050 - 1200°C, keeping it at this temperature until the dissolution of carbides occur and then cooling in water. An example of properly performed duplex cast steel processing is shown in chapter 9. The problem of complete dissolution of carbides may affect corrosion-resistant castings with a wall thickness above 30 - 50 mm in combination with a wall thickness of 8 - 15 mm (pumps for the mining industry, castings for the power industry or marine applications). In these conditions, additional time and temperature may be required to dissolve carbides

(carbides dissolve slowly). In the conventional cast grades, the carbon content is usually higher than in other steels. For this reason, but also in cast steel with molybdenum, castings should be heated at higher temperatures than 1050 - 1100°C. However, in corrosion-resistant cast steel stabilised with titanium or niobium, solution treatment can be performed at lower temperatures (950 - 1050°C) [8].

A second problem may occur during cooling of the castings: slow cooling within the range of 450 - 900°C [8] leads to precipitation of not only carbides, but also sigma (σ), chi (χ), Laves (η) and other phases [2, 8, 9]. In molybdenum-containing cast steels (18%Cr-11%Ni-2-4%Mo) and in cast duplex steels, there is a tendency towards sensitisation. The effect of molybdenum on the formation and precipitation of σ – or χ – phases is well known (and dangerous in practice, (Figs. **17** and **18** in chapter 9). In steels containing 2 - 5% Mo, the first precipitations of the phases already occur within one hour. Therefore, their application should be avoided in the range of 550 - 760°C.

Another example of the application of solution treatment concerns Hadfield cast steel. Its chemical composition (in general 12-14% Mn + 1-1.4%C; in other grades: 17%Mn – 2-3%Cr, 14% Mn - 1-2% Mo, 12-14% Mn – 2-4%Cr – 2-4% Ni) is austenitic at all temperatures, but in as-cast conditions, in addition to austenite, carbides (Fe, Mn_3C) precipitate at grain boundaries, (Fig. (**5**). Carbide precipitates increase the tendency to cracking in the castings, while decreasing the impact toughness, elongation and reduction of area. Since Hadfield cast steel is difficult to work by mechanical tooling, risers are removed after solution annealing.

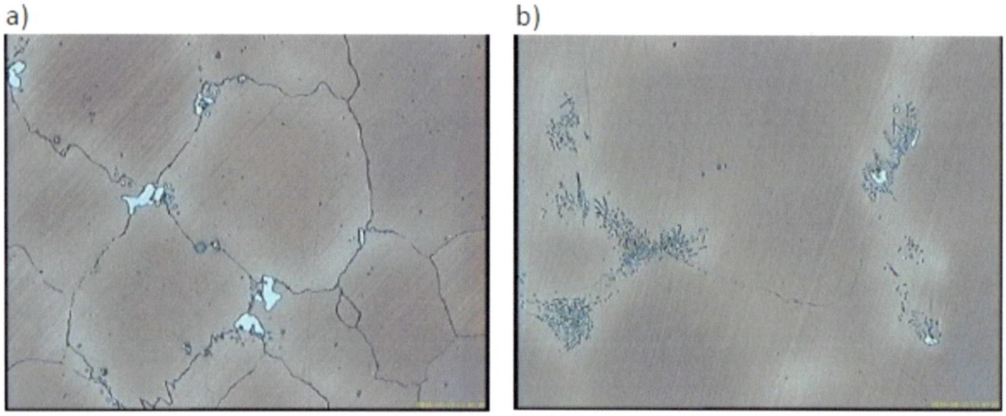

Fig. (5). Microstructure of Hadfield after solution annealing 1050°C/water (a – Magn.400X, white areas – phosphorous eutectic) and after too short a time of solution annealing (b – Magn.500X) [10].

The austenite temperature ranges from 1050-1150°C; it depends, among other factors, on the ratio of Mn:C. At a lower ratio of Mn:C (*e.g.* Mn = 12 - 12.5% and a carbon content of 1.3 - 1.4%), the austenite temperature should be higher [11]. The growth in the amount of carbides at the grain boundaries increases with the wall thickness of castings, their carbon content and chromium alloying. Therefore, in massive castings, which are of interest to the industry (for railway crossings, crawler pads, crusher mantles and jaw plates), the danger of formation of carbides is very high. Quick cooling in water with rapid quenching from the annealing temperature, *e.g.* from 1050°C, suppresses the formation of carbide in castings.

The austenitic manganese cast steels have the ability to harden in depth during wear applications. Reheating these grades of cast steels can result in embrittlement and decomposition of the microstructure, (Fig. **6**). In the temperature range of 380 to 450°C, heat generated by friction also causes precipitation of carbides. This condition leads to rapid lowering of the elongation [12, 13].

Fig. (6). Formation of carbides in reheated high manganese cast steel; service conditions: 400°C -25 hrs, Magn.800X [10].

3. QUENCHING OF CAST STEELS

Quenching of cast steel aims at increasing the hardness on the casting surface or throughout its volume. Quenching comprises the following steps:

1. heating rate to the austenite temperature in [°C/h],
2. holding castings at the austenite temperature (Fig. **1**), for hypo-eutectoid cast steel – 30-50°C above Ac_3 and for hyper-eutectoid cast steel 30 - 50°C above Ac_1,
3. cooling the castings in water, oil or air.

The heating rate has a very important influence on the final quality of castings. Depending on the shape, weight and thickness of casting walls, the heating rate can lead to stress formation and, consequently, to cracks. Castings with various wall thicknesses should be heated slower, taking into account the initial temperature of the furnace. The example of the gradual heating of a massive casting (9 tons) made of cast steel with a 0.7%C content is given in Fig. (**7**).

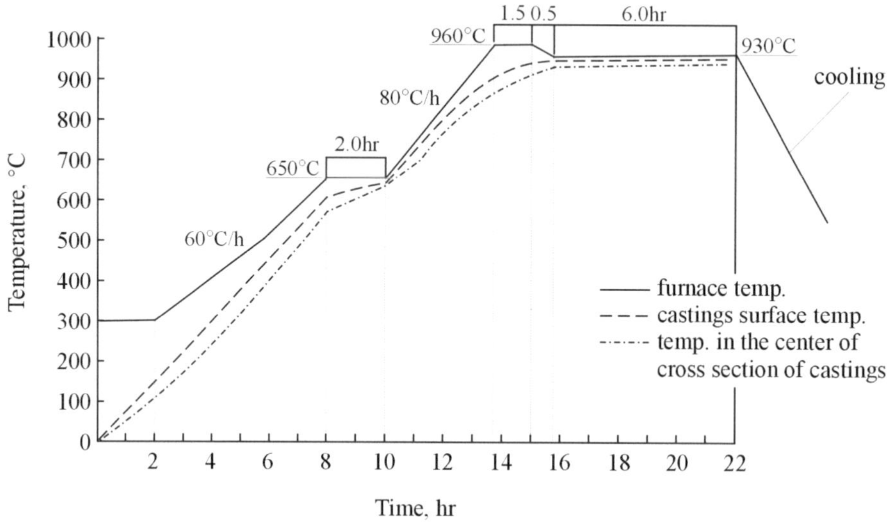

Fig. (7). The heating of heavy steel castings to quenching temperature [4].

It is assumed that the heating rate of massive castings made of alloy cast steel equals about 40°C/h, and for castings made of carbon cast steel with wall thickness up to 100 mm, 150°C/h [4]. These values are often diverse: up to 650°C, the rate is slightly lower at 50 - 80°C/h, and for complex high-alloy castings, it can equal even 20 - 30°C/h.

There are practical indications for holding castings for 1 hour per each 25 mm of the wall. However, in practice, these values must be considered individually, as they depend not only on the chemical composition of the cast steel and the structure of the casting, but also on the furnace type, efficiency of heating (arrangement of gas burners and their performance) and on the arrangement of castings on the furnace hearth.

The method of cooling determines the type of microstructure obtained, martensite or bainite. For a given cast steel, the cooling rate and the T-T-T diagram (time-temperature-transformation diagram) characterise the obtained structure, (Fig. **8**). In the temperature range of 450 - 650°C, there is a critical cooling rate, which determines the austenite - martensite transformation. For low-alloy Mn – Ni cast steel, the martensite structure is obtained at a cooling rate above 130°C/s.

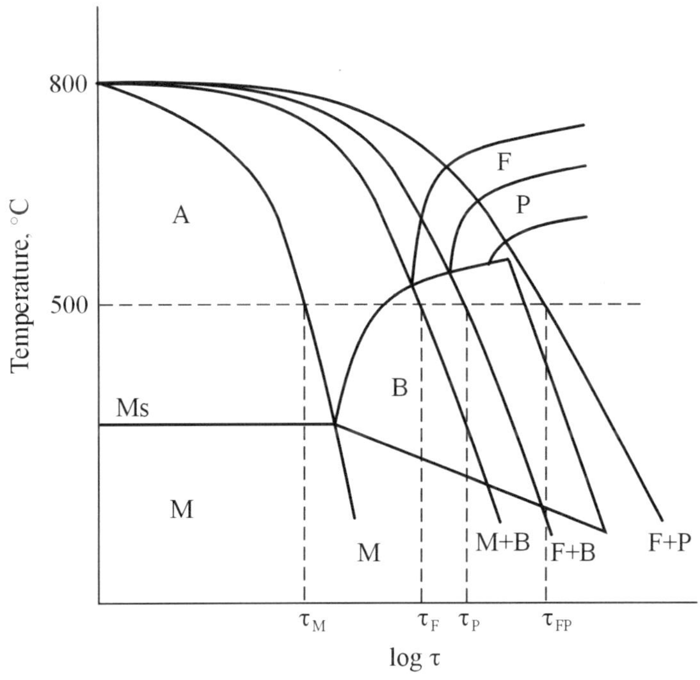

Fig. (8). General TTT-diagram for the quenching of cast steels.

Cooling media in practice give very diverse cooling rates (Table **2**).

Table 2. Cooling rates in different media [4].

Cooling medium	**Cooling rate [° C/s] in the temperature range**	
	650 – 550 [°C]	300 – 200[°C]
Water 18°C	600	270
Water 28°C	500	270
Water 50°C	100	270

(Table 2) contd.....

Cooling medium	Cooling rate [° C/s] in the temperature range	
Water 74°C	30	200
Pressured air	70	8
Oil	100 – 130	≈ 25
Water + polymer (12-20%)	120 - 200	30

The values of the cooling rate mentioned may be different in industrial practice. The reasons are due to: gritty casting, burnt-on surfaces, too much time elapsing from taking the casting out of the furnace until immersion in water, contamination of water with oil or quartz dust, or the amount of movement of the cooling medium in the water tank.

Quenching of castings can be divided into:

1. direct quenching,
2. surface quenching.

Each of these groups has their own methods [14]. For example, depending on the cooling method, direct quenching can be performed:

- with continuous cooling in one medium (the most common method of quenching of castings in water, oil or aqueous polymer solutions). In this method, cooling occurs directly from the austenite temperature to room temperature;
- from the so-called interrupted quenching: depending on rapid quenching from the austenite temperature to the temperature above the M_s temperature, where it is held for a time sufficient to ensure the desired transformation. This quenching technique comprises either martempering, austempering or isothermal annealing;

martempering [3] - a method similar to direct quenching, but has the advantage that when holding in the salt bath at a temperature above M_s, there is a balancing of temperature throughout the wall thickness and of hot spots throughout the casting. Thereby, residual stresses are minimised, which brings very beneficial effects in castings of diverse wall thicknesses. Castings are kept in molten salt baths or technical oils, and after taking out of the bath are cooled in air;

austempering - differs from martempering of steels only by holding for longer at a temperature above M_s, so that the bainite transformation occurs;

isothermal annealing - consists in austenitisation at higher temperatures, so that the austenite transformation into pearlite can take place.

A schematic course of the cooling techniques mentioned is presented in Fig. (**8**).

Heat treatment of steel castings results from the characteristics of the crystallisation process. The effect of segregation causes variable concentration of elements in the middle of grains, compared to their boundaries (micro-segregation). The three zones of crystals in castings (frozen grains at the chill zone, columnar and equiaxed) create the effect of macro-segregation, where the largest concentration of elements is in the equiaxed zone. The presence of structural heterogeneity requires homogenisation annealing. Since such annealing is performed at a high temperature (1000 - 1150°C) for a long period of time (8 to 15 hours), the costs of the operation are high. Therefore, it is used for massive (several dozen tons) castings requiring homogeneous structure and reduction of the effects of segregation (carbon, chromium) in walls with a thickness of more than 100 mm. An example of complex heat treatment of cast steel containing 0.35%C - 1.5%Cr - 3.5% Ni is shown in Fig. (**16**), chapter 8. Application of double deoxidation of steel, introduction of modifications and application of one of the secondary metallurgy processes (*e.g.* vacuum degassing) allows for a reduction in segregation in castings, thus avoiding a costly diffusion procedure. The phenomenon of segregation is revealed during cooling in two ways:

1. fine-grain structure at grain boundaries in relation to the matrix (Fig. **9**),
2. precipitates of carbides at grain boundaries.

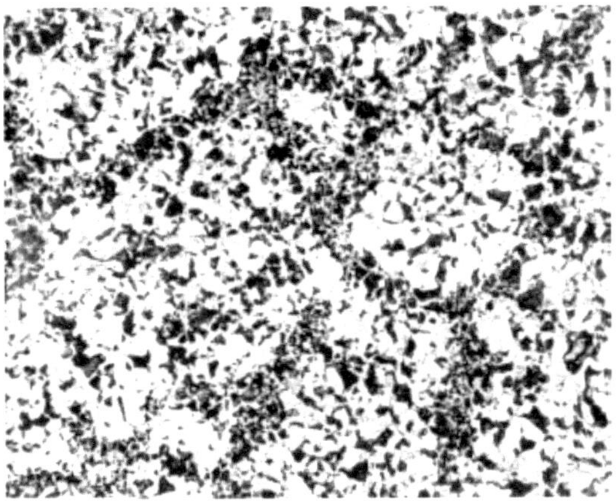

Fig. (9). Effect of segregation on the grain size on the prior austenite grain boundary in carbon cast steel, Magn.100X.

Heterogeneity of the structure can be met in castings during the transformation of austenite into pearlite, as a result of the segregation of carbon and slow cooling between the temperatures Ar_3 and Ar_1 [9].

Transformations during austempering differ from the process of quenching and tempering:

1. after austempering, there is no tempering,
2. the final structure is bainite, not martensite,
3. there is only one transformation of austenite into bainite.

Austempering process technology consists in:

1. heating to the austenite temperature above the Ac_3, holding the casting at this temperature until a homogenous structure is obtained at the cross-section of the casting wall (about 1 hour/25 mm),
2. quenching into a hot bath (250 - 450°C) and holding at this temperature until the transformation of austenite into bainite is finished,
3. removing the casting from the molten bath and cooling it to room temperature in air.

The austempering process brings measurable benefits. Compared to direct quenching, it reduces the tendency to crack, because the difference between the austenite temperature and austempering transformation (hot bath) is 200 - 400°C lower than the difference given by direct quenching in water [9]. These smaller temperature differences also bring smaller differences in temperature between the surface of the workpiece and its centre. In the case of castings, including massive castings (with a wall thickness of 80 to 150 mm), smaller thermal stresses are formed. This is a very important effect of austempering. Another one is that there are no changes in volume: the austenite - martensite transformation is characterised by an increase in volume (6% [5, 15]), while the austenite - bainite transformation takes place gradually, both in thin and thick walls of the casting. The two beneficial effects reduce the number of cracks in castings, repairs by welding and re-grinding, which eliminate a certain percentage of scrapped castings.

A comparison of both processes used in steels is very common. It shows that they lead to an improvement in impact toughness, but also in the fatigue life [16]. The major difference is the temperature in which maximum impact toughness is obtained: 350°C in the austempering process and 600°C after quenching and tempering.

Introduction of austempering to steel castings is limited because:

- in carbon steels, the complete transformation of austenite into bainite occurs only in the walls with a thickness of up to 5 mm,

- in low-alloy steels (*e.g.* 0.5%C - 0.8%Cr), this transformation takes place in walls with a thickness of up to 25 mm [9].

Therefore, heat treatment of castings takes place in two stages: quenching in water or oil, and then tempering. If, after quenching, tempering is performed at high temperatures (450 - 650°C), the heat treatment is called toughening. Such heat treatment is used for castings intended for the construction of machinery and equipment. It provides optimal mechanical properties (UTS, YS) in relation to the plastic parameters in castings (EI, RA, KCV). Due to the hardenability of cast steel containing 0.3 - 0.4% of carbon, the preferred set of properties can be only obtained in castings with a wall thickness of 50 - 60 mm. Ju.A. Szulte revealed [8] that toughening of carbon cast steel with a content of 0.35%C gives the best set of mechanical properties, (Table **3**).

Table 3. Mechanical properties of carbon cast steel (0.35%C) after quenching and tempering [8].

Heat treatment	**UTS [MPa]**	**YS [MPa]**	**RA [%]**	**El [%]**	**KCU [kJ/m²]**	**HB**
as cast	608	310	16.3	14.5	190	179
Normalisation 930°C/air	678	430	18.5	15.0	400	183
Double norm.930°C + 830°C/air	650	415	21.0	18.8	530	175
Norm.930°C+temp.650°C/air	588	385	29.0	25.3	580	175
Quenching 900°C+temp.650°C/air	660	553	32.8	21.8	620	202

In cast steel containing 0.45% C, the strength parameters are much higher with low-temperature tempering, but similarly to cast steel with a content of 0.35%C, an optimal set of properties is obtained after quenching and tempering in the temperature range of 500°C - 550°C.

Many machine constructions require much higher yield strength, YS, at appropriate values of elongation. For such conditions, low-alloy cast steel and its heat treatment is more complex. The presence of alloy elements (mainly chromium, nickel and molybdenum) allows for quenching in air. Low-alloy cast steel with the chemical composition 0.35%C-1.1%Cr-1.9Ni-0.3% Mo, subject to normalisation + quenching + tempering in a temperature range between 350° - 650°C, allows a ratio of YS/UTS above 0.9 to be obtained, (Fig. **10**). The obtained values of YS exceed 1200 MPa at an elongation of 6-8%. Cast steel with such a chemical composition is used for castings working in the coal mine industry (*e.g.* yokes) and in the machine industry as vehicle elements, military equipment [17] and elements of clutches or spindles.

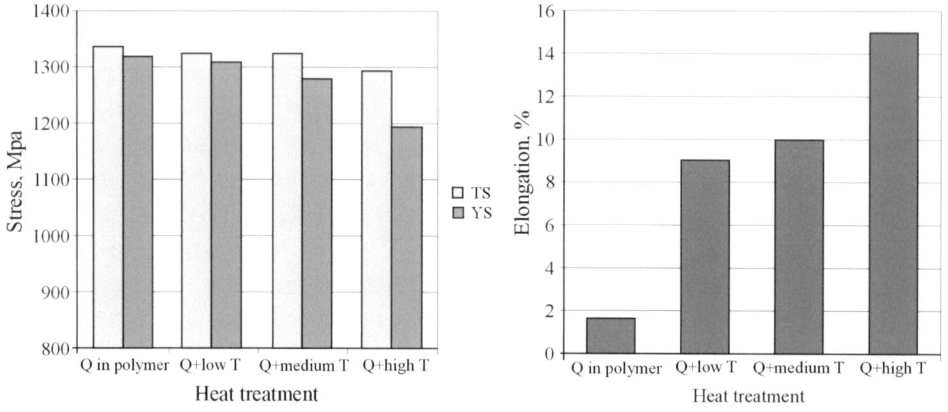

Fig. (10). Effect of different methods of quenching on the mechanical properties of low-alloy cast steel [18].

Stress-Relief Annealing

The cooling of castings during solidification and quenching cooling causes temperature gradients and, consequently, thermal stresses and structural changes, leading to volume changes (transformational stresses). These effects are caused by the earlier shrinking of thin casting walls, which act forcibly on thicker casting walls, the latter still being in the liquid-solid range of the 'mushy-zone'. The greater the differences in wall thicknesses and the greater the temperature differences between them, the larger the stress. The value of these stresses decides whether there are deformations in the casting (if they are smaller than the matrix strength of the cast steel), or cracking - if they exceed the YS. Time also plays a big role here, because only additional stress causes cracks.

Elimination of these stresses can be performed by the annealing of castings at a temperature below A_1 for several hours (at least 5 hrs [15],). Holding at this temperature, however, depends on the complexity of the casting (its construction, wall connections, reinforcing ribs), maximum wall thickness, chemical composition (carbon or alloy cast steel, content of sulphur, phosphorus and non-metallic inclusions). The cooling rate of castings from the temperature of stress-relief annealing should be less than 15°C - 25°C/hr. In order to maintain the mechanical properties after the final heat treatment, the temperature of stress-relief annealing should be lower by 30°C - 40°C.

4. HARDENING PROCESSES

A wide set of various heat treatment techniques also includes another concept for increasing the hardness. The hardening procedure assumes maintaining the plastic matrix of steel (cast steel), wear resistance and fatigue resistance. There are several groups of these methods:

- surface hardening,
- carburising,
- nitriding and carbonitriding,
- ion implantation.

The aim of **surface hardening** of castings is obtaining a hard surface at the plastic matrix of cast steel. This heat treatment is applied to castings that work in abrasive conditions with a carbon content of 0.35% to 0.65%. Before surface hardening, castings are subjected to normalising or toughening. Rapid cooling of the outer film of the casting from the austenite temperature leads to the formation of a martensite structure within it. The surface layer usually has a hardness of 50 to 60 HRC at a depth of several millimetres (up to 4-6 mm).

There are two methods of surface hardening: by induction heating the casting's outer surface or by flame hardening.

Induction hardening: the depth of obtained martensite depends on the frequency of the current (at 25 kHz – "d" ≈ 6-10 mm; at 440 kHz - ≈ 1 mm)

Flame hardening: is applied when the shape or area of the hardened surface hinder the use of another method. Construction of gas burners and construction of the quenching machine make the cost of this method high.

Carburising: is applied for carburising of the surface from 0.15%C to 0.9-1%C. Before carburising, the surface is machined and then exposed to gas or liquid at a temperature of up to 920-930°C. The carburising materials are:

1. coke, charcoal with barium carbonate,
2. molten salt bath (sodium cyanide + barium chloride or sodium chloride with accelerators),
3. hydrocarbon gases (methane, propane, alcohol and alcohol derivatives, other organic carbon compounds).

Carbonitriding: gives a tougher hardened layer, but not as deep as after carburising (up to 0.1 mm at 760°C within 1 hr, or 0.75 mm at 840°C within 6 hrs). Nitriding may be achieved by heating tool steel in a cyanide bath or gaseous nitrogen at 510 – 565°C. Carbonitriding uses a mixture of ammonia and hydrocarbons. Gas nitriding is used for high-alloy steels (stainless steels) and hot work die steels (dies, gears, seals and camshafts).

Ion Implantation: is applied in modifying metal surfaces to improve the properties of tools. This process involves embedding atoms into the substrate surface using a high-energy beam of ionised particles. Its advantages include low-temperature processing, good reproducibility of the penetration depth of atoms and good solubility of atoms without precipitation.

All methods of hardening are generally connected with the use of high-alloy steels for tools, and have very little use with castings. They are described in reference [9].

CONFLICT OF INTEREST

The author (editor) declares no conflict of interest, financial or otherwise.

REFERENCES

[1] G.E. Totten, *Steel Heat Treatment,* G.E. Totten, Ed., vol. 1. 2nd ed.. CRC Press: Boca Raton, 2007.

[2] D. Koshal, Ed., *Manufacturing Engineer's Reference Book* Butterworth-Heinemann Ltd.: Oxford, 1993.

[3] C.R. Brooks, *Principles of the Heat Treatment of Plain Carbon and Low Alloy Steels.* Chapter 7. ASM Intern, 1996.

[4] S. Gawlik, Obróbka cieplna Odlewów, Poradnik Odlewnictwa, Ed., *Polish Foundrymen`s Association, t.1, chapter II.*, 2013, pp. 397-421. (in Polish)

[5] K. Roesch, H. Zeuner, and K. Zimmermann, *Stahlguss.* Verlag Stahleisen GmbH: Düsseldorf, 1982.

[6] G. Krauss, *Steels: Heat Treatment and Processing Principles.* ASM Intern: Material Park, Ohio, 1990.

[7] Ju.A Szulte, *Elektrometalurgija Stalnowo Litja.*Metalurg, Moskwa, 1970. (in Russian)

[8] P.R. Rios, and A.F. Padilha, Precipitation from Austenite.*Encyclopedia of Materials Science and Technology Updates.,* K.H. Buschow, Ed., Elsevier: Amsterdam, 2005. [http://dx.doi.org/10.1016/B0-08-043152-6/01408-X]

[9] B. Liscic, Steel Heat Treatment:Metallurgy and Technologies*Steel Heat Treatment,* G.E. Totten, Ed., CRC Press.2nd ed.

[10] G. Tęcza, and A. Garbacz-Kledmpka, Microstructure of cast high-manganese steel containing titanium, Archives of Foundry Engineering*Polish Academy of Sciences, Commision of Foundry Engineering* vol. 16. , 2016, no. 4, pp. 163-168.

[11] G. Kniaginin, Śląsk Ed., *Staliwo-Metalurgia i Odlewnictwo*, 1972.

[12] H.A. Fabert, Manganese Steels in Crushing and Grinding Service, Symp, *Materials for the Mining Industry.* Climax Molybdenum Co., 1974, pp. 166-167.

[13] Steel Founders` Soc. of America, *Steel Castings Handbook* Chapter 16. 5th ed. , 1980.

[14] C.R. Brooks, *Principles of the Heat Treatment of Plain Carbon and Low Alloy Steels, Chapter 5-Tempering, Chapter 6-Austenitization of steels* ASM Intern., The Materials Information Society, 1996.

[15] E.J. Mittermeier, J. Grosch, Ed., *Progress in Heat Treatment and Surface Engineering* ASM Intern., Materials Park: Ohio, 2000.

[16] F.W. Eysell, "Zeit-Temperatur-Umwandlungs-Schaubild (ZTU) fuer kontinuierliches Abkuehlen des Stahls", *Z. Prakt. Bearbeitung,* no. 27, p. 3, 1972.

[17] M.A. Layton, K.O. Yu, S. Guleyupoglu, and G.K. Upadhya, "Casting process development for a steel, low-loss, launch-valve plug", In: *AFS Trans.* vol. 97-127. , 1997, pp. 273-278.

[18] S. Sobula, M. Rapala, G. Tecza, and J. Glownia, "Cast steel with a yield strength above 1300MPa-comparable to forgings", In: *Przegląd Odlewnictwa* vol. 3. , 2009, pp. 102-106.

SUBJECT INDEX

A

abrasion 147, 148, 154, 219, 231, 247, 248, 250, 251, 253, 254, 263, 265, 266, 268, 269, 271
acid 5, 6, 11, 18, 45, 48, 54, 58, 63, 95, 194, 201, 211, 213
active domain 194, 201, 210
advantages 80, 113, 114, 124, 126, 128, 130, 157, 213, 280, 307
advantages 80, 113, 114, 124, 126, 128, 130, 157, 213, 280, 307
anions 21, 29, 42, 210, 226
annealing 2, 63, 106, 181, 183, 187, 206, 213, 233, 234, 253, 263, 268, 303, 306, 307
AOD 7, 11, 73, 131, 132, 147, 157, 183, 208, 210
aqueous corrosion 192
argon blowing 112, 120

B

basic 5, 6, 18, 20, 21, 25, 28, 44, 45, 47, 48, 54, 63, 85, 93, 95, 107, 113, 117, 122, 126, 129, 145, 168, 176, 180, 210, 219, 220, 263, 268, 269, 278, 294, 295
basicity 16, 18, 20, 21, 25, 26, 29, 38, 41, 42, 45, 122, 248
blowing of oxygen 15, 47
boron 62, 145, 160, 176, 177, 235, 258, 259
bottom ladle 277, 280
breakdown of passive film 196
brittle fracture 6, 61, 70, 165, 166, 168, 169, 180, 184, 185, 190
by complex deoxidisers 98

C

carbide 30, 47, 93, 97, 145, 155, 160, 204, 205, 207, 208, 235, 253, 271, 299, 300
carbides 97, 139, 142, 145, 154, 165, 170, 173, 176, 177, 183, 184, 188, 194, 200, 213, 214, 221, 231, 232, 239, 240, 253, 258, 299, 300, 304
carbon 6, 7, 15, 29, 31, 41, 45, 46, 54, 55, 58, 59, 73, 74, 79, 81, 83, 84, 99, 109, 113, 115, 120, 127, 128, 131, 132, 139, 140, 148, 149, 151, 152, 154, 155, 158, 159, 169, 190, 192, 200, 214, 227, 231, 240, 249, 253, 261, 272, 304-308
carbon 17, 66
carburising conditions 229, 239, 240
cast steels ii, 5, 41, 46, 60, 63, 64, 81, 84, 86, 106, 109, 139, 140, 144, 148, 159, 165, 176, 181, 182, 186, 190, 192, 196, 200, 204, 219, 229, 230, 240, 247, 250, 254, 261, 263, 268, 271, 295, 297, 308
cast steels ii, 5, 41, 46, 60, 63, 64, 81, 84, 86, 106, 109, 139, 140, 144, 148, 159, 165, 176, 181, 182, 186, 190, 192, 196, 200, 204, 219, 229, 230, 240, 247, 250, 254, 261, 263, 268, 271, 295, 297, 308
cations 21, 42, 216, 226
charge 18, 22, 24, 47, 53, 63, 92, 93, 97, 115, 121, 123, 126, 131
chlorides 194, 200
chromium 13, 31, 39, 46, 47, 62, 64, 65, 67, 75, 78, 97, 128, 131, 139, 149, 152, 172, 175, 182, 183, 186, 202, 211, 214, 215, 220, 221, 226, 247, 250, 253, 254, 256, 258, 259, 270, 272, 284, 296, 299, 300, 304, 306

Chromosil 259, 260, 262
compounds 12, 17, 28, 41, 43, 57, 63, 78, 79, 203, 206, 220, 222, 223, 225, 232, 234, 307
computer methods 290
conditions ii, iii, 7, 12, 16, 24, 26, 29, 41, 44, 53, 67, 81, 83, 92, 95, 96, 102, 112, 114, 115, 117, 122, 126, 127, 139, 151, 152, 154, 177, 186, 197, 206, 207, 213, 214, 223, 259, 263, 266, 268, 269, 277, 283, 284, 306, 307
consumption of electric energy 14
contraction 283, 284
converter 5, 6, 11, 38, 43, 48, 122, 123, 125, 130, 132, 183, 276
cooling rate 72, 142, 149, 152, 155, 159, 184, 186, 187, 205, 265, 270, 285, 302, 303, 306
cooling rates 139, 155, 214, 302
corrosion ii, 1, 5, 7, 12, 22, 24, 26, 41, 46, 49, 64, 65, 73, 79, 81, 97, 106, 113, 122, 124, 130, 145, 157, 224, 228, 250, 251, 255, 256, 270, 271, 294, 299
crevice corrosion 202, 204
critical 24, 71, 83, 139, 144, 145, 154, 170, 194, 197, 199, 200, 204, 215, 216, 234, 302
current density 194, 196, 197, 200, 209

D

damp materials 14
decarburisation 11, 15, 24, 26, 127, 129, 130
dendrites 100, 101, 104, 106, 295
deoxidation ii, 2, 6, 11, 15, 17, 18, 23, 24, 30, 38, 39, 45, 47, 48, 55, 56, 58, 59, 75, 76, 84, 85, 118, 124, 126, 132, 147, 149, 155, 160, 168, 175, 176, 187, 190, 258, 259, 304
deoxidation ii, 2, 6,11, 15, 17, 18, 23, 24,30, 38, 39, 45,47, 48, 55, 56, 58, 59,75, 76, 84, 85, 118, 124, 126, 132, 147, 149, 155, 160, 168, 175, 176, 187, 190, 258, 259, 304
deoxidisers 72, 73, 75, 76, 78, 80, 92, 94, 104, 106, 107, 114, 124, 131, 178
dephosphorisation 113, 155
desulphurisation 11, 15, 26, 27, 39, 43, 47, 48, 70, 80, 92, 99, 128, 130, 132, 147, 155, 238
DETEM 7, 128-130
diameter of inclusions 68
diffusion 15, 24, 38, 39, 44, 45, 55, 59, 63, 73, 93, 122, 207, 208, 235, 240, 254, 304
duplex structure 192

E

effect of argon stirring 118
effect of elements 140
effect of rare earth 80
element 13, 24, 46, 47, 54, 58, 64, 65, 68, 70, 73, 94, 125, 143, 145, 173, 175, 176, 178, 184, 185, 196, 201, 209, 210, 219, 220, 224, 229, 233, 237, 255, 290
elevated temperature 219
equilibrium 17, 18, 21, 27, 31, 38, 44, 60, 63, 73, 84, 105, 108, 124, 160, 266, 294

F

feeding 78, 119, 120, 122, 276, 290-292
filling velocity 276
first castings 2, 3, 5
freeboard 114, 115, 127, 129
friction 206, 247, 248, 251, 252, 270, 271, 283, 300

G

gases 7, 11, 14, 59, 66, 67, 104, 112, 113, 116, 126, 130, 133, 157, 170, 180, 186, 192, 219, 220, 230, 231, 237, 254, 276, 307

gating systems 170, 283
general corrosion 201, 202

H

Hadfield cast steel 142, 168, 299
Hadfield cast steels 308
hardenability 152, 153, 156, 157, 182, 263, 266, 267, 305
hardening 145, 161, 178, 182, 253, 281, 307, 308
hardness 147, 167, 251, 263, 266, 268, 269, 294, 301, 307
harmful elements 39, 46, 112
HAZ 206, 299
heat resistance 92, 145, 219, 220, 222, 229, 231, 232, 235
heat treatment ii, 11, 61, 106, 148, 150, 165, 168, 181, 182, 184, 185, 188, 190, 206, 210, 211, 214, 216, 234, 253, 255, 257, 259, 263, 266, 268, 269, 271, 294, 295, 297, 303-308
high manganese 247, 256, 268, 269, 272, 300
high strength 85, 147, 150, 151, 153, 173, 215, 253
hydrated ions 194
hydrogen, nitrogen 115, 124
hydrogen in steel 53, 54

I

immunity 193
impact toughness 1, 13, 19, 54, 69, 70, 76, 80, 81, 83, 84, 86, 92, 106, 107, 112, 132, 139, 143, 145, 147, 149, 150, 152, 154, 155, 157, 160, 234, 250, 253, 259, 263, 268, 269, 296, 297, 299, 305
impact toughness 1, 13, 19, 54, 69, 70, 76, 80, 81, 83, 84, 86, 92, 106, 107, 112, 132, 139, 143, 145, 147, 149, 150, 152, 154, 155, 157, 160, 234, 250, 253, 259, 263, 268, 269, 296, 297, 299, 305
inclusions 7, 11, 17, 38, 39, 48, 53, 54, 58, 59, 75, 94, 112, 120, 122, 124, 126, 128, 130, 142, 147, 155, 158, 165, 175, 186, 192, 194, 203, 204, 221, 240, 254, 259, 276, 278, 295, 306
induction furnace 5, 38, 43, 107, 129, 130, 150, 276
ingates 277, 278, 280, 283
in steel castings 53, 54, 60, 63, 66, 69, 92, 130, 165, 295, 298
in temperature 15, 28, 38, 60, 94, 229, 230, 238, 305
intergranular corrosion 200, 202, 204, 205, 208, 210, 211, 213, 214, 216
in vacuum 105, 106, 112, 115, 125, 126, 130-133
ionic theory 21, 29, 41, 42
iron castings 2-4

K

kinetics of oxidation 221

L

layer 24, 155, 193, 200, 210, 220, 221, 223, 224, 228, 229, 251, 252, 254, 255, 276, 307
LD 45, 55, 122, 214
LF 73, 106, 119, 121, 122, 128, 130, 147, 214
lining 5, 6, 15, 20, 23, 24, 38, 41, 48, 53, 59, 66, 112, 113, 119, 122, 126, 127, 129, 277
LTS process 122

M

magnetite 56, 226, 230
manganese 6, 7, 13, 22, 27, 38, 46, 47, 57, 58, 62, 67, 69, 70, 83, 84, 94, 96, 99, 106, 124, 125, 131, 139, 152, 154, 160,

168, 172, 173, 175, 176, 186, 190, 199, 221, 232, 247, 256, 258, 263, 266, 272, 283, 296, 300, 308
manganese 6, 7, 13, 22, 27, 38, 46, 47, 57, 58, 62, 67, 69, 70, 83, 84, 94, 96, 99, 106, 124, 125, 131, 139, 152, 154, 160, 168, 172, 173,175, 176, 186, 190, 199, 221, 232, 247,256, 258, 263, 266, 272, 283, 296, 300, 308
melting and evaporation 46
melting of steel in EAF 115
MHD process 122
mixture 21, 26, 30, 41, 47, 55, 56, 63, 93, 115, 220, 223, 240, 255, 256, 307
Mn 6, 12, 41, 46, 47, 53, 54, 56, 66, 67, 69, 70, 83, 86, 92, 94, 96, 98, 107, 108, 115, 125, 130, 140, 141, 144, 160, 166, 186, 190, 208, 211, 222, 242, 249, 250, 253, 261, 262, 265, 271, 282, 283, 299, 300, 302
Mn 6, 12, 41, 46, 47, 53, 54, 56, 66, 67, 69, 70, 83, 86, 92, 94, 96, 98, 107, 108, 115, 125, 130, 140, 141, 144, 160, 166, 186, 190, 208, 211, 222, 242, 249, 250, 253, 261, 262, 265, 271, 282, 283, 299, 300, 302
modulus method 289
moulds 5, 11, 13, 54, 102, 104, 112, 117, 130, 131, 236, 269, 276, 277, 279-281

N

nitrogen in molten steel 68
Niyama 285, 292
normalisation 19, 106, 107, 150, 152, 154, 158, 259, 294, 297, 305, 306

O

O 7, 8, 12, 68, 70, 72, 73, 76, 79, 80, 102, 103, 105, 108, 117, 119, 141, 157, 160, 180, 193, 195, 197, 199, 200, 202, 204, 206, 209, 211, 215, 220, 221, 240, 254, 255, 272, 297-308
of carbon 3, 6, 13, 18, 19, 22, 29, 45, 55, 66, 70, 71, 74, 79, 81, 86, 94, 105, 107, 109, 113, 123, 124, 128, 131, 132, 139, 140, 142, 143, 146, 149, 154, 159, 172, 173, 175, 180, 190, 200, 205, 207, 209, 231, 234, 240, 249, 253, 264, 266, 267, 269, 270, 284, 295, 297, 301, 304, 305
of cast steels 81, 106, 139, 152, 154, 165, 181, 190, 213, 214, 219, 271, 300-302
of charge 13
of chromium 47, 62, 64, 65, 75, 78, 126, 128, 144, 145, 147, 152, 154, 155, 175, 182, 183, 197, 200, 205, 220, 226, 230, 232, 238, 240, 256, 258, 259, 263, 264, 272
of hot spots 284, 303
of low alloyed cast steel 187
of the bath 15, 16, 22, 26, 29, 63, 66, 93, 94, 96, 100, 106, 124, 130, 131, 303
of the bath 15, 16, 22, 26, 29, 63, 66, 93, 94, 96, 100, 106, 124, 130, 131, 303
on properties 242
outflow 23, 66, 68, 80, 94, 106, 123, 128, 278, 280, 282
oxidation 6, 17, 18, 38, 41, 58, 63, 66, 67, 69, 83, 92, 95, 98, 107, 114, 120, 126, 127, 129, 192, 201, 202, 250
oxide film 221, 222, 224, 225, 236
oxygen activity 29, 58, 72, 85, 93, 94, 175, 180

P

P 16, 28, 31, 38, 45, 46, 48, 49, 56, 60, 62, 64, 66, 72, 94, 99, 105, 108, 122, 133, 141, 153, 155, 156, 160, 185, 189, 190, 200, 202, 203, 211, 215, 216, 222, 223, 228, 241, 256, 257, 261, 268, 291, 308
passive film 210, 255
passivity 192, 193, 215

period 1, 11, 17, 18, 21, 22, 26, 27, 30, 39, 70, 92, 93, 97, 104, 113, 123, 124, 127, 128, 180, 204, 207, 214, 221, 226, 229, 231, 270, 303
period 1, 11, 17, 18, 21, 22, 26, 27, 30, 39, 70, 92, 93, 97,104, 113, 123, 124, 127, 128, 180, 204, 207, 214, 221, 226, 229, 231, 270, 303
phases in stainless steels 206
phosphorus 2, 15, 16, 24, 39, 46, 84, 94, 112, 147, 172, 178, 187, 203, 258, 269, 270, 306
phosphorus 2, 15, 16, 24, 39, 46, 84, 94, 112, 147, 172, 178, 187, 203, 258, 269, 270, 306
pitting corrosion 86, 195, 197, 198, 202, 204, 210, 215
pitting temperature 199
polarisation curve 194, 195, 197, 198, 202, 203
potential 197, 198, 200, 202, 203, 210, 211, 216, 239
Pourbaix diagrams 192
pouring systems ii, 276
PRE 47, 119, 198
precipitation 18, 26, 39, 55, 59, 78, 93, 142, 155, 161, 231, 234, 235, 263, 264, 269, 299, 300, 308
preliminary 15
primary steelmaking 113
process 20, 23, 24, 26, 28, 30, 38, 58, 59, 63, 66, 67, 69, 72, 73, 92, 93, 97, 118, 119, 156, 175, 178, 186, 204, 206, 214, 220, 221, 236, 238, 239, 251, 254, 255, 264, 266, 267, 269, 285, 286, 290, 298, 299, 307, 308
Processes ii, 27, 41, 43, 44, 46, 48, 55, 67, 68, 73, 76, 80, 84, 85, 106, 108, 116, 124, 125, 130, 132, 142, 147, 155, 157, 170, 178, 183, 205, 208, 211, 214, 215, 224, 228, 229, 252, 254, 263, 270, 295, 304, 305, 307
processes ii, 27, 41,43, 44, 46, 48, 55, 67, 68, 73, 76, 80, 84, 85, 106, 108, 116,124, 125, 130, 132, 142, 147, 155, 157, 170, 178, 183,205, 208, 211, 214, 215, 224, 228, 229, 252, 254, 263, 270, 295, 304, 305, 307
process 20, 23, 24, 26, 28, 30, 38, 58, 59, 63, 66, 67, 69, 72, 73, 92, 93, 97, 118, 119, 156, 175, 178, 186, 204, 206, 214, 220, 221, 236, 238,239, 251, 254, 255, 264, 266, 267, 269, 285, 286, 290, 298, 299, 307, 308
protective film 201, 204, 220, 222, 223, 236
purge plug 115-117

Q

quenching 61, 106, 139, 141, 185, 186, 234, 249, 263, 265, 266, 268, 294, 297, 298, 300-307

R

rate 6, 13, 24, 39, 63, 68, 72, 80, 93, 94, 100, 101, 103, 104, 119, 120, 127, 139, 142, 149, 152, 154, 155, 159, 168, 184, 198, 199, 201, 202, 205, 207, 208, 219, 221, 241, 259, 265, 266, 269, 270, 272, 276, 280, 282, 285, 297, 299, 306
rate 6, 13, 24, 39, 63, 68, 72, 80, 93, 94, 100, 101, 103, 104, 119, 120, 127, 139, 142, 149, 152, 154, 155, 159, 168, 184, 198, 199, 201, 202, 205, 207, 208, 219, 221, 241, 259, 265, 266, 269, 270, 272, 276, 280, 282, 285, 297, 299, 306
Re 80, 81, 128, 141, 203, 226, 305
reaction 17, 18, 44, 60, 63, 72, 78, 79, 99, 100, 113, 114, 116, 119, 122, 124, 131, 192, 194, 200, 204, 220, 221, 223, 225, 227, 228, 236, 240, 247, 251, 266
reaction 17, 18, 44, 60, 63, 72, 78, 79,

99, 100, 113, 114, 116, 119, 122, 124, 131, 192, 194, 200, 204, 220, 221, 223, 225, 227, 228, 236, 240, 247, 251, 266
reduction of heat transfer 15
refining 11, 12, 14, 21, 30, 44, 45, 47, 55, 63, 70, 92, 93, 113, 115, 118, 121, 123, 124, 128, 149, 175, 180
relief annealing 298, 306, 307
removal 15, 20, 23, 38, 63, 66, 92, 100, 104, 113, 116, 123, 124, 126, 131, 133, 143, 254, 257, 297, 298
removal 15, 20, 23, 38, 63, 66, 92, 100, 104, 113, 116, 123,124, 126, 131, 133, 143, 254, 257, 297, 298
Rennfeuer 4, 5
resistance 1, 5, 7, 65, 73, 81, 83, 85, 92, 106, 112, 150, 151, 154, 157, 160, 168, 170, 176, 184, 197, 198, 200, 202, 222, 226, 228, 229, 231, 232, 238, 240, 253, 263, 282, 294, 299, 307
retained austenite 144, 145, 182, 263-268
reversion of phosphorus 21
risers 13, 46, 131, 143, 257, 269, 276, 281, 290, 298, 299
risers 13, 46, 131, 143, 257, 269, 276, 281, 290, 298, 299
rising ii, 276

S

S iii, 2, 3, 17, 18, 21, 23, 24, 39, 42, 46, 49, 55, 60, 66, 69, 70, 84, 92, 94, 101, 102, 105, 123, 124, 128, 133, 141, 145, 149, 160, 167, 170, 179, 180, 189, 192, 202, 211, 216, 219, 225, 231, 241, 242, 255, 268, 298, 302, 303, 307, 308
scale 3, 27, 73, 82, 232, 234, 251
scales 226, 237, 238
Schaeffler diagram 211, 212
schematic layout 123
Schoefer diagram 211-213
scrap 22, 46, 53, 54, 93, 113, 124, 128, 131, 178
secondary metallurgy ii, 7, 39, 73, 121, 125, 129, 130, 132, 133, 147, 170, 178, 304
segregation of elements 72, 73, 106, 142, 155, 186, 265, 296
service conditions ii, 92, 139, 168, 172, 192, 206, 228, 230, 231, 240, 247, 253, 255, 300
Si 6, 16, 17, 42, 47, 53, 56, 58, 67, 70, 72, 77, 103, 107, 115, 128, 140, 141, 144, 153, 172, 185, 211, 222, 223, 237, 242, 249, 258, 259, 268, 284
Si 6, 16, 17, 42, 47, 53, 56, 58, 67, 70, 72, 77, 103, 107, 115, 128, 140, 141, 144, 153, 172, 185, 211, 222, 223, 237, 242, 249, 258, 259, 268, 284
silicon 2, 6, 7, 13, 21, 22, 31, 38, 46, 57, 58, 64, 67, 70, 79, 85, 99, 103, 120, 122, 123, 127, 131, 139, 144, 146, 149, 172, 175, 180, 186, 209, 210, 220, 221, 228, 229, 232, 233, 236, 238, 240, 242, 259, 284
slag 55, 58, 63, 66, 67, 69, 102, 106, 124, 219, 276, 280
slags ii, 15, 18, 20, 22, 47, 48, 58, 63, 84, 85, 93, 95, 96, 113, 115, 117, 118, 220, 276
smelting 1, 17, 22, 23, 30, 38, 41, 43, 45, 46, 48, 63, 73, 83, 93, 94, 106, 118, 127, 130, 131, 147, 157, 175, 180, 183, 186, 188, 210, 213, 214, 258, 269
solubility of oxygen 53, 56, 57, 59, 71, 72, 75, 92, 108
spalling 254
stabilisers 208
steel castings ii, 1, 5, 7, 13, 53, 54, 63, 66, 69, 70, 78, 81, 86, 92, 108, 130, 154, 169, 171, 186, 189, 192, 214, 229, 240,

241, 247, 271, 276, 285, 298, 301, 303, 305, 308
stirring 15, 39, 100, 104, 118, 133
stress intensity factor 169, 170
structure 20, 38, 61, 67, 76, 78, 139, 146, 148, 149, 153, 165, 168, 170, 192, 197, 210, 211, 213, 223, 231, 232, 238, 239, 249, 251, 253, 254, 263, 294, 302, 304, 307
structure 20, 38, 41
sulphides 27, 28, 43, 97, 99, 108, 139, 142, 155, 165, 178, 194, 204, 226, 258, 259
sulphides 27, 28, 43, 97, 99, 108, 139, 142, 155, 165, 178, 194, 204, 226, 258, 259
sulphur 2, 27, 29, 30, 39, 48, 79, 80, 85, 94, 97, 99, 112, 113, 115, 119, 120, 122, 123, 126, 128, 132, 142, 147, 155, 178, 179, 187, 203, 205, 219, 220, 230, 258, 259, 296, 306
synthetic 22, 38, 48, 113, 122
systems ii, 13, 38, 39, 43, 44, 49, 55, 73, 112, 166, 170, 266, 271, 283

T

tap hole 280, 282, 283
tapping 6, 21, 22, 27, 38, 63, 68, 69, 83, 96, 100, 112, 113, 115, 119, 121, 122, 124, 127, 129, 131, 236, 276
technology of smelting 7
temperature ii, 5, 6, 12, 22, 43, 44, 53, 63, 64, 66, 76, 77, 80, 83, 85, 99, 100, 119, 131, 132, 139, 151, 152, 157, 159, 165, 167, 168, 199, 206, 207, 209, 212, 238, 239, 263, 269, 270, 272, 276, 279, 284, 285, 294, 297-307
thermal conductivity 41, 142, 248, 269
thermodynamics of aqueous corrosion 192
Ti 16, 20, 46, 47, 58, 67, 69, 70, 74, 75, 77, 78, 83, 92, 94, 97, 112, 115, 118, 120, 130, 145, 153, 176, 180, 183, 208, 223, 236, 238, 242, 268
transition temperature 59, 132, 159, 165, 167, 168, 179, 180, 182, 184, 186
transpassive 194, 197, 210, 211
Tropenas 6, 55
TTT diagram 156, 260
type 12, 13, 58, 61, 63, 92, 94, 100, 102, 107, 122, 139, 154, 155, 157, 159, 160, 175, 177, 178, 184, 194, 201, 206, 219, 223, 226, 234, 235, 238, 248, 249, 251, 259, 270, 286, 294, 298, 302
types of corrosion 192, 200, 201, 209

U

unit 12, 113, 115, 122, 124, 221, 228, 229, 254
unit 12, 113, 115, 122, 124, 221, 228, 229, 254
Utaloy 262, 263

V

Vacuum Degassing 7, 11, 125, 304
VAD 7, 11, 48, 73, 106, 125, 128, 147, 157, 170, 183, 208, 210, 214
viscosity 29, 39, 40, 44, 48, 49, 96, 100, 101, 248
viscosity 29, 39, 40, 44, 48, 49, 96, 100, 101, 248
VOD 7, 11, 48, 55, 73, 122, 130, 132, 133, 147, 157, 170, 183, 208, 214

W

wear ii, 5, 7, 12, 15, 117, 152, 154, 160, 161, 206, 220, 261, 263, 294, 300, 307
wear ii, 5, 7, 12, 15, 117, 152, 154, 160, 161, 206, 220, 261, 263, 294, 300, 307
white 2, 213, 300
with Al 78, 79, 114, 120, 122

Made in the USA
Las Vegas, NV
03 November 2026

57422193R00197